Studies in Classification, Data Analysis, and Knowledge Organization

More information about this series at http://www.springer.com/series/1564

Domenica Fioredistella Iezzi ·
Damon Mayaffre · Michelangelo Misuraca
Editors

Text Analytics

Advances and Challenges

Springer

Editors
Domenica Fioredistella Iezzi
Department of Enterprise Engineering Mario Lucertini
Tor Vergata University
Rome, Italy

Damon Mayaffre
BCL
University of Côte d'Azur and CNRS
Nice, France

Michelangelo Misuraca
Department of Business Administration and Law
University of Calabria
Rende, Italy

ISSN 1431-8814 ISSN 2198-3321 (electronic)
Studies in Classification, Data Analysis, and Knowledge Organization
ISBN 978-3-030-52679-5 ISBN 978-3-030-52680-1 (eBook)
https://doi.org/10.1007/978-3-030-52680-1

Mathematics Subject Classification: 97K80

This Springer imprint is published by the registered company Springer Nature Switzerland AG
The registered company address is: Gewerbestrasse 11, 6330 Cham, Switzerland

Preface

The Statistical Analysis of Textual Data is a broad field of research that has been developed since the 1950s. The subjects that significantly contributed to its development are Linguistics, Mathematics, Statistics, and Computer Science. Over time, the methodologies have been refined, and the applications have been enriched with new proposals from the segmentation of texts, to the development of linguistic resources, the creation of lexicons, concordance analysis, text classification, sentiment analysis. The fields of application are the most varied, ranging from Psychology to Sociology, Marketing, Economics, Medicine, and Politics. Noteworthy, the Internet has become an inexhaustible source of data, also providing through social media a cross section of the changing society. Big data is the word that most echoes among the data scientists.

This volume aims to collect methodological and applicative contributions to text analysis, previously discussed during the JADT18 conference which took place in Rome from 12 to 15 June 2018. This biennial conference, which has continuously gained importance since its first occurrence in Barcelona (1990), is open to all scholars and researchers working in the field of textual data analysis; ranging from lexicography to the analysis of political discourse, from information retrieval to marketing research, from computational linguistics to sociolinguistics, from text mining to content analysis. After the success of the previous meetings, the three-day conference in Rome continued providing a workshop-style forum through technical paper sessions, invited talks, and panel discussions.

This book, composed of 23 papers, is divided into four macro-parts: (1) techniques, methods, and models parts, (2) dictionaries and specific languages, (3) multilingual text analysis, and (4) applications.

As Umberto Eco said in The Name of the Rose (1980) “The beauty of the cosmos is given not only by unity in variety but also by variety in unity.” This latter claim outlines the traits of the present book, a variety in unity of analysis, techniques, and methods for the interpretation of the textual phenomenon in a variety of applications to several domains.

Those papers represent a virtual showcase of the wealth of this research field, a classical music concert where each instrument has its role and is indispensable for the general harmony.

The first part (*Techniques, Methods and Models*) is composed of six papers.

Iezzi and Celardo traced the timeline of text analytics, illustrating indices and techniques that have had a strong impact on this field of research. They showed the long way from the past to the present demarcating what could be the future scenarios.

Misuraca and Spano explained how to prepare a set of documents for quantitative analyses and compared different approaches widely used to extract information automatically, discussing their advantages and disadvantages.

Felaco presented a joint use of text analysis and network text analysis in order to study the narrations.

Vanni, Corneli, Longree Mayaffre, and Precioso compared statistical analysis and deep learning approaches to textual data.

Fronzetti and Naldi described concentration indices for dialogue dominance phenomena in TV series.

Naldi introduced the methods to measure interactions in personal finance forums.

The second macro-area (*Dictionaries and Specific Languages*) consists of six papers focusing the attention on dictionaries in particular areas and on the search for specific forms.

Iezzi e Bertè analyzed the judicial measures issued by the Court of Audit, from 2010 to 2018 in matters of responsibility and pension. At this aim, the authors proposed a dictionary to support the accounting magistracy in drafting the final measures the judge's decision-making process.

Revelli reviewed the scholastic Italian modeled by teachers in the first 150 years after unification, with a view to assessing the strengths and weaknesses of applying lexicometric parameters to a linguistic variety that was targeted at an inexpert audience. Battisti and Dolcetti analyzed emotional textual analysis based on the study of the association of dense words that convey most of the emotional components of texts.

Romano, Baldasserini, and Pavone proposed a specific dictionary for public administration. They presented the preliminary results on 308 sentences of the Court of Cassation.

Pincemin, Lavrentiev, and Guillot-Barbance designed several methodological paths to investigate the gradient as computed by the correspondence analysis (CA) first axis. Rossari, Dolamic, Hütsch, Ricci, and Wandel analyzed the discursive functions of a set of French modal forms by establishing their combinatory profiles based on their co-occurrence with different connectors.

Multilingual Text Analysis part is composed of four papers.

Moreau examined the access to the signs in French Sign Language (LSF) within a corpus taken from the collaborative platform Ocelles, from a multilingual French bijective/LSF perspective.

Farina and Billero described the work done to exploit the LBC database for the purpose of translation analysis as a resource to edit the bilingual lexical sections of our dictionaries of cultural heritage (in nine languages).

Henkel observed the frequency of the conditional perfect in English and French in a corpus of almost 12 million words corpus consisting of four 2.9 million-word comparable and parallel subcorpora, tagged by POS and lemma, and analyzed using regular expressions.

Shen underlined the need for multilingual monitoring and on the current or future developments of text mining, for three major languages (French, English, and Chinese), in crucial areas for the world's future, and to describe the specificity and efficiency of anaphora.

Applications part is composed of seven papers.

Lebart presented a brief review of several endeavors to identify latent variables (axes or clusters) from an empirical point of view.

Gerli examined the extended abstracts of the EU-funded research projects (2007–2013) realized within the broad domain of the Social Sciences and Humanities (SSH). The aim is to verify how the emergence of a European research funding affects the directions and processes of scientific knowledge production.

Sanandres proposed a latent Dirichlet allocation (LDA) topic model of Twitter conversations to determine the topics shared on Twitter about the financial crisis in the National University of Colombia.

Celardo, Vallerotonda, De Santis, Scarici, and Leva presented a pilot project of the Italian Institute of Insurance against Accidents at Work (INAIL) about mass media monitoring in order to find out how the press deals with the culture of safety and health at work.

Santelli, Ragozini, and Musella analyzed the information included in the open-ended question section of Istat survey "Multiscopo, Aspetti della vita quotidiana" (Multi-purposes survey, daily life aspects), released in the year 2013, regarding the description of the tasks performed individually as volunteers.

Bitetto and Bollani reflected on the valorization of the wide availability of clinical documentation stored in electronic form to track the patient's health status during his care path.

Cordella, Greco, Meoli, Palermo, and Grasso explored teachers' and students' training culture in clinical psychology in an Italian university to understand whether the educational context supports the development of useful professional skills.

Rome, Italy — Domenica Fioredistella Iezzi
Nice, France — Damon Mayaffre
Rende, Italy — Michelangelo Misuraca

Contents

Techniques, Methods and Models

Text Analytics: Present, Past and Future

Domenica Fioredistella Iezzi and Livia Celardo

Abstract Text analytics is a large umbrella under which it is possible to report countless techniques, models, methods for automatic and quantitative analysis of textual data. Its development can be traced back the introduction of the computer, but the prodromes date back, the importance of text analysis has grown over time and has been greatly enriched with the spread of the Internet and social media, which constitute an important flow of information also in support of official statistics. This paper aims to describe, through a timeline the past, the present and the possible future scenario of text analysis. Moreover, the main macro-steps for a practical study are illustrated.

1 Introduction

Data and statistics represent a key for decision-making processes. Nowadays, the amount of data produced by human activities generated the age called "data revolution" [1, 2]. Currently, unstructured data (documents, posts, tweets, video, photos, open questions in a survey, product complains and reviews) represent the most important part of the information present, but not available, to companies and decision-makers. As early as 1995, a Forrester Research report estimated that 80% of useful information in a company is hidden in unstructured documents and therefore not available for the same company [3]. Currently, the exponential growth in the use of social media and social networking sites such as Facebook, Twitter and Instagram have created a new information flow. According to Global Digital Report of "We Are Social and Hootsuite" (2020), about half of the world's population (3.8 billion

D. F. Iezzi (✉) · L. Celardo
Department of Enterprise Engineering, Tor Vergata University,
Via del Politecnico, 1, 00133 Rome, Italy
e-mail: stella.iezzi@uniroma2.it

L. Celardo
e-mail: livia.celardo@uniroma2.it

D. F. Iezzi et al. (eds.), *Text Analytics*, Studies in Classification,
Data Analysis, and Knowledge Organization,
https://doi.org/10.1007/978-3-030-52680-1_1

people) regularly use social media, while 4.54 billion people are connected to the Internet, with almost 300 million users who had access to the Internet for the first time in 2019.[1]

Textual analytics is an automatic and quantitative approach for collecting, processing and interpreting text data. Generating new knowledge from a text analytics process is very complex. It is a chain of operations strong link to research goals, that can allow to have information in real time, and help to make a decision in uncertainty conditions.

This field of research represents an extraordinary challenge for the analysis of many research fields (artificial intelligence, social media, linguistics, statistics, mathematics, marketing, medicine, ...). During the Coronavirus Disease (COVID-19) pandemic, e.g. Google users searched countless times the expressions "I can't smell". A sort of a "truth machine" on one's health. Indeed, anosmia, or loss of smell, has been proven to be a symptom of COVID-19, and some estimates suggest that between 30 and 60 per cent of people with the disease experience this symptom. In this way, it was possible to identify new outbreaks of pandemic propagation[2] [4].

This paper aims is to examine the main changes from the origin of text analytics to the present and with a look towards the future. A zoom on scientific works on these issues in the past 20 years is explored.

This article is structured as follows. In Sect. 2, the prodromes of the text analysis are described; in Sect. 3, origins and developments are discussed; in Sect. 4, the main steps of an text analysis are explored; in Sect. 5, the main applications developed in the last 20 years are explored, and conclusions are also drawn on possible future scenarios.

2 The Text Analysis Pioneers

The quantitative approach to the study of language originates from experimental psychology and the need to demarcate the differences between this new science and philosophy. In 1888, in a research on the expression of emotions through words, Benjamin Bourdon analysed the *Exodus* of the Bible and calculated the frequencies by rearranging and classifying them, eliminating the stop words [5].

A real interest in the quantitative analysis of the lexicon (Lexicometry) is due to stenographers such as Keading and Estoup. Keading coordinated a research in 1898 on the frequencies of graphemes, syllables and words in the German language. Estoup, one of the major exponents French shorthand, in a study published in 1907, defined the notion of rank as a position occupied by a word in a list of words sorted according to decreasing frequencies [6].

[1] See the website https://wearesocial.com/it/blog/2020/01/report-digital-2020-i-dati-global.

[2] Stephens-Davidowitz, in an article published in the New York Times (April 5, 2020), suggested that this methodology could be adopted to search for places where dissemination has escaped the reporting of official data as in the case of the state of Ecuador.

Table 1 Ranks and frequencies in Joyce's Ulysse

Rank	Frequency
10	2653
100	265
1,000	26
10,000	2

An acceleration to grow to text analysis comes from psycholinguistic studies on children's language. Bouseman (1925) demonstrated the existence of two distinct linguistic styles. Emotionally unstable children used verbs preponderantly [7].

In the 1940s, Boder obtained a grammatical measure called "adjective–verb ratio", using different literary genres and discovering that in the drama, the genre closest to spoken language, verbs are five times more than adjectives in 85% of cases [8].

Zipf's work, which takes up what Estoup had started, states "the principle of relative frequency", better known as "Zipf's Law". In a sufficiently long corpus, by classifying words by decreasing ranks, there is a fundamental relationship between the frequency (f) and the rank (r) of the words (Zipf's law): $fr = c$, where c is a constant [9, 10].

Zipf himself carries out an experiment on *Joyce's Ulysse* [27]. From the vocabulary of a corpus of 260,000 occurrences, Zipf noted stating from the highest rank, the frequency of words is inversely related and that the product of frequency (f) and rank (r) is approximately a constant (see Table 1).

Zipf's contribution to linguistics, albeit ingenious, is quite controversial and has given a strong impetus to other theoretical and empirical contributions [11, 26] that only with subsequent development of calculating have found wider fields of application, e.g. the frequency of access to web pages. From this debate, it arose that it is more appropriate to express the law as $fr^a = c$, which on a logarithmic scale expresses itself as $alog(r) + log(f) = c$. This formula can also be written as follows:

$$log(f) = c + alog(r) \tag{1}$$

This is the equation of a straight line, where a indicates the slope or *richness of the vocabulary*, which, in turn, depends on the size of the text [21].

Many indices on the richness of the vocabulary are based on the measure of the relationship between type (V) and token (N): V/N (type/token ratio). This measure allows to connect only texts of equal size. It is well known that the ratio between types and tokens (TTR) decreases systematically with increasing text length because speakers/writers have to repeat themselves. This makes it impossible to compare texts with different lengths.

The Yule index (K) is a measure to identify an author, assuming that it would differ for texts written by different authors. Let N be the total number of words in a text, $V(N)$ be the number of distinct words (dictionary of a corpus), $V(m, N)$ be the number of words appearing m times in the text, and $m_{\max}$ be the largest frequency of a word. Yule's K is then defined as follows, through the first and second moments of the vocabulary population. Yule defines the two quantities S_1 and S_2 (two first moments about zero of the distribution of words), such that

$$S_1 = N = \sum_m mV(m, N) \tag{2}$$

$$S_2 = N = \sum_m m^2 V(m, N) \tag{3}$$

$$K = C\frac{S_2 - S_1}{S_1^2} = C[-\frac{1}{N} + \sum_m^{m_{max}} m^2 V(m, N)\frac{m^2}{N^2}] \tag{4}$$

where C is a constant enlarging of the value of K, defined by Yule as $C = 10^4$. The theme of word frequencies in a text, treated by Zipf and Yule, returns to the studies of Guiraud , who identifies the relationships between the length of the language and the extension of the vocabulary. Guiraud argues that there are very few words that cover more than half of the occurrences that make up the majority of words. Guiraud underlines that concentration of the most frequent words (thematic words) and the dispersion of the less frequent words represented a measure of the richness of the vocabulary (Guiraud, 1954):

$$G = \frac{V}{\sqrt{N}} \tag{5}$$

Guiraud showed empirically that his index is stable over texts between 1000 and 100,000 words [29]. The richness of the vocabulary finds a mathematical formulation with Herdan (1956), which proposes the use of a measure based on the exponential function [28].

3 Timeline Representation Framework

Computer science has a leading role in the development of text analysis. It presupposes, in fact, the automatic treatment of a text without reading by the researcher. The introduction of calculators from the mid-1940s and then the widespread use allowed their growth and application in all disciplinary areas.

Its fortune is originated from the fruitful encounter between computer science, linguistics, statistics and mathematics [23, 27].

Towards the end of the 1950s, the Besançon study centre began a mechanographic examination of Coraille's works, on which Muller's studies on lexicometric analyses were based (Muller, 1967, 1977). The perspectives change and the texts become a representative sample of the language. In which specific expressions are searched for Indices and specific vocabulary measures are born (Tounier, 1980; Lafon, 1984),

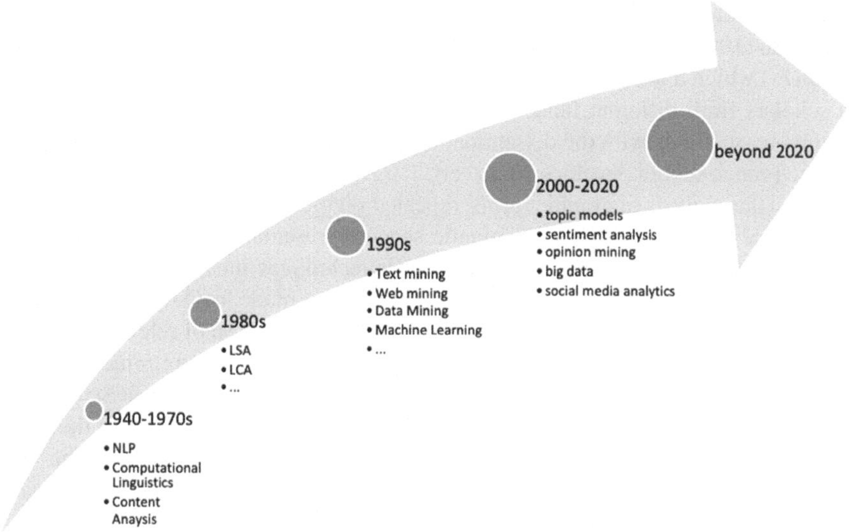

Fig. 1 Text analysis timeline

which soon become widely used by scholars thanks to the implementation of the HyperBase software (Brunet, 1993) [27].

Figure 1 presents a timeline of text analysis from 1940 to the present day. In the first phase (1940–1970's) text analysis studies focus on natural language processing (NPL), computational linguistics and content analysis. Among the most significant works are the first quantitative studies by Guiraud (1954, 1960) and Hedan (1964). The analysis of the content, which has its prodrome in the chiropractic culture of 4000 B.C., has known hybridizations with the computer [31] (Krippendorff, 2004).

The new objective is no longer just to describe the language of a specific work (e.g. The Divine Comedy) or an author (Corneille or Shakespeare), but to describe the written or spoken language through a representative sample of different textual genres (letters, newspaper articles, theatre,...). In the 1960s, "Corpus Linguistics" was born, a new method of investigating language which takes as its reference a finite set of linguistic executions that can be from spoken or written language. This approach is not universally shared by the scientific community, since the language is alive, therefore potentially infinite.

A corpus can only represent an incomplete way of all possible statements. According to Noam Chomsky, the object of linguistics is constituted by an inevitably partial set of performances but by the implicit and unconscious structures of natural language from which it is necessary to reconstruct the rules. If a corpus is a representative sample of the reference language, although it cannot claim to exhaust the complexity of the linguistic analysis, it can provide reliable, it is generalizable, and controllable information. In the 1960s, Jean-Paul Benzecrí introduced a new and innovative approach based on graphic forms. This approach of orientation is opposite to that of Chomsky, with the creation of the French School. Ludovic Lebart and Alain Morineau (1985) develop the SPAD software (*Système Portable pour l'Analyse des Donneés*) which allows a wide application of these techniques to a large community of scholars from different fields of knowledge. Andrè Salem produced innovative textometry methods with the dissemination through the Lexico software proposal.

The French school develops (Benzecrí, 1977; Lebart 1982) which bases its originality on the analysis of graphic forms, repeated segments, analysis of the correspondence of lexical tables (LCA). Semantic space representation and latent semantic analysis (LSA) are the most relevant analysis techniques introduced in the 1980s [27]. In the 1990s, the widespread use of the Internet made available an enormous circulation of textual documents. Techniques for the extraction of relevant information are developed, such as data mining, text mining and web mining, which together with the techniques and machine learning models have allowed the dissemination and use of textual data in the following corporate and institutional areas. At the dawn of the development of the Internet, Etienne Brunet recognizes the role of the web as a reference for corpora linguistics: "*The Internet has become a huge storehouse of information and humanities researchers can find the statistics they need [...]. Statistics have always had ample space, large amounts of data and the law of large numbers. The Internet opens up his wealth, without control, without delay, without expenses and without limits*" [21].

At the beginning of 2000, there were many proposals for the identification of topics, such as the Latent Dirichlet allocation or analyses based on the mood of users as sentiment analysis [25]. Starting from 2014, big data and social media have been used to support official statistics, and the United Nations Global Working Group (GWG)[3] on big data for official statistics was created under the UN Statistical Commission. The GWG provides strategic vision, direction and the coordination of a global program on the use of new data sources and new technologies, which is essential for national statistical systems to remain relevant in a fast-moving data landscape. It is evident that from a methodological point of view, big data sources, the social data generation mechanism does not fall under the direct control of the statistician and is not known. This radically distinguishes this information from the usual statistical outputs derived both from traditional sample surveys, whose sampling design is designed and controlled by statistical institutes, and from administrative sources, whose data generation mechanism is usually known to survey designers. As a result, there is no rigorous methodology to date that guarantees the general validity of statistical information derived from textual social media data, customer reviews and posts. In particular, users cannot be considered as a representative sample of the population. It follows that these data can be explored, but they cannot guarantee the accuracy of any inferences on the population. In Italy, the Italian Institute of Statistics (Istat) proposed a statistical tool capable of assessing the mood of Italians on well-defined topics or aspects of life, such as the economic situation.[4]

[3]See the website: https://unstats.un.org/bigdata/.

[4]See the website https://www.istat.it/it/archivio/219585.

4 Text Analytics Process

The steps of a text analysis process, albeit strongly linked to the objectives of the analysis, elaborate a process of chain analysis that is expressed in different macro-phases (Fig. 2).

The first step is characterized by the definition of the research problem. The project, in fact, should be clear, well-focused and flexible. In this phase, it is necessary to specify the population under study, defined deadlines and goals. If an inductive approach is applied, it is appropriate to examine any external influences such as backgrounds and theories. In the deductive approach, the existence of data or information can be used to test hypotheses, benchmark and build models.

In the second step, the corpus is generated, collecting a set of documents. Therefore, many aspects related to the objectives and the sample collection will be clarified at this stage.

The third step is dedicated to pre-processing, which is extrinsic in a set of sub-phases dependent on the type of document and the aims of the analysis. A corpus that comes from a social survey with open questions, or a corpus that presented a collection of Tweets will get "the need" for a very different treatment. In a corpus from a survey, pretreatment requires a fairly traditional simple standardization, with the removal of punctuation, numbers and so on, and then it will be chosen whether to carry out a morphological normalization such as lemmatization [21]. Moreover, any meta information from survey can support eventually the building of a lexical table [27]. In the case of a corpus coming from a social media, there will be emojis and emoticons and many special characters that can be removed or incorporated the analysis [22].

In the previous step, the pre-processed text document is first transformed into an inverted index, that, in the fourth step, is transformed into a term-document matrix (TDM) or document-term matrix (DTM). In a TDM, the rows correspond to terms, and the columns correspond to documents. Alternatively, in a DTM, the

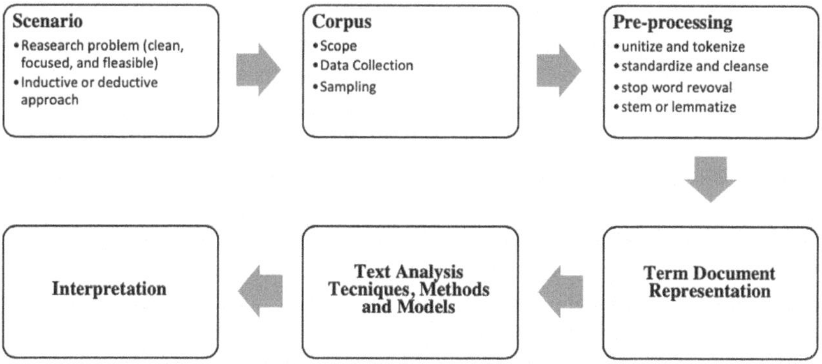

Fig. 2 Steps of a text analysis process

rows correspond to documents, and the columns correspond to terms. Local, global and combinatorial weighting can be applied to TDM or DTM.

In the fifth phase, methods, techniques and models based on goals are applied. The most used techniques in the exploration phase are the latent semantic analysis (LSA) and the correspondence analysis of lexical tables (LCA), and clustering techniques. For the research of the topics, a widespread probabilistic approach is the application of the probabilistic topic models. In recent years, machine learning classification techniques and sentiment analysis models have developed.

The analysis phase is followed by the interpretative phase, which allows to generate knowledge. This sixth phase of the supply chain can end the process or start another one if the results are not satisfactory.

5 Scientific Research and Collaborations

The world's scientific community has been publishing an enormous amount of papers and books in several scientific fields, in the way that it is essential to know which databases are efficient for literature searches. Many studies confirmed that today, the two most extensive databases are Web of Science (WoS) and Scopus [30]. Many articles are comparing Scopus and WoS databases, trying to find which one is the most representative of the existing literature [12, 13, 15, 16, 19, 20, 24]. These comparative studies concluded that even if these two databases are constantly improving, Scopus could be preferred due to its quality of outcomes, time savings and ease of use and because Scopus provides about 20% more coverage than WoS. For these reasons, we decided to implement the research about text analytics subject on Scopus. We selected all the articles in English, with no limits in time, including in their abstract, using the keyword "text analytics". The selected articles from the database were 580 (excluding the manuscripts published in 2020). In Fig. 3 we displayed, by year, the number of published articles.

We noticed that, using the keyword "text analytics" the first manuscript was published in 2003. Under this umbrella name, in 2019 the number of published articles was about 120 while until 2012 there were less than 25 manuscripts published by year. Moreover, from 2018 to 2019, there was a large jump in the number of published articles. After we had analysed the most active authors in the text analytics field, we investigated the co-authorship network [18] of the selected articles (Fig. 5), in order to describe the interactions between authors in this area. For a better readability of the graph, we constructed the network keeping only those nodes with a centrality degree score higher than 4. Among the authors, Liu S. has the highest centrality degree ($D = 19$), showing that it is the one who works more together with the other authors. On the other hand, Atasu K. and Polig R. have the highest number of links connecting each other. If we look to which subject area these articles belong, we can observe in Fig. 4 that most of the manuscripts are related to computer science area (41%), but also to many other areas, such as engineering (10.8%), social sciences (9.7%) and mathematics (8.6%). It means that, even if text analytics seems to be strongly related

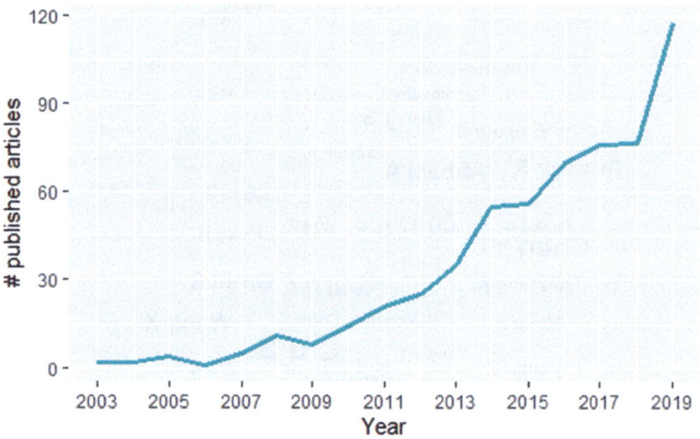

Fig. 3 Time plot of published articles related to text analytics subject from 2000 to 2020

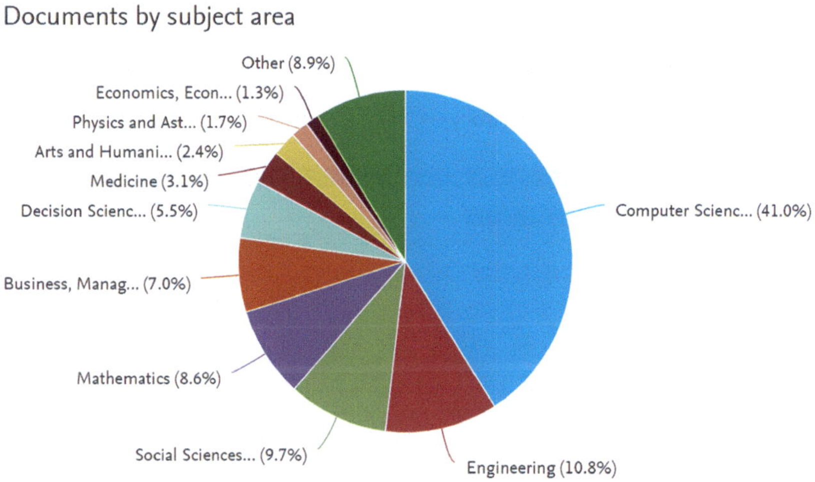

Fig. 4 Published articles related to text analytics subject, by subject area (2003–2019)

to the information technologies, it is a multidisciplinary topic, since it appeared in articles belonging to many areas (Fig. 6).

Table 2 shows the total number of published manuscripts listed by top eight authors. The most active authors in this field, from the Scopus query, are Raphael Polig (IBM Research), Kubilay Atasu (IBM Research) and Shixia Liu (Tsinghua University). The articles year of publication by these authors ranges from 2003 to 2019; 39 out of 41 (almost all) manuscripts belong to computer science area, confirming that top authors in the field of text analytics work in the information technology sector. The most productive researchers work in the USA (48%), India (13%) and Germany (5%).

Fig. 5 Co-authorship network

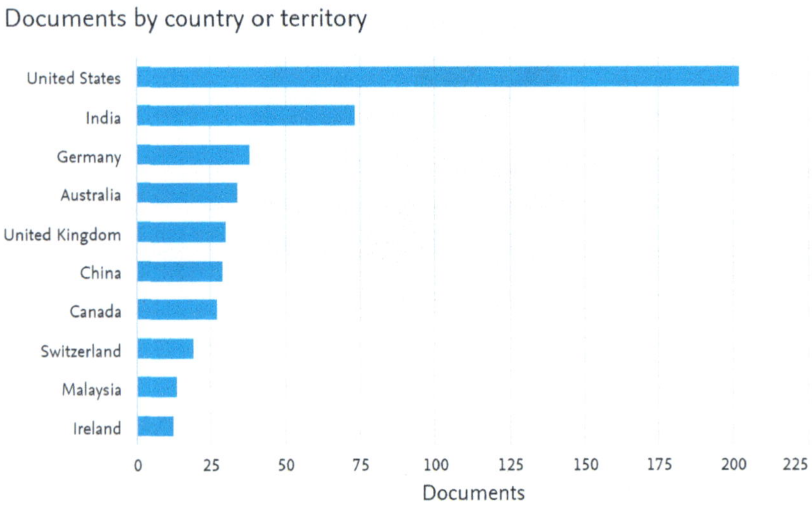

Fig. 6 Articles (no.) by Country

Table 2 Top eight authors, by number of published articles and country

Author	Article (no.)	Country
Polig R.	10	Switzerland
Atasu K.	8	Switzerland
Liu S.	7	China
Hagleitner C.	6	Switzerland
Chiticariu L.	5	USA
Dey L.	5	India
Palmer S.	5	Australia
Zhou Y.	5	USA

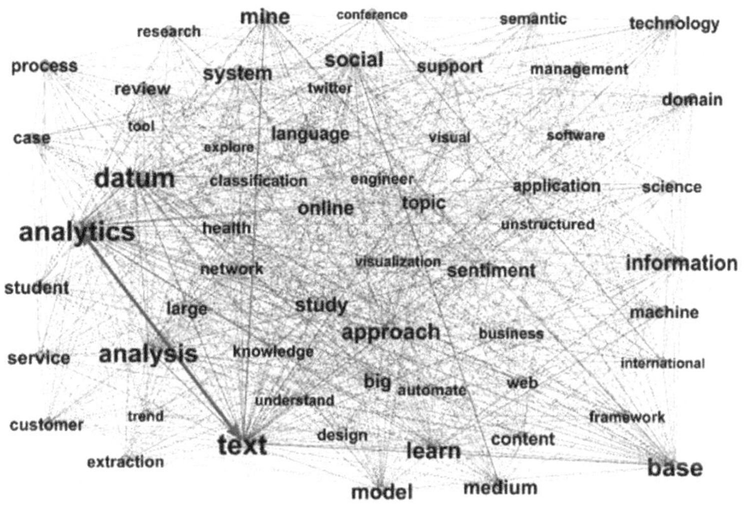

Fig. 7 Word co-occurrences network

In order to better understand the topics covered in the articles, we constructed a one-mode network of the words [14, 17] from the articles' titles. Figure 7 shows a graph, collecting all the titles of the selected manuscripts in order to construct a corpus composed of these short texts. This corpus had, before pre-processing, 6,149 occurrences and 1806 forms. We lemmatized the corpus, and we removed hapax forms and stop words (articles, prepositions, auxiliary verbs, conjunctions and pronouns), obtaining a term-document matrix of size (550 × 580). From the term-document matrix, we obtained the adjacent matrix of terms (550 × 550); for a better readability of the graph, we constructed the word network keeping only those terms with a centrality degree score higher than 100.

From the network emerged that text analytics subject is mostly connected to topics belonging to computer science area, e.g. text mining, data mining, social media but is also connected to more general terms which regard issues connected to many different areas, e.g. information, system, student, customer, learn, model, approach, service, confirming the multidisciplinary nature of this topic. We have reviewed the research topics addressed in the past 20 years from work on this topic, and we have found that although the expression "text analysis" is relatively new (the first articles were published in 2003), it currently incorporates all the techniques, methods and models for automatic text analysis. Networks of scientific collaborations should be increased, also between different sectors. If the birth and development of the text analysis are due to the hybridization of knowledge, today more than collaborations between different research fields can ensure its future.

References

1. Giovannini E (2014) Scegliere il futuro, il Mulino., Bologna
2. United Nations (2014) Independent expert advisory group on a data revolution for sustainable development. A world that counts: mobilising the data revolution for sustainable development. Independent Advisory Group Secretariat. Available from http://www.undatarevolution.org/report/
3. Forrester Research (1995) Coping with complex data. The Forrest Report, April
4. Stephens-Davidowitz S, Pinker S (2017) Everybody lies : big data, new data, and what the Internet can tell us about who we really are, New York, NY
5. Bourdon B (1892) L'espression des émotions et des tendence dans le language. Alcan, Paris
6. Estoup JB (1916) Gammes sténografiques. Institut sténografiques de France, Paris
7. Busemann (1925) Die Sprache der Jugend als Ausdruck des Entwicklungsbrbythmuds. Fischer, Jena
8. The adjective-verb quotient: a contribution to the psychology of language. Psychol Rec 3:310–343
9. Zipf GK (1929) Relative frequency as a determinant of phonetic change. Harvard Stud Classical Philology 40: l–95
10. Zipf GK (1935) The psycho-biology of language. Houghton Mifflin, Boston, MA
11. Mandelbrot B (1954) Structure formelle des textes et communication. Word lO:l-27
12. Bakkalbasi N, Bauer K, Glover J, Wang L (2006) Three options for citation tracking: Google Scholar, Scopus and Web of science. Biomed Digital Libraries 3(1):7
13. Boyle F, Sherman D (2006) Scopus: the product and its development. The Serials Librarian 49(3)
14. Celardo L, Everett MG (2020) Network text analysis: a two-way classification approach. Int J Inf Manage 51:102009
15. Malagas ME, Pitsouni EI, Malietzis GA, Pappas G (2008) Comparison of PubMed, Scopus, web of science, and Google scholar: strengths and weaknesses. FASEB J 22(2):338–342
16. Harzing AW, Alakangas S (2016) Google Scholar, Scopus and the Web of Science: a longitudinal and cross-disciplinary comparison. Scientometrics 106(2):787–804
17. Iezzi DF (2012) Centrality measures for text clustering. Commun Statistics-Theor Methods 41(16–17):3179–3197
18. Liu X, Bollen J, Nelson ML, Van de Sompel H (2005) Co-authorship networks in the digital library research community. Information Process Manage 41(6):1462–1480
19. Mongeon P, Paul-Hus A (2016) The journal coverage of Web of Science and Scopus: a comparative analysis. Scientometrics 106(1):213–228
20. Vieira ES, Gomes JANF (2009) A comparison of Scopus and Web of Science for a typical university. Scientometrics 81(2):587–600
21. Bolasco S (2013) L'analisi automatica dei testi: fare ricerca con il text mining. Carocci, Roma

22. Celardo L, Iezzi DF (2020) Combining words, emoticons and emojis to measure sentiment in Italian tweet speeches. In: JADT 2020 : 15th international conference on statistical analysis of textual data, 16–19 Jun 2020, TOULOUSE (France)
23. Bolasco S, Iezzi DF (2012) Advances in textual data analysis and text mining-special issue Statistica Applicata Italian. J Appl Statistics 21(1):9–21
24. Aria M, Misuraca M, Spano M (2020) Mapping the evolution of social research and data science on 30 years of social indicators research
25. Bing L (2012) Sentiment analysis and opinion mining. Synthesis Lect Human Language Technol 5(1):1–167
26. Witten IH, Bell TC (1991) The zero-frequency problem: estimating the probabilities of novel events in adaptive text compression. IEEE Trans Information Theor 37:1085–1094
27. Lebart L, Salem A (1988) Analyse statistique des données textuelle. Dunod, Paris
28. Herdan G (1964) Quantitative linguistics or generative grammar? Linguistics 4:56–65
29. Guiraud P (1954) Les caractheres statistiques du vocabulaire, Paris, P.U.F
30. Aghaei Chadegani A, Salehi H, Yunus M, Farhadi H, Fooladi M, Farhadi M, Ale Ebrahim N (2013) A comparison between two main academic literature collections: web of science and Scopus databases. Asian Soc Sci 9(5):18–26
31. Krippendorff K (2012) Content analysis: an introduction to its methodology, 3rd edn. Sage, Thousand Oaks, CA, p 441

Unsupervised Analytic Strategies to Explore Large Document Collections

Michelangelo Misuraca and Maria Spano

Abstract The technological revolution of the last years allowed to process different kinds of data to study several real-world phenomena. Together with the traditional source of data, textual data became more and more critical in many research domains, proposing new challenges to scholars working with documents written in natural language. In this paper, we explain how to prepare a set of documents for quantitative analyses and compare the different approaches widely used to extract information automatically, discussing their advantages and disadvantages.

1 Introduction

Reading a collection of documents to discover recurring meaning patterns is a time-consuming activity. Besides the words used to write the document content, it is necessary to consider high-level latent entities that express the different concepts underlying the text. As stated by cognitive psychologists, these concepts are "the glue that holds our mental world together" [39]. They are an abstraction of a particular set of instances retained in the human mind from experience, reasoning and/or imagination. Highlighting the primary topics running through a document can increase its understanding, but this task is challenging when there is not any other prior knowledge about the document itself. A topic can be represented as a set of meaningful words with a *syntagmatic* relatedness [47].

Identifying the topic of large document collections has become more and more important in several domains. The increased use of computer technologies allowed drawing up automatic or semi-automatic strategies to extract meaning from texts written in natural language. Because of the multifaceted nature of the linguistic

M. Misuraca (✉)
DiScAG—University of Calabria, Rende, Italy
e-mail: michelangelo.misuraca@unical.it

M. Spano
DiSeS—University of Naples Federico II, Naples, Italy
e-mail: maria.spano@unina.it

D. F. Iezzi et al. (eds.), *Text Analytics*, Studies in Classification, Data Analysis, and Knowledge Organization,
https://doi.org/10.1007/978-3-030-52680-1_2

phenomenon, scholars belonging to diverse fields investigated this research issue founding their proposals on sometimes distant theoretical frameworks, with proper terminology and notation. Nevertheless, the mathematical (and statistical) base of quantitative natural language processing often led to very similar solutions.

Aim of this paper is briefly explaining how to prepare and organise a set of documents for quantitative analyses (Sect. 2), comparing the leading solutions proposed in the literature to extract a knowledge base as a set of topics (Sect. 3) and discussing advantages and disadvantages of the different approaches (Sect. 4).

2 From Unstructured to Structured Data

Let us consider a collection of documents written in natural language. Texts encode information in a form difficult to analyse from a quantitative viewpoint because their content does not follow a given data model. In the framework of an information mining process, it is then necessary to process the texts and obtain a set of structured data that can be handled with statistical techniques.

2.1 Data Acquisition and Pre-treatment

A document belonging to a given collection can be seen as a sequence of characters. Documents are parsed and tokenised, obtaining a set of distinct strings (*tokens*) separated by blanks, punctuation marks or—according to the particular analysed phenomenon—other kinds of special characters (e.g. hashtags, at-signs, ampersands). These tokens correspond to the words used in the documents. However, since the expression "word" is too generic and does not encompass other valuable combinations of words—e.g. collocations and multiword expressions—*term* is conventionally used instead of word. The particular scheme obtained with tokenisation is commonly known as *bag-of-words* (BoW), viewing each document as a multiset of its tokens, disregarding grammatical and syntactical roles but keeping multiplicity.

Once the documents have been atomised into their basic components, a pre-treatment is necessary to reduce language variability and avoid possible sources of noise, improving in this way the effectiveness of the next analytical steps [48, 55]. First of all, because of case sensitivity, it is often necessary to change all terms' characters to lower case. To keep track of the variety of language and proceed uniformly in the whole collection, it is then possible to perform other normalising operations such as fixing misspelt terms or deleting numbers. A more complex operation considers the inflexion of the terms, reducing variability at a morphological level. Two alternative solutions are commonly considered, *stemming* and *lemmatisation*. The main difference between the two approaches is that the first reduces inflected terms to their roots by removing their affixes, whereas the second uses terms' lemma transforming each inflected term in its canonical form. Even if stemmers are easier and faster to implement, it is preferable to use lemmatisers to preserve the syntactic role of each term and improve the readability of the results (see [27, 28, 54]). When the texts have been pre-treated, it is possible to build the so-called *vocabulary* by stacking identical tokens/terms and counting the number of occurrences of each vocabulary entry (*type*) in the collection of documents.

Additional vocabulary pruning can be carried on to avoid non-informative terms, removing the so-called *stop words*, i.e. the most common terms used in the language and in the specific analysed domain. For the same reason, rare terms with a low number of occurrences are usually removed from the vocabulary. At the end of this process, the documents' content is ready to be processed as structured data.

2.2 Data Representation and Organisation

Vector space model [45] is commonly used to represent a document from an algebraic viewpoint. Let us consider a collection $\mathcal{D}$ containing n different documents. Each document d_i can be seen as a vector $(w_{i1}, w_{i2}, \ldots, w_{ij}, \ldots, w_{ip})$ in a p-dimensional vector space $\mathcal{R}^p$ spanned by the terms belonging to the vocabulary. When a term j occurs in the document, its value w_{ij} in the vector is nonzero. This latter quantity can be seen as a weight representing the importance of the term j in d_i, i.e. how much the term contributes to explain the content of the document.

Several ways of computing these *term weights* have been proposed in the literature, mainly in information retrieval and text categorisation domains. According to Salton and Buckley [44], three components have to be taken into account in a weighting scheme:

- *term frequency*, used to express the relative importance of the terms in each document (local weight);
- *collection frequency*, used to capture the discrimination power of the terms with respect to all the documents (global weight);
- *normalisation*, used to avoid biases introduced by unequal document lengths, where the length is represented as the total number of tokens used in a document.

A primary distinction can be done between *unsupervised* and *supervised* term weighting schemes, depending on the use of available information on document membership. Unsupervised term weighting schemes include *binary* weights, *raw frequency* weights (*tf*), *normalised frequency* weights and *term frequency—inverse document frequency* weights (*tf-idf*). Several studies showed the influence of these schemes in a text mining domain (e.g. [2, 40]). Binary weights just consider the presence or the absence of a term j in a document d_i by assigning to w_{ij} a value of 1 or 0, respectively. Raw frequency weights are calculated as the number of occurrences of a term in a document and correspond to the absolute frequency. Normalised frequency weights incorporate $1/(\max tf_{ij})$ as normalisation factor of the raw frequencies, where $max\ tf_{ij}$ is the highest number of occurrences observed in the document d_i. In all these schemes, the global weight and the length normalisation factor are set to 1. To include the discrimination power of the terms into the weighs, it is possible to consider a *tf-idf* scheme. The raw frequencies or the normalised frequencies are multiplied by a global weight $log(n/n_j)$, where n/n_j is the reciprocal of the fraction of documents containing the term j [50]. Often the *tf-idf* weights are normalised by the Euclidean vector length of the document, obtaining a so-called *best fully weight* [44]:

$$w_{ij} = \frac{tf_{ij} \cdot log(\frac{n}{n_i})}{\sqrt{\sum_{d_i} (tf_{ij} \cdot log(\frac{n}{n_i}))^2}}$$

Supervised term weighting schemes follow the same rationale of the *tf-idf*, using as global weights other measures developed for feature selection, such as χ^2, *gain ratio* or *information gain* [15, 17].

The different document vectors can be juxtaposed to form a *document-term* matrix $\mathbf{T}$ with n rows and p columns. This matrix is a particular contingency table whose marginal distributions provide different information depending on the chosen weighting scheme. The transposition of the matrix in a term-document matrix does not change data reading. In a binary document-term matrix $\mathbf{T_b}$, the marginal row totals state how many terms of the vocabulary are used in each document, whereas the marginal column totals state to the number of documents in which each term appears. In a raw frequency document-term matrix $\mathbf{T_{tf}}$, the marginal row totals state the lengths of the different documents, whereas the marginal column totals represent the term distribution in the overall collection, i.e. the vocabulary. When other weighting schemes are used, the marginal totals are not directly interpretable. As an example, the two marginal distributions of a tf-idf document-term matrix represent the total tf-idf per document and the total tf-idf per term, respectively.

When the research interest concerns the association between terms, it is possible to build a p-dimensional *term-term* square matrix $\mathbf{G}$. This matrix is commonly called *co-occurrence* matrix, because the generic element $g_{ii'}$ $(i \neq i')$ represents the number of times two terms co-occur together. The easiest way to obtain $\mathbf{G}$ is from the matrix product $\mathbf{T_b}^{\mathsf{T}}\mathbf{T_b}$, where $g_{jj'}$ $(j \neq j')$ is the number of documents in which the term i and the term i' co-occur. The diagonal elements g_{jj} correspond to the marginal column totals of $\mathbf{T_b}$. The co-occurrence matrix can be also obtained by splitting in sentences the document of the collection and building a binary sentence-term matrix $\mathbf{S_b}$. The generic element $\hat{g}_{jj'}$ of $\hat{\mathbf{G}} = \mathbf{S_b}^{\mathsf{T}}\mathbf{S_b}$ can be read as the number of sentences in which the term i and the term i' co-occur and more properly interpreted as the co-occurrence of the two terms in the collection. It is possible to obtain a different granularity of term-term co-occurrences changing the way documents are split, e.g. breaking documents in correspondence of stronger punctuation marks as full stops or splitting each document in clauses. Other measure can be used to express term association, such as *simple matching coefficient*, *Jaccard similarity*, *cosine similarity* [52, 53]. Following the same logic of term-term matrix, a n-dimensional *document-document* square matrix $\mathbf{H}$ can be derived to represent the pairwise document asso-

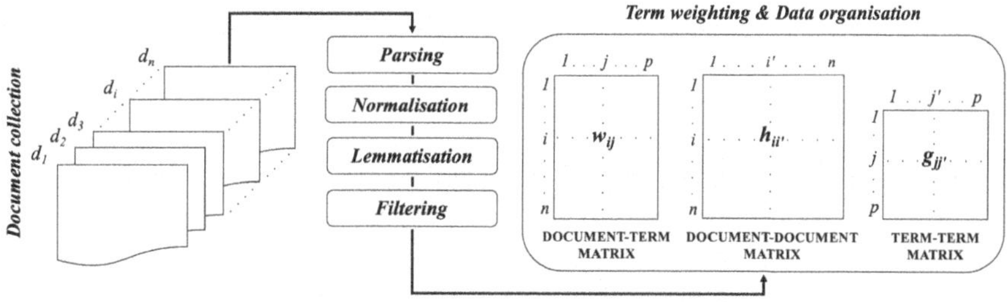

Fig. 1 From unstructured data to structured data: a workflow model

ciation in the collection. The document-document matrix is obtained from the matrix product $\mathbf{T_b}\mathbf{T_b}^\top$, where $h_{ii'}$ is the number of terms shared by documents d_i and $d_{i'}$. The diagonal elements h_{ii} correspond to the marginal row totals of $\mathbf{T_b}$. As above for term-term matrix **G**, different measure can be used to express pairwise association in a document-document matrix **H** [51].

A schematic drawing of the different steps above described is shown in Fig. 1.

3 Looking for Knowledge: A Comparison Between Text Analytics

Taking into account the different textual data representations and the distributional assumptions that can be made about the data themselves, it is possible to set up a general classification of the approaches mostly used in recent years to extract information from document collections automatically.

3.1 Factorial-Based Approach

Factorial techniques aim at minimising the number of terms required to describe a collection of documents, representing the most relevant information in a low-rank vector space. The dimensionality reduction is reached by considering linear combinations of the original terms expressing the content of the documents and visualising these latent association structures onto factorial maps.

Methods like *lexical correspondence analysis* (LCA:[5, 32]) and *latent semantic analysis* (LSA:[16]) are commonly used. LCA was developed in the framework of the French school of data analysis and is generally used to identify the association structure in the document-term matrix [4], using a raw frequency term weighting scheme. LSA became popular in the information retrieval domain to represent the semantic structures of a collection of documents, starting from a document-term matrix with a raw frequency or a tf-idf term weighting scheme. Although the two techniques have been proposed in different contexts and with different aims, they rely on a common algebraic base. The document-term matrix is decomposed in both cases via a *generalised singular value decomposition* [20]. The differences between LCA and LSA are related to the weights assigned to the rows (documents) and the columns (terms), introducing proper orthonormalising constraints [3]. In the first case, a weighted-Euclidean metric based on χ^2 statistics is introduced, giving the same importance to all the terms (normalising their number of occurrences) and to all the documents (normalising their lengths), respectively. In the second case, a Euclidean metric is instead considered both for terms and documents, giving more importance to the most occurring terms and the longest documents. Starting from LCA and LSA, several other related techniques were developed and applied to extract information from document collections (e.g.[13, 37, 56]).

An obvious drawback of the factorial-based approaches is that the new elements may not have a definite meaning, then the results are quite difficult to interpret. To overcome this problem, the French school proposed a two-step sequential approach known as *tandem analysis* [1].

The original data are firstly transformed by performing a factorial method, then a clustering onto the transformed data is carried out. This strategy has the advantage of working on new features that are orthogonal and ordered with respect to the borrowed information. From a statistical point of view, the tandem approach is then very useful for solving some difficulties in reading and interpreting the factorial analysis. The choice of the factorial method is an important and critical phase because it may affect the final results, but taking into account this caveat an improvement of the overall readability is achieved. A satisfying way for obtaining non-overlapping clusters is to consider a hierarchical clustering, in which different levels of aggregation are investigated at the same time [19]. In practical applications, the agglomerative methods are the most used since they allow to construct a hierarchy of partitions with a significantly reduced computational cost. A partition with a higher aggregation index means that the distance between the two closest clusters is large, so they are well separated. Cutting the tree at a level corresponding to a significant "jump" in the index level leads to a good partition. A possible alternative is a clustering step that implements a hierarchical clustering, i.e. Euclidean distance and *Ward linkage* method [59], followed by *k-means* [35], using the cluster centres obtained from the hierarchical algorithm to initialise the k-means algorithm. This joint use is known as *consolidation approach* [31] and can balance the advantages and disadvantages of both methods.

3.2 Network-Based Approach

The term-term matrix—as well as the document-document matrix—can also be seen as an adjacency matrix and depicted as a graph $G(\mathcal{V}, \mathcal{E}, \mathcal{W})$, where $\mathcal{V}$ is a finite set of nodes and $\mathcal{E}$ is a set of undirected edges [12, 42]. If two nodes v_i and v_j are connected, there is an undirected edge with a weight $w_{ij} \subset \mathcal{W}$. The underline

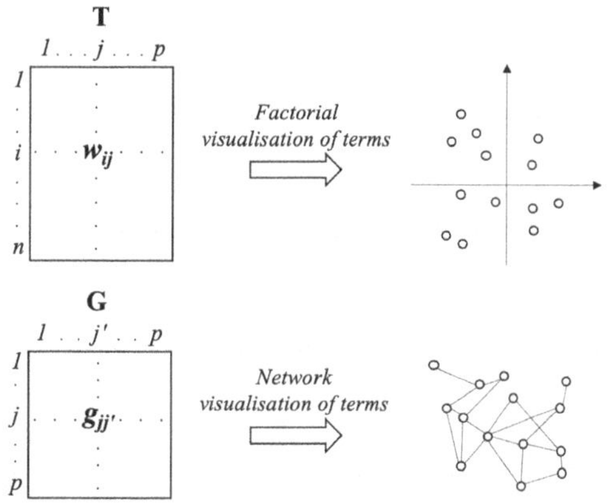

Fig. 2 Visualisation of terms in factorial-based approaches and network-based approaches

assumption is that language can be effectively modelled as a network of terms [49]. If the nodes are the terms, the edges express the co-occurrence or the similarity between paired terms (Fig. 2). If the nodes are the documents, the edges represent the strength of the similarity between linked documents. Several methodological proposals (e.g.[11, 24]) have been produced in this framework, known as *network analysis* [41].

One of the advantages of this approach is to overcome the limits of BoW encoding, where the context of use of each term is lost. In a network of terms, it is possible to visualise both single terms and subsets of terms frequently co-occurring together, recovering the context and the syntagmatic relations of the terms. The task of grouping nodes sharing common characteristics, and/or playing similar roles within the graph, is usually performed by referring to *community detection* methods. Communities are usually thought as subgraphs densely interconnected and sparsely connected to other parts of the graph [60]. From a theoretical viewpoint, communities are then not very different from clusters [18]. As a consequence, there is an overlap among the techniques used in multidimensional data analysis and in network anaysis.

Several recent works addressed the problem of extracting information from textual networks to the identification of high-level structures, considering the combination of terms as concepts/topics occurring in the collection of documents. Sayyadi and Raschid [46] proposed an approach called *KeyGraph* that applies a *betweenness centrality* community detection algorithm on the term co-occurrence graph. Analogously, Lim et al. [34] developed *ClusTop* a clustering-based topic model that applies *Louvain* community detection algorithm [8]. Jia et al. [26] proposed *WordCom* algorithm, using the *k-rank-D* algorithm [33] on the term co-occurrence graph weighted by pointwise mutual information. Misuraca et al. [38] proposed instead a twofold strategy based on a *fast greedy* community detection algorithm [14] and hierarchical clustering to detect the main concepts listed in a collection of short texts, using then the knowledge emerging from the community structure to categorise the different documents.

3.3 Probabilistic Approach

The so-called *topic modelling* offers a probabilistic viewpoint of the information extraction problem. This approach includes a family of generative statistical models used to discover semantic patterns reflecting the underlying (not observable) topics of a document collection. With these models, it is possible to discover the topics of a document collection and at the same time to assign a probabilistic membership of documents to the latent topic space, allowing to consider a lower-dimensional space. Moreover, the topics extracted with topic models are easily human-interpretable because each topic is characterised by the terms most strongly associated with.

The seminal idea of this approach was directly drawn from LSA in [22] when Hoffmann [22] introduced his *probabilistic-LSA* algorithm (pLSA). Differently from techniques based on singular value decomposition, pLSA considers a probabilistic mixture decomposition of the document-term matrix [23]. The probability that a term occurs in a document is modelled as a mixture of conditionally independent multinomial distributions. In contrast to classical LSA, the representation of terms in topics and the representations of topics in documents are non-negative and sum up to one. This latter means that the probabilities associated with the topics are nonzero because each document has a certain probability of being generated by any topic. Each document contains all the topics with different proportions. *Latent Dirichlet allocation* (LDA: [7]) was proposed as a more effective Bayesian model overcoming some limits of pLSA, improving the way mixtures capture the exchangeability of both terms and documents. The rationale of pLSA and LDA is quite similar, but in the second one the topic distribution is drawn by a Dirichlet. Topics are distributions over the vocabulary so that each topic contains each term belonging to the entire collection, but some of them have very low probability. Documents are randomly generated and can be seen as mixtures over the topics.

Several LDA-based models have been developed and proposed [25]. One of the most important improvement of LDA is the *correlated topic model* (CTM:[6, 29]). LDA imposes a mutual independence among topics, CTM takes into account and highlights the correlations between topics by using a logistic normal to draw the topic proportions. An interesting line of research involves topic models and term co-occurrence matrices, as an alternative to the techniques based on the classic document-term matrices. The *biterm topic model* proposed by [61] directly models the term co-occurrence patterns to enhance the topic learning in the document collection. Zuo et al. [62] argued that BTM is a special form of the unigram mixtures rather than a proper LDA-based technique, proposing as an alternative a LDA *word network topic model* based on the term co-occurrence graph.

4 Discussion

Many times, the scientific background of the scholar may substantially influence the choice of an approach. In practical applications, each of the three approaches shows both advantages and disadvantages, considering the computational complexity, the readability of the results and their interpretability, the visualisation capability.

Factorial-based techniques are well consolidated in textual data analysis literature and extensively implemented in various software, some of which are free and user-friendly. This approach aims at reducing the dimensionality of the term space, extracting information about the association relationship and visualising the emerging patterns onto factorial maps. As discussed above, the latent factors could be challenging to interpret. Usually, a two-dimensional map is considered to represent terms originally spanned into a $\mathscr{R}^n$ space. Many scholars not familiar with factorial methods look at the amount of variability represented in the lower-rank space disregarding the quality of the representation. The visualisation and the interpretation of the results are even harder with more extensive document collections. To improve the readability, it is possible to group the terms by performing a subsequent clustering step. Choosing a higher- or lower-order partition allows obtaining topics with different granularity. The drawback of these integrated strategies is that they could be as a black box, since the preliminary data reduction may mask or distort the original relations of data. Each step of the tandem approach involves a different optimisation criterion, and these criteria are addressed separately [58]. A possible alternative is jointly performing both dimensionality reduction and clustering (e.g.[57]), but this solution has been poorly investigated for textual data.

In recent years, network-based techniques were the most applied ones in the field of textual data analysis, despite the fact that the analysis of term-term matrices has a long history in other scientific domains (e.g.[9, 43]). One of the main advantages of this approach is the possibility of preserving the context of use of the different terms. The original terms are ever retained in the analyses, consequently the results are easier to interpret. The choice of a threshold on the co-occurrence value allows highlighting more connected structure representing the core of the document collection. On the other hand, graphs are non-metric representations and do not allow to consider the association between terms as distances. Lebart [30] showed that correspondence analysis could be satisfactorily used to map a graph-structured matrix in a vector space, reconstituting both the relative positions of the nodes and an acceptable order of magnitude for the lengths of the edges. This means that it would be possible to combine network and factorial approaches. As concern community detection, a noteworthy aspect is that knowing in advance the number of topics embodied in the document collection and consequently setting the input parameters is not required, leading to a completely automatic extraction.

Topic modelling is becoming increasingly popular in many applicative domains. The probabilistic framework of these techniques requires to scholars a more profound statistical background. A clear advantage is the possibility to discover topics and categorise documents on the base of their topic distribution. Differently from factorial techniques and network-based analysis, there is a soft assignment of each term to the topics, allowing to take into account the semantic relatedness of the terms. There are different ways to visualise the topics and the related terms, including the possibility of considering the graph underlying a *topic-topic* co-occurrence matrices [36]. Unlike the other described approaches, it is necessary to set up different parameters related to the distribution used in the analysis. Furthermore, when no prior information on the documents is available or considered, there is not a well-established procedure to identify the number of topics. Some proposals tried to cope with this limit (e.g. [10, 21]) but they are not widely shared in the reference literature.

References

1. Arabie P, Hubert L (1994) Cluster analysis in marketing research. In: Bagozzi RP (ed) Advanced methods of marketing research. Blackwell, Oxford, pp 160–189
2. Balbi S, Misuraca M (2005) Visualization techniques for non symmetrical relations. In: Sirmakessis S (ed) Knowledge Mining. Proceedings of the NEMIS final conference. Springer, Heidelberg, pp 23–29
3. Balbi S, Misuraca M (2006) Procrustes techniques for text mining. In: Zani S, Cerioli A, Riani M, Vichi M (eds) Data analysis, classification and the forward search. Springer, Heidelberg, pp 227–234
4. Beaudouin V (2016) Statistical analysis of textual data: Benzécri and the french school of data analysis. Glottometrics 33:56–72
5. Benzécri JP (1982) Histoire et préhistoire de l'analyse des données. Dunod, Paris
6. Blei DM, Lafferty JD (2007) A correlated topic model of science. Ann Appl Statistics 1(1):17–35
7. Blei DM, Ng AY, Jordan MI (2003) Latent dirichlet allocation. J Mach Learn Res 3:993–1022
8. Blondel VD, Guillaume JL, Lambiotte R, Lefebvre E (2008) Fast unfolding of communities in large networks. J Statistical Mech Theor Exp (10)
9. Callon MJPC, Turner WA, Bauin S (1983) From translations to problematic networks: an introduction to co-word analysis. Soc Sci Information 22(2):191–235
10. Cao J, Xia T, Li J, Zhang Y, Tang S (2009) A density-based method for adaptive lDA model selection. Neurocomput Int J 72(7–9):1775–1781
11. Carley KM (1988) Formalizing the social expert's knowledge. Sociol Methods Res 17(2):165–232
12. Carley KM (1997) Network text analysis: the network position of concepts. In: Roberts CW (ed) Text analysis for the social sciences. Lawrence Erlbaum Associates, pp 79–102
13. Choulakian V, Kasparian S, Miyake M, Akama H, Makoshi N, Nakagawa M (2006) A statistical analysis of synoptic gospels. In: Viprey JR (ed) Proceedings of 8th international conference on textual data. Presses Universitaires de Franche-Comté, pp 281–288
14. Clauset A, Newman MEJ, Cristopher M (2004) Finding community structure in very large networks. Phys Rev E 70(066111)
15. Debole F, Sebastiani F (2003) Supervised term weighting for automated text categorization. In: Sirmakessis S (ed) Text mining and its application. Springer, Heidelberg pp 81–97
16. Deerwester S, Dumais ST, Furnas GW, Landauer TK, Harshman R (1990) Indexing by latent semantic analysis. J Am Soc Information Sci 41(6):391–407
17. Deng ZH, Tang SW, Yang DQ, Zhang M, Li LY, Xie KQ (2004) A comparative study on feature weight in text categorization. In: Lin X, Lu H, Zhang Y, Yu JX (eds) Advanced web technologies and applications. Springer, Heidelberg, pp 588–597
18. Fortunato S, Hric D (2016) Community detection in networks: a user guide. Phys Rep 659

19. Gordon AD (1999) Classification. Chapman & Hall/CRC
20. Greenacre M (1983) Theory and applications of correspondence analysis. Academic Press, London
21. Griffiths TL, Steyvers M (2004) Finding scientific topics. Proceedings of the National Academy of Sciences 101(suppl. 1):5228–5235
22. Hofmann T (1999) Probabilistic latent semantic indexing. In: Proceedings of the 22nd annual international ACM SIGIR conference on research and development in information retrieval, pp 50–57
23. Hofmann T (2001) Unsupervised learning by probabilistic latent semantic analysis. Mach Learn 42(1–2):177–196
24. James P (1992) Knowledge graphs. In: Meersman R, van der Riet R (eds) Linguistic instruments in knowledge engineering. Elsevier, pp 97–117
25. Jelodar H, Wang Y, Yuan C, Feng X, Jiang X, Li Y, Zhao L (2018) Latent dirichlet allocation (lDA) and topic modeling: models, applications, a survey. Multimedia Tools Appl 78:15169–15211
26. Jia C, Carson MB, Wang X, Yu J (2018) Concept decompositions for short text clustering by identifying word communities. Pattern Recogn 76(C):691–703
27. Kettunen K, Kunttu T, Järvelin K (2005) To stem or lemmatize a highly inflectional language in a probabilistic IR environment. J Documentation 61(4):476–496
28. Konkol M, Konopík M (2014) Named entity recognition for highly inflectional languages: effects of various lemmatization and stemming approaches. In: Sojka P, Horák A, Kopeček I, Pala K (eds) Proceedings of the 17th international conference on text, speech and dialogue, Lecture Notes in Computer Science, pp 267–274
29. Lafferty JD, Blei DM (2006) Correlated topic models. In: Weiss Y, Schölkopf B, Platt J (eds) Advances in neural information processing aystems, 18. MIT Press, chap 147–154
30. Lebart L (2000) Contiguity analysis and classification. In: Gaul W, Opitz O, Schader M (eds) Data Analysis. Scientific modeling and practical application. Springer, Heidelberg pp 233–244
31. Lebart L, Morineau A, Warwick KM (1984) Multivariate descriptive statistical analysis: correspondence analysis and related techniques for large matrices. Wiley, New York
32. Lebart L, Salem A, Berry L (1988) Exploring textual data. Kluwer, Dordrecht
33. Li Y, Jia C, Yu J (2015) A parameter-free community detection method based on centrality and dispersion of nodes in complex networks. Phys A Statistical Mech Appl 438(C):321–334
34. Lim KH, Karunasekera S, Harwood A (2017) Clustop: a clustering-based topic modelling algorithm for twitter using word networks. In: Nie JY, Obradovic Z, Suzumura T, Ghosh R, Nambiar R, Wang C, Zang H, Baeza-Yates RA, Hu X, Kepner J, Cuzzocrea A, Tang J, Toyoda M (eds) Proceedings of the 2017 IEEE international conference on big data, pp 2009–2018
35. MacQueen JB (1967) Some methods for classification and analysis of multivariate observations. In: Le Cam LM, Neyman J (eds) Proceedings of 5th Berkeley symposium on mathematical statistics and probability, vol 1, pp 281–297
36. Maiya AS, Rolfe RM (2014) Topic similarity networks: visual analytics for large document sets. In: Proceedings of 2014 IEEE international conference on big data, vol 1, pp 364–372
37. Misuraca M, Scepi G, Grassia MG (2005) Extracting and classifying keywords in textual data analysis. Italian J Appl Statistics 17(4):517–528
38. Misuraca M, Scepi G, Spano M (2020) A network-based concept extraction for managing customer requests in a social media care context. Int J Information Manage 51(101956)
39. Murphy G (2004) The big book of concepts. MIT press
40. Nakov P, Popova A, Mateev P (2001) Weight functions impact on lsa performance. In: Proceedings of the 2001 conference on recent advances in natural language processing, pp 187–193
41. Newman M (2010) Networks: An introduction. Oxford University Press
42. Popping R (2000) Computer-assisted text analysis. Sage Publications, London
43. Popping R (2003) Knowledge graphs and network text analysis. Soc Sci Information 42(1):91–106
44. Salton G, Buckley C (1988) Term-weighting approaches in automatic text retrieval. Information Process Manage 24(5):513–523

45. Salton G, Wong A, Yang CS (1975) A vector space model for automatic indexing. Commun ACM 18(11):613–620
46. Sayyadi H, Raschid L (2013) A graph analytical approach for topic detection. ACM Trans Internet Technol 13(2):1–23
47. Schutze H, Pedersen J (1993) A vector model for syntagmatic and paradigmatic relatedness. In: Making sense of words : 9th annual conference of the UW Centre for the New OED and Text Research, Oxford, pp 104–113
48. Song F, Liu S, Yang J (2005) A comparative study on text representation schemes in text categorization. Pattern Anal Appl 8(1–2):199–209
49. Sowa J (1984) Conceptual structures: information processing in mind and machine. Addison-Wesley, Boston
50. Spärck Jones K (1972) A statistical interpretation of term specificity and its application in retrieval. J Documentation 28(1):11–21
51. Sternitzke C, Bergmann I (2009) Similarity measures for document mapping: a comparative study on the level of an individual scientist. Scientometrics 78(1):113–130
52. Tan PN, Kumar V, Srivastava J (2002) Selecting the right interestingness measure for association patterns. In: Proceedings of the 8th ACM SIGKDD conference on knowledge discovery and data mining, pp 32–41
53. Terra EL, Clarke CLA (2003) Frequency estimates for statistical word similarity measures. In: Hearst M, Ostendorf M (eds) Proceedings of the 2003 human language technology conference of the North American Chapter of the Association for Computational Linguistics, pp 244–251
54. Toman M, Tesar R, Jezek K (2006) Influence of word normalization on text classification. In: Proceedings of the 1st international conference on multidisciplinary information sciences & technologies, vol II, pp 354–358
55. Uysal A, Gunal S (2014) The impact of preprocessing on text classification. Information Process Manage 50(1):104–112
56. Valle-Lisboa JC, Mizraji E (2007) The uncovering of hidden structures by latent semantic analysis. Information Sci 177(19):4122–4147
57. van de Velden M, Iodice D'Enza A, Palumbo F (2017) Cluster correspondence analysis. Psychometrika 82(1):158–185
58. Vichi M, Kiers HAL (2001) Factorial k-means analysis for two-way data. Comput Statistics Data Anal 37(1):49–64
59. Ward JH (1963) Hierarchical grouping to optimize an objective function. J Am Statistical Assoc 58:236–244
60. Wasserman S, Faust K (1994) Social network analysis: methods and applications. Cambridge University Press, New York
61. Yan X, Guo J, Lan Y, Cheng X (2013) A biterm topic model for short texts. In: Proceedings of the 22nd international conference on World Wide Web, pp 1445–1456
62. Zuo Y, Zhao J, Xu K (2016) Word network topic model: a simple but general solution for short and imbalanced texts. Knowl Information Syst 48(2):379–398

Studying Narrative Flows by Text Analysis and Network Text Analysis

The Case Study of Italian Young People's Perception of Work in Digital Contexts

Cristiano Felaco and Anna Parola

Abstract The paper presents a joint use of text analysis and network text analysis in order to study the narrations. Text analysis allows to detect the main themes subjects in the narrations and hence the processes of signification, network text analysis permits to track down the relations between linguistic expressions of text, identifying, therefore, the path of flow of thoughts. Using jointly the two methods is possible not only to explore the content of narrations, but starting from the words and concepts with higher semantic strength, also to identify the processes of signification. To this purpose, we will present a research aiming to understand high school students' perception of employment precariousness in Italy.

1 Introduction

The narration, and more specifically the exposition, is a process that constitutes a textual texture endowed with significance and conveying meanings. Analyzing the texts allows us to grasp on one hand the perception of those who narrate on a given topic and the process of significance attributed to the narrated experience, but on the other, to understand the flows of thought, entering specifically the words used and their sequence. The narration is a crucial moment in the contemporary society because of tendency and, at the same time, the need of people to express feelings, opinions, acts in order to affirm their own identity [1]. The increasingly development of the Internet and ICT and the spread of new digital channels, such as blogs, social networks, and forums, have enhanced the ways through which individuals may express the narratives of their life experiences [2]. Using the statistical text analysis to study the narrations, it is possible to detect in-depth meaning of the words and the process of signification of the experiences [3].

C. Felaco (✉) · A. Parola
University of Naples Federico II, Naples, Italy
e-mail: cristiano.felaco@unina.it

A. Parola
e-mail: anna.parola@unina.it

D. F. Iezzi et al. (eds.), *Text Analytics*, Studies in Classification, Data Analysis, and Knowledge Organization,
https://doi.org/10.1007/978-3-030-52680-1_3

Aim of the paper is to show a methodology to reconstruct narrative processes from which run an automatic extraction of the meaning through the jointly use of text analysis (TA) and network text analysis (NTA). If, on the one hand, TA enables to pinpoint the relevant themes and the words used, as well as the meaning perceived [4] in other words "what" it is told, on the other hand, the use of means of TNA may shed light on "how" it is told, then on the process of narration.

The joint use of textual data analysis and network analysis is not new. Many researches focused on semantic relations between the words, when representing texts as graphs and applying network text analysis to figure out text's meanings and structures. This approach is commonly used to uncover social interaction structures from semantic structures of texts [5], to identify "plot units" within a text [6], graphical representations of affect states, or to establish affect as a moving force behind narrative structures [7]. Later work, payed more attention to the use of map analysis [8] and the construction of spatial visualization techniques for textual analysis [9]. This approach assumes that the structure of relations between the words of a text may correspond to the mental or model map [10]—also usually defined as knowledge networks [11], semantic networks or meta-networks [12, 13]—and the cognitive connections put into action by the authors of the text. Such method enables to model language as network of words and relations [14]: starting to the words and concepts with higher semantic strength is possible to detect key concepts which function as junctions for meaning circulation within a text, the main pathways for meaning circulation [15] and to identify the processes of signification [16].

Specifically, this approach is used to study the students' perception of labor market in Italy paying particularly attention to psycho-social and cognitive aspects, with the aim at identifying keywords able to shed light on mental processes and experiences that precede the choice of words. The lexical corpus was built by narrations collected from 2013 to 2016 in blog of Repubblica "Microfono Aperto." Furthermore, in our approach, according to [17], empirical research techniques have been adapted for the digital context.

This paper is structured as follows: Sect. 2 shows the method and procedures used; Sect. 3 illustrates the case study of young people's job perceptions, paying attention to youth employment condition in Italy; while Sect. 4 more specifically discusses the results of the research. The last section concludes with some remarks about the undertaken work.

2 A Joint Approach: Method and Procedures

Our approach is built upon two integrated phases: An explorative study of the content of narrations through a statistical textual analysis [4, 18] by T-LAB software, and an analysis of the relationship between the words and the paths of meaning by the means of social network analysis [19–21] using Gephi software.

The textual material is previously handled through the text normalization and the dictionary building. The normalization phase allows to correctly detect the word as raw forms entailing a transformation of the corpus (blank space in excess elimination, apostrophe marking, capital letter reduction, etc.); along with this, a set of strings are recognized as proper nouns and the sequences of row forms recognized are converted as multiword in unitary strings. The dictionary building was performed through the phases of lemmatization and disambiguation of words with the same graphic form but different meanings which permit to rename the types into lemmas. The disambiguation is an operation by which homograph cases (when graphic form is the same but with a different meaning) are solved, while in the lemmatization operation, the forms of the verbs are brought back to their present infinitive forms, the nouns and adjectives to their singular masculine form and articulated prepositions to their article-less form. After this stage, lexical characteristics of the corpus are checked by calculating measure of lexical richness: Type/token ratio with values no lower than 0.2, and hapax amount with a threshold no lower than 50% of the corpus. Along with this, it needs to check the list of keywords created through an automatic procedure, and their occurrence in the corpus; moreover, occurrence threshold is set, ruling out all the lemmas which appeared less than n times in the corpus. The level of threshold depends of lexical characteristics and dimensions of the corpus.

In order to detect the main thematic groups and paths of meanings, we jointly used textual data analysis and network analysis.

Thematic analysis of elementary contexts was performed; this is a cluster technique that allows to construct and explore the content of the corpus [22]. Each cluster consists of a set of elementary contexts (they approximately have the length of a sentence) which is defined by the same pattern of keywords and described by those lexical units most characteristics of context units from which is composed. Cluster analysis was carried out by an unsupervised ascendant hierarchical method (bisecting K-means algorithm), characterized by the co-occurrence of semantic features. More specifically, the procedure analysis consists of: Co-occurrences analysis through the creation of a data table context units*lexical units with presence/absence values; data pre-treatment by TF-IDF procedure and scaling of row vectors to unit length (Euclidean norm); clustering of the context units using cosine measure and bisecting K-means method; construction of a contingency table (lexical units*clusters); chi-square test applied to each intersection of lexical units*clusters. The cluster number is determined by an algorithm which uses the relationship between intercluster variance and total variance, and it takes as optimal partition the one in which this relationship exceeds the threshold of 50%.

The interpretation of the position of the clusters in factorial space and the interpretation of the words which characterize the clusters permit to identify implicit relations which organize the subjects thought allowing to understand the narrator's point of view with regard to the event narrated. The latter includes also a series of evaluative factors, thoughts, meanings, value judgments, and emotional projections.

The lexical corpus obtained was imported in Gephi software which organize the lemmas in an adjacency matrix (lemmas*lemmas); the network structure of lemmas is formalized as a 1-mode network, a useful tool to spatially visualize these relations by means of a graph, whose points (or nodes) represent the single lemmas, whereas the lines (or edges) represent the ties (that is, the co-occurrences) which link them together. This method allows to understand how the nodes are connected, fostering the identification of the neighborhood, and the identification of those nodes that are strategically located in different sets or in the whole network. To this aim, indices associated with the centrality are used to identify important concepts, associations among concepts, themes, and so on. Firstly, the degree centrality that indicates the words used with greater frequency in relation to other words in the accounts and the various contexts of meaning. In more detail, the incidence for each node can be expressed both as in-degree, that is, the number of edges incoming in a node (identifying in this way the so-called predecessors of each lexical unit) and as out-degree, that is, the number of edges outgoing from the node (showing in this case the so-called successors). The relation between predecessors and successors within the textual network is interesting because it helps to understand the semantic variety produced by the nodes. Betweenness centrality is an additional index used, which expresses the degree of a node to being located 'between' other nodes of the graph. The node showing a high degree of betweenness carries out a role of 'mediator' or 'guardian' of information, that is, it has the control over the information flow, allowing the link between two or more set of networks [23]. In terms of textual analysis, these lemmas play a central role in relation to the flow of meanings in the network, acting as bridge that allow the link between different parts of text and from which specific path of meaning originate, defining in this way the semantic variety of the narrations.

3 Young People's Job Perceptions: A Case Study

We present a case study conducted through the combined use of techniques in order to understand students' perception of the work world in the Italian context.

The latest available data show that Italy is the first European country with the highest percentage of NEET (Not in Education, Employment or Training, 30.7%) [24]. The Italian context is characterized by a particularly difficult job market for young people and is conceived as a prototype of countries of Southern Europe where the chances of young people to develop coherent and satisfying future careers and life plans are influenced by the current socio-economic difficulties [25]. Most of young Italians, after concluding their prolonged programs of study go through periods of strong instability and uncertainty [26–28], with consequences on transition to adulthood [29] and psychological well-being [30].

The economic crisis does not spare the coping strategies of young people [31]. In fact, the lack of a future planning and the feeling of anxiety and insecurity affects the career transition, that is, the maintenance of high aspirations, the crystallization of career goals, and the intensive behavior of job search [32]. In a psychological point of view, the adolescence is a crucial period for construct one's career. According to life span developmental psychology [33, 34], career development constitutes a life-long process from childhood through adolescence and adulthood. In the career choices, in fact, individuals have a more active role, compared to the past, in the construction of their own professional paths [35].

In this sense, the narrative's own story is central in the careers field. According to [36], a career story helps a person to define who is and how should act within a career context creating and providing meaning and direction [37] and by constructing a sense of causality and continuity about one's career path [38, 39].

For the study, we used a source of textual data from the Repubblica's blog "Microfono Aperto" in which high school students, in the period from 2013 to 2016, answered to the prompt "Four young people out of ten without work. And what do you think? Whose faults is that? What would you like to see done as soon as possible to guarantee you a dignified future?". According to [40], the digital technologies lead to the creation of new environments through which individuals can narrate and share their life experiences. Talking through the Web facilitates the process of self-reflection, on one's own role and on the relationship with what happens in the context in which the young person is inserted. In a situation of unease due to precarious employment, the Web can be a useful container for sharing the experience of precariousness, constituting an environment for sharing and socializing one's own experiences [41].

4 Results and Discussion

The lexical corpus constitutes of 130 narrations (10,110 occurrences (tokens), 2484 types, 821 lemmas, 1590 hapax) and two descriptive variables, macro-region of residence (north, center, south, and islands) and type of secondary school (technical or professional school or high school). The thematic analysis of elementary contexts produced four clusters (Fig. 1), named CL1 "Looking at opportunities" (14.6%); CL2 "And the government?" (19.8%); CL3 "From dreams to crisis" (38.5%); CL4 "The job search, where?" (27.1%).

The narrations of the cluster "Looking at the opportunities" refer to the analysis of sacrifices and opportunities; the need for an "activity," as put into practice actions in the present in view of a better future emerges clearly. For this reason, the crisis is, at the same time, also an opportunity that young people must seize to demonstrate their abilities:

> at this point, from what you hear, anyone wonders about their future. To ensure that one day there is more work opportunities, one must act NOW.[...]. Even those looking for work, however, must fly low and be content, for the moment, with little instead of staying home in a surrendered state. In my opinion young people must have the opportunity to demonstrate what they are worth, to show the world what they know, who they are and to make everyone understand that they are capable to do "if they commit" any work, from the simplest to the most complex.

The texts of the second cluster are more oriented toward the search for "guilt" and a request for solutions mainly from the government:

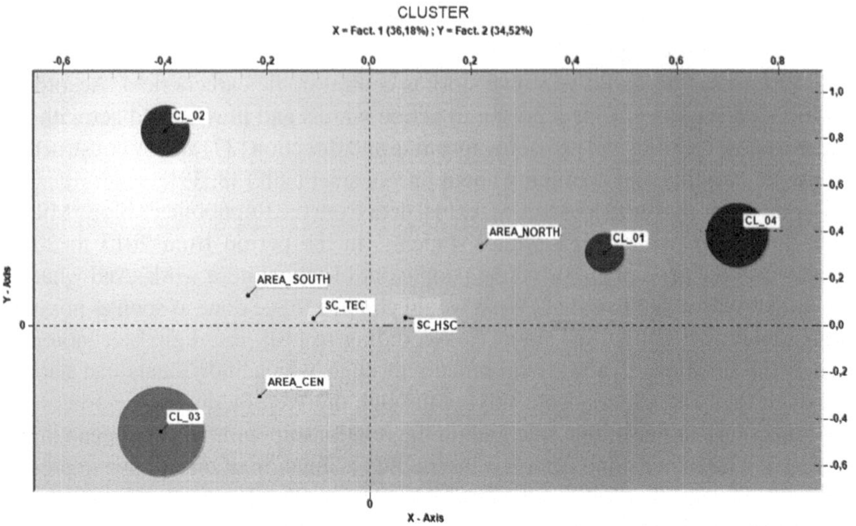

Fig. 1 Cluster analysis of Young' narrations

> I think the state should give more space to young people by ensuring them protection and safeguard. Parliamentarians must preserve the rights and possibilities of every young person, as we are the future of this state and, as such, we need opportunities.

The cluster "From dreams to crisis" refers to the more internal dimension of being immersed in a society that is going through a time of economic crisis. Students point out that the lack of work nullifies dreams:

> I am really worried, we all dream about what to do when we grow up and knowing that 38.7% of young people can't find work makes me outraged. Young people are the future, the progress, they commit [...] We all know what the first article of our splendid constitution says, yet it seems to be left ignored. We need to give more opportunities to young people, and we need to take our constitution into consideration, to open the doors to the future and make Italy a better place.

The narratives of the last cluster relate transversally to all the difficulties of looking for work (the breathless research, the companies that do not hire because of too many taxes) and the need to go abroad:

> Italy finds itself in a period of profound crisis and if it does not recover economically by giving back to us, young people, the chance to make it clear to those who are responsible that we have the ability and willingness to work, Italy will lose all these young workers but above all, those minds that will go abroad in search of more favorable living conditions and specifically more job opportunities.

The position of the descriptive variables shows a difference for the territory's origin variable and no difference for the institution attended one. If, in fact, attending a school rather than another does not seem to affect the perception of the work world and the feelings of distrust, which are instead common, territorial origin has its weight. The northern areas mode is, in terms of proximity, located near clusters 1 and 4, the center to number 3, and the south to cluster 2. This indicates that students in the northern areas tend more to problematize the phenomenon of precariousness and the difficulties linked to the searching for work, with an emphasis on the opportunities young people have to demonstrate their worth; the themes of those in the southern areas tend to blame the context, in line with a greater resonance of the topic of discussion due to a high incidence of youth unemployment; the narrations of the students of the center, instead, mostly recall their own internal experiences.

As we said, SNA' tools were used to better understand the meanings that flow in the narrations. The network obtained presents 259 nodes and 414 arcs. A first in-depth visualization form of the relations structure between the various lemmas shows the highest values of degree centrality in which "job," "young," "future," "problem" are the nodes with more connections.[1] These nodes indicate, therefore, words used with greater frequency in relation to other words in the narrations and the various contexts of meaning. In more detail, the incidence for each node can be expressed both as in-degree, that is, the number of edges incoming in a node (identifying in this way the so-called predecessors of each lexical unit) and as out-degree, that is, the number of edges outgoing from the node (showing in this case the so-called successors). In this case, the nodes with higher values of degree centrality show also the highest in-degree centrality: Furthermore, they are "well" nodes, presenting more incoming edges than outgoing edges compared to all the other nodes. In this view, students tend to channel their speeches and, more in general, the flow of thoughts toward job-related issues, focusing both on future possibilities and current problematics (Fig. 2).

On the other hand, "make an effort" and "conditions" are *source* nodes, those lemmas with the highest out-degree centrality: In other words, conditions as regards the employment opportunities and the need to use energies, in trying to make an effort to represent the fulcrum from which narration shifts toward other words (Fig. 3).

Lemmas which refer to students' experiences, their feelings about current condition and job insecurity scenario play a central role in the circulation of flows of meanings in the network, showing the highest values of betweenness centrality. In particular, "unemployed", "obligate", "stay" (intended as remain in own your city or country) and "discourage" are those words with the greater semantic weight that enable the link between network sets: Different parts of the narrations are connected by those lemmas that refer to job insecurity, a forced and daunting situation (Fig. 4).

5 Conclusions

From methodological point of view, the mixed use of TA and NTA allows us to represent a synthetic picture of the semantic structure, to understand what we are

[1]In Figs. 2, 3, and 4, centrality measures are represented by those nodes and labels with greater size.

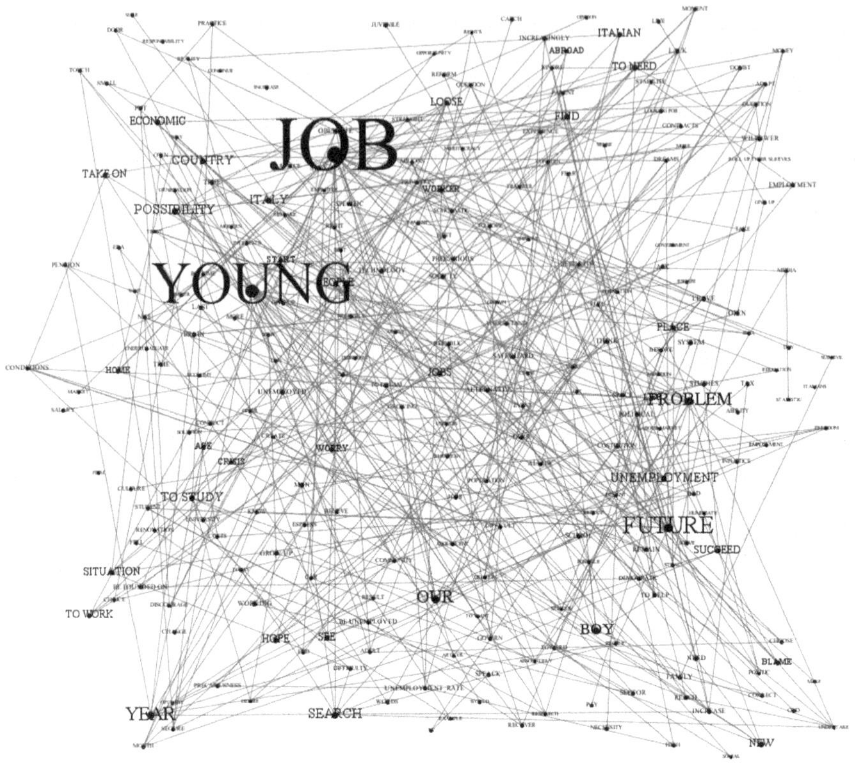

Fig. 2 Network of lemmas (size and label of nodes according to values of in-degree centrality)

talking about, but also how we do it: The choice of words and the order of presentation of an idea or opinion in relation to the subject in question.

The joint use of the two techniques provides: (a) A summary of the information contained in the narratives; (b) analysis of the topics addressed; (c) a focus on structuring sentences in terms of relationships between headwords. In this way, it permits us to relate thematic and content categories as a latent structure, reconstructing the discursive process in reverse.

Moreover, the analyzes also show the potential of the narrative. In fact, in a psychological point of view, the narrative fosters a mediation between subject and experience, sets space between the self and the violence of emotions. Therefore, the blogs act as a tool leading toward awareness of personal distress [42].

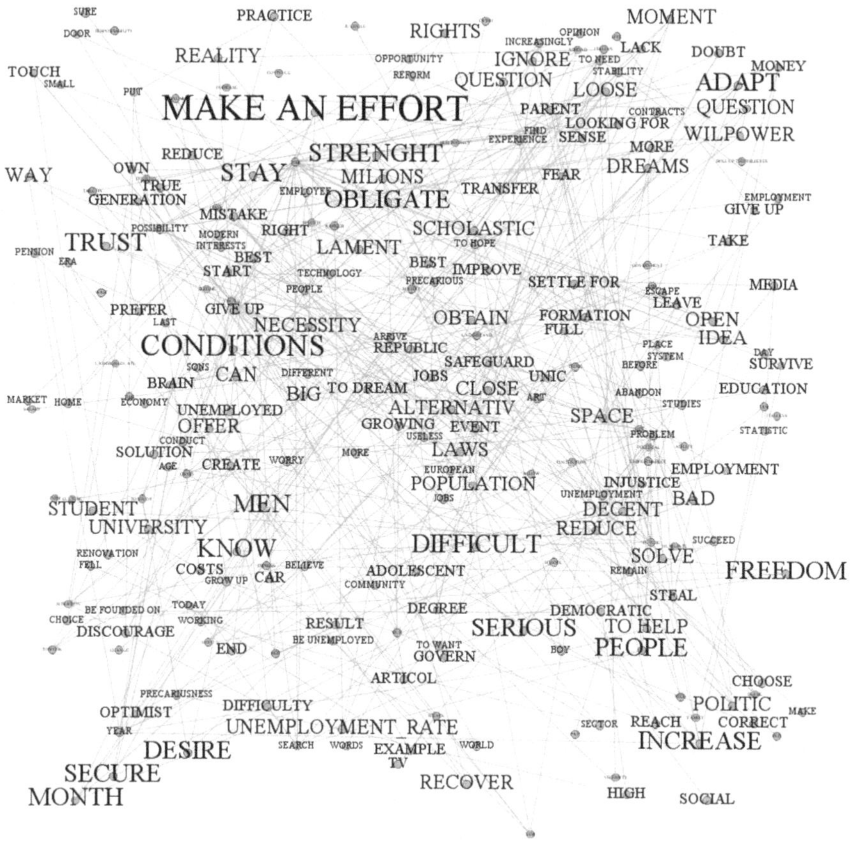

Fig. 3 Network of lemmas (size and label of nodes according to values of out-degree centrality)

Nevertheless, the study is not free from limitations. Youth has voluntarily responded to an online inquiry and the narrative products could be affected by social desirability. Furthermore, contents could be influenced by the previous narrations left by other users.

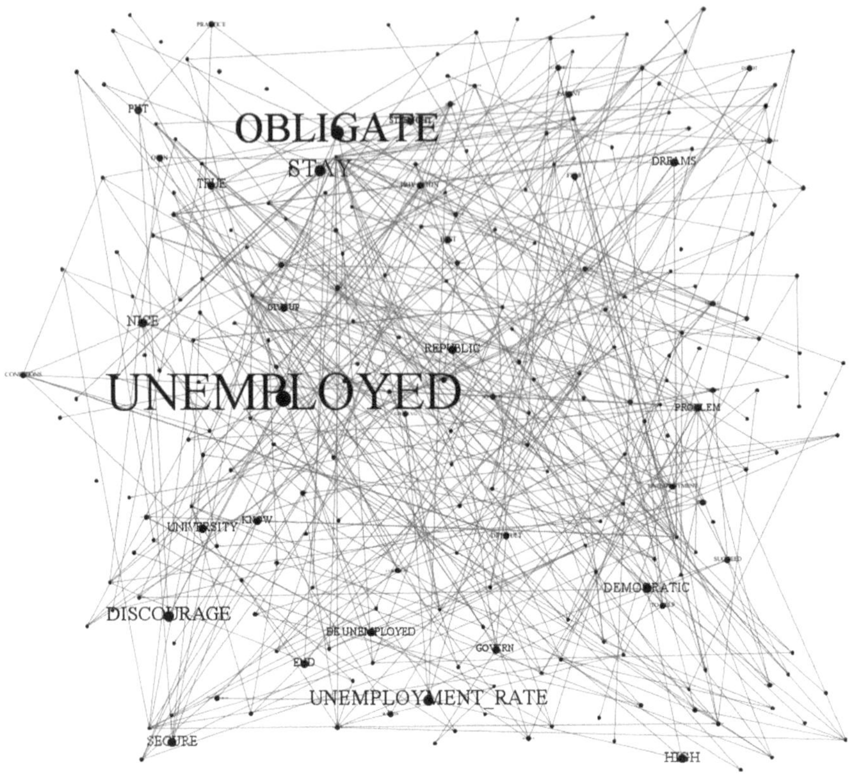

Fig. 4 Network of lemmas (size and label of nodes according to values of betweenness centrality)

References

1. Lupton D (2014) Digital sociology. Routledge, London
2. Catone MC, Diana P (2019) Narrative in social research: between tradition and innovation. Ital J Sociol Educ 11(2):14–33
3. Bolasco S (2005) Statistica testuale e text mining: alcuni paradigmi applicativi. Quaderni di Statistica 7:1–37
4. Lebart L, Salem A, Berry L (1998) Exploring textual data. Kluwer Academic Publishers, Dordrecht
5. Bruce D, Newman D (1978) Interacting plans. Cogn Sci 2(3):195–233
6. Lehnert W (1981) Plot units and narrative summarization. Cogn Sci 5(4):293–331
7. Dyer G (1983) The role of affect in narratives. Cogn Sci 7:211–242
8. Carley KM (1993) Coding choices for textual analysis: a comparison of content analysis and map analysis. Sociol Methodol 23:75–126
9. Popping R (2000) Computer-assisted text analysis. Sage, London
10. Carley KM (1997) Extracting team mental models through textual analysis. J Organ Behav 18(1):533–558
11. Popping R (2003) Knowledge graphs and network text analysis. Soc Sci Inf 42(1):91–106
12. Diesner J, Carley KM (2005) Revealing social structure from texts: meta-matrix text analysis as a novel method for network text analysis. In: Narayanan VK, Armstrong DJ (eds) Causal

mapping for information systems and technology research. Idea Group Publishing, Harrisburg, PA, USA
13. Diesner J, Carley KM (2008) Conditional random fields for entity extraction and ontological text coding. Comput and Math Organ Theory 14:248–262
14. Popping R, Roberts CW (1997) Network approaches in text analysis. In: Klar R, Opitz O (eds) Classification and knowledge organization. Springer, Berlin, New York
15. Paranyushkin D (2011) Identifying the pathways for meaning circulation using text network analysis. Venture Fiction Pract 2(4):2–26
16. Hunter S (2014) A novel method of network text analysis. Open J Mod Linguist 4(2):350–366
17. Rogers R (2013) Digital methods. MIT press, Cambridge
18. Lebart L, Salem A (1994) Statistique Textuelle. Dunod, Paris
19. Scott J (1991) Social network analysis: a handbook. Sage, London
20. Wasserman S, Faust K (1994) Social network analysis: Methods and applications. Cambridge University Press, New York
21. Borgatti S, Everett M, Johnson J (2013) Analyzing social networks. Sage, London
22. Lancia F (2004) Strumenti per l'analisi dei testi. Franco Angeli, Milano
23. Freeman LC (1979) Centrality in social networks conceptual clarification. Social Networks 1:215–239
24. Eurostat (2019) Statistics on young people neither in employment nor in education or training. Report.
25. Leccardi C (2006) Redefining the future: youthful biographical constructions in the 21st century. New Dir Child Adolesc Dev 113:37–48
26. Berton F, Richiardi M, Sacchi S (2009) Flex-insecurity: perché in Italia la flessibilità diventa precarietà. Bologna, Il Mulino.
27. Boeri T, Galasso V (2007) Contro i giovani. Milano, Mondadori
28. Iezzi M, Mastrobuoni T (2010) Gioventù sprecata. Perchè in Italia si fatica a diventare grande. Editori Laterza, Bari
29. Aleni SL, Sica LS, Ragozini G, Porfeli E, Weisblat G, Di Palma T (2015) Vocational and overall identity: a person-centered approach in Italian university students. J Vocat Behav 91:157–169
30. Parola A, Donsì L (2018) Sospesi nel tempo. Inattività e malessere percepito in giovani adulti NEET. Psicologia della Salute 3:44–73
31. Santilli S, Marcionett J, Rochat S, Rossier J, Nota L (2017) Career adaptability, hope, optimism, and life satisfaction in Italian and Swiss adolescents. J Career Dev 44(1):62–76
32. Vuolo M, Staff J, Mortimer JT (2012) Weathering the great recession: psychological and behavioral trajectories in the transition from school to work. Dev Psychol 48(6):1759
33. Baltes PB, Lindenberger U, Staudinger UM (1998) Life-span theory in developmental psychology. In Damon W, Lerner RM (eds) Handbook of child psychology: Vol. 1. Theoretical models of human development, 5th edn. Wiley, Hoboken, NJ, pp. 1029–1143
34. Bynner J, Parsons S (2002) Social exclusion and the transition from school to work: the case of young people not in Education, employment or training. J Vocat Behav 60:289–309
35. Duarte ME (2004) O indivíduo e a organização: Perspectivas de desenvolvimento. [The individual and the organization: Perspectives of development]. Psychologica (Extra-Série), 549–557.
36. Meijers F, Lengelle R (2012) Narratives at work: the development of a career identity. Br J Guid Couns 40:157–177
37. Wijers G, Meijers F (1996) Career guidance in the knowledge society. Br J Guid Couns 24(2):185–198
38. Linde C (1993) Life stories. The creation of coherence. Oxford University Press, New York
39. Savickas ML (2011) Career counselling. American Psychological Association, Washington, DC

40. Marres N (2017) Digital sociology: the reinvention of social research. Polity Press, Cambridge
41. Di Fraia G (ed) (2007) Il fenomeno blog. Blog-grafie: identità narrative in rete. Guerini e Associati, Milano
42. Margherita G, Gargiulo A (2018) A comparison between pro-anorexia and non-suicidal self-injury blogs: from symptom-based identity to sharing of emotions. Psychodynamic Practice 24(4):346–363

Key Passages : From Statistics to Deep Learning

Laurent Vanni, Marco Corneli, Dominique Longrée, Damon Mayaffre, and Frédéric Precioso

Abstract This contribution compares statistical analysis and deep learning approaches to textual data. The extraction of *key passages* using statistics and deep learning is implemented using the Hyperbase software. An evaluation of the underlying calculations is given by using examples from two different languages—French and Latin. Our hypothesis is that deep learning is not only sensitive to word frequency but also to more complex phenomena containing linguistic features that pose problems for statistical approaches. These linguistic patterns, also known as *motives* Mellet and Longrée (Belg J Linguist 23:161–173, 2009 [9]), are essential for highlighting key passages. If confirmed, this hypothesis would provide us with a better understanding of the deep learning *black box*. Moreover, it would bring new ways of understanding and interpreting texts. Thus, this paper introduces a novel approach to explore the hidden layers of a convolutional neural network, trying to explain which are the relevant linguistic features used by the network to perform the classification task. This explanation attempt is the major contribution of this work. Finally, in order to show the potential of our deep learning approach, when testing it on the two corpora (French and Latin), we compare the obtained linguistic features with those highlighted by a standard text mining technique (z-score computing).

L. Vanni (✉) · D. Mayaffre
Université Côte d'Azur, CNRS, BCL (UMR 7320), Nice, France
e-mail: laurent.vanni@univ-cotedazur.fr

D. Mayaffre
e-mail: damon.mayaffre@univ-cotedazur.fr

M. Corneli
Université Côte d'Azur, Center of Modeling, Simulation & Interaction, Nice, France
e-mail: Marco.Corneli@univ-cotedazur.fr

D. Longrée
Université de Liège - L.A.S.L.A., Liège, Belgium
e-mail: dominique.longree@uliege.be

F. Precioso
Université Côte d'Azur, CNRS, I3S (UMR 7271), Nice, France
e-mail: Frederic.Precioso@univ-cotedazur.fr

D. F. Iezzi et al. (eds.), *Text Analytics*, Studies in Classification, Data Analysis, and Knowledge Organization,
https://doi.org/10.1007/978-3-030-52680-1_4

1 Introduction

Due mainly to technical reasons, until the 1960s textual analysis developed around the "token," which in computer science denotes the string between two white spaces.[1] Since then, practitioners have continued to widen the set of observable features, believing that the complexity of a text cannot be reflected by tokens alone. Thus, although the tokenization, along with computing word specificity,[2] is still the preliminary steps in the statistical analysis of textual data, research into wider, and more complex phraseological units characterizing the structure of texts has now become the focus of the discipline. Indeed, since 1987, the calculation of the *repeated segment* [16] or n-grams has marked a breakthrough, since significant segments of the text, with no predetermined size, were automatically detected. Further, the automatic and unsupervised detection of *patterns*[3] [7, 9, 10, 13] is a key issue. From this perspective, the present contribution focuses on the notion of *key passages*, which we define in Sect. 2.1 of the text, as they are implemented in the Hyperbase[4] software. The outlined approach consists of two steps. First, we propose a statistical extraction of key features, with an evaluation of their interpretative relevance on two corpora, a French and a Latin one. Then, a methodological comparison with a deep learning approach is implemented. Thus, by using the Text Deconvolution Saliency (TDS) method [17], we identify within each corpus areas of activation (Sect. 3.2.2) and observe that, from a linguistic point of view, they constitute relevant *patterns*.

The paper is organized as follows. Section 2 introduces the notion of key passage and reviews a standard statistical approach. Section 3 outlines an original deep learning approach for key passage detection and illustrates how TDS can be used to highlight some key *patterns* in the text. Section 4 concludes the paper.

2 Key Passages and Statistics

2.1 Definition

In the following, the word "passage" will be used to denote any text portion *not* necessarily being a sentence of a paragraph. Indeed, the approaches outlined in this paper do not have a syntactic model. Thus, strong punctuation delimiting a sentence, although useful, is not needed for our processing. As theorized by [14]

[1]This contribution has been founded by the French government, Agence Nationale de la Recherche, project Investissement d'Avenir UCA^{JEDI} n° ANR-15-IDEX-01.

[2]Also known as z-score, specificity in textual data analysis since [5] is based on hypergeometric distribution.

[3]Also known as motives: complex linguistic objects with variable and discontinuous spans.

[4]The software is used by the UMR Bases, Corpus, Language and was developed in collaboration with the LASLA. A first local version was coded by Etienne Brunet. A recent web development was realized by Laurent Vanni http://hyperbase.unice.fr/.

in an eponymous article, the notion of *passage* refers to the interpretability of a part of a text. A passage is therefore a piece of text judged meaningful enough to characterize the whole text, independently of its size (one word, one phrase, etc.). Based on this idea, we can formally define a key passage as a text portion containing a high number of specificities [5]. Textometric software such as Hyperbase, Dtm-Vic, Iramuteq implement key passage extraction. The more specificities a passage contains, the more remarkable it is considered. The next section illustrates the given definition with two examples.

2.2 Implementations

2.2.1 Unfiltered Processing

Let us denote the ith text portion (henceforth text) by y_i, that can be seen as a vector of length N_i, where N_i is the number of words in that text. The corpus considered for the experiments in Sect. 3.3 is constituted by texts sharing the same length, therefore $N_i = N$ for all i. The jth entry of y_j, namely $y_{ij} \in \mathbb{N}$, identifies the jth token (of the ith text) via a natural number, corresponding to the ID of the token in a dictionary containing V vocables. The dictionary is obtained by collecting all the different vocables appearing in the corpus. So, for instance, if the first token of the fifth passage is "nation" and the vocable "nation" is the 25th word of the dictionary, then $y_{51} = 25$. In a standard statistical analysis, when inspecting a text, the specificity indexes of words can be summed in order to obtain a score characterizing the whole text. Moreover, each word has several specificity indexes, corresponding (for instance) to different authors. More formally, let us denote by K the number of different authors. A real matrix Z with K rows and V columns is introduced such that Z_{kv} is the specificity index of the vth vocable with respect to the kth author. Roughly speaking, Z_{kv} measures (in terms of numbers of standard deviations) how much the word v is overused or underused by the kth author with respect to the average in the whole corpus. Then, a score S_{ik} can be computed for y_i with respect to the kth author as

$$S_{ik} = \sum_{j=1}^{N} Z_{k y_{ij}}. \tag{1}$$

Note that the specificity indexes in Z can be either positive or negative. A positive index corresponds to a word v that is overused by an author. Thus, that word contributes toward the identification of a passage as being representative of that author. Conversely, a negative index marks a word as underused by the author. Thus, negative indexes contribute to a "non-identification" of the passage. Once the scores S_{ik} have been assigned to all texts for all the authors, Hyperbase sorts the texts from the more representative (highest S_{ik}) to the least representative (lowest S_{ik}) for each author. In the unfiltered version, the scores S_{ik} are computed as in Eq. (1).

2.2.2 Filtered Processing

With our linguistic and statistical backgrounds, the approach described in the previous section can be refined and improved. For instance, only positive specificities—and among them, the strongest ones—might be considered, based on the assumption that an object is better identified by its strengths than by its weaknesses. Thus, modified scores S_{ik} can be computed as

$$S_{ik} = \sum_{j=1}^{N} Z_{ky_{ij}} \mathbf{1}_{[c,+\infty[}(Z_{ky_{ij}}), \tag{2}$$

where c is a positive constant, corresponding to the desired threshold and $\mathbf{1}_A(\cdot)$ denotes the indicator function on a set A.

Additional filters might be considered. For instance, the tool words (conjunctions, determinants) could be discarded. Indeed, they bring the double disadvantage of having very high frequencies (potentially crucial when computing the specificities) and of not being very meaningful, from a semantic/thematic point of view. Finally, the grammatical category can be considered. For example, using only proper and common names can be more efficient than using the entire set of words.

2.3 *Examples*

Key passage detection via the Hyperbase software is performed on two different data sets. The first contains the speeches of the French presidents of the 5th republic and the second is a Latin corpus that regroups texts from the principle classical authors. See Table 1 for more details.

2.3.1 French Corpus

With a view to characterizing contemporary French political discourse, the statistical extraction of Emmanuel Macron's key passages (versus his predecessors at the Elysée: de Gaulle, Pompidou, Giscard, Mitterrand, Chirac, Sarkozy, and Holland) represents obvious added value for the linguist and historian. In a corpus of several tens of thousands of passages, the machine selects the following extract:

Table 1 French and Latin data sets

Data set	Authors	Vocab	Occurrences
French	8	46,978	2,738,652
Latin	22	56,242	2,035,176

> [...] l'engagement de l'Europe pour réussir justement cette aventure de l'intelligence artificielle. Donc, vous l'avez compris, je souhaite que la France soit l'un des leaders de cette intelligence artificielle, je souhaite que l'Europe soit l'un des leaders de cette intelligence artificielle [...]

(Macron, the 29th March 2018, speech "AI for Humanity" at the collège de France)

Everything in this passage is typical of Macron's speeches. From an enunciative point of view, *je / vous* is a persuasive rhetorical device where the polite but repeated affirmation of the president's personality (*je souhaite*, *je souhaite*) versus people in a position of inferiority (*vous avez compris*). In France, this politico-enunciative posture marked by the over-use of *je* and of *vous* has been described by commentators as a "Jupiterian" or dominant posture. Especially from a lexical point of view, two major themes of macronism seem to be highlighted by the machine: Europe on the one hand, and post-industrial modernity on the other. Europe has been a theme throughout the French presidential corpus since 1958. But Macron, more than anyone else has used the word *Europe* with such recurrence (Fig. 1).

And in this excerpt, as is often the case in Macron's speech, if the President still speaks of *France*, it is indeed *Europe*, twice repeated, that is the subject of the speech. Post-industrial productionism, referred to here as *intelligence artificielle*, represents another facet of Macron's fundamental message since he was elected President: digital modernity, digital sciences, and artificial intelligence represent the economic future. And the terms *réussir* or *leaders* end up characterizing this modern discourse on the *start-up nation*.

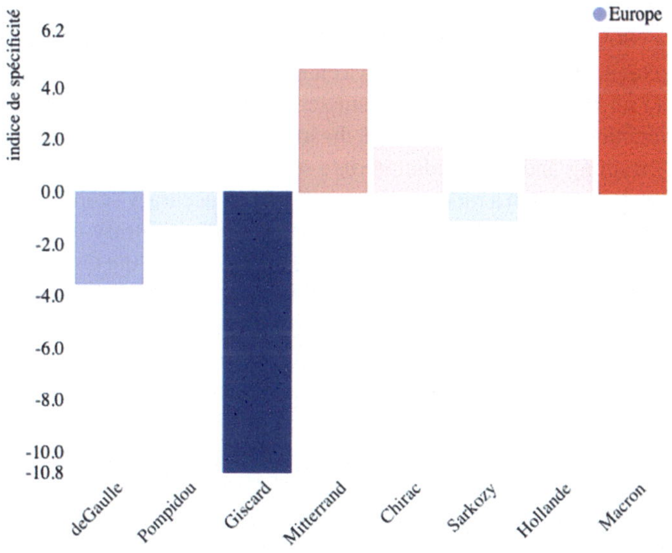

Fig. 1 Distribution of word "Europe" in the french presidential corpus

2.3.2 Latin Corpus

Focusing on the Latin corpus, we consider the following key passage from Julius Caesar, in contrast with many authors contained in the LASLA database:

> [...] partes Galliae uenire audere quas Caesar possideret neque exercitum sine magno commeatu atque molimento in unum locum contrahereposse sibi autem mirum uideri quid in sua Gallia quam bello uicisset aut Caesari aut omnino populo Romano negotii esset his responsis ad Caesarem relatis iterum ad eum Caesar [...]

(César, Bellum Gallicum, 1, 34)

This passage from the *Gallic Wars* can indeed be considered as very representative of Caesar's work. There are known proper names (*Galliae*, *Caesar*, *Gallia*) or common names reflecting the military context of the time (*bello*, *commeatu*). However, the method does not allow the identification of structures characteristic of Caesar's language and style, such as a specific participial clause marking the transition between episodes in a negotiation: *His responsis ad Caesarem relatis*, "These answers having been reported to Caesar."

3 Key Passages and Deep Learning

3.1 Most Relevant Passages Detected by the Neural Network

In machine learning, the observed data sets are usually split into three parts: a *training* data set (e.g., 80% of all the data), a *validation* data set (e.g., 10% of all the data) and a *test* data set (e.g., 10% of all the data). In a classification framework, for instance, the training data set is used to learn the parameters of a classifier, via minimization of a given loss function. During optimization, the loss function is progressively monitored on both the training and the validation data sets, in order to avoid *overfitting*. This phenomenon occurs when a model obtains a very high accuracy score on the training data set but cannot generalize to the rest of the data. On the contrary, a smart classifier is asked to work reasonably well on new, unseen data sets like the test data set. Thus, a reasonable goal is to reach the highest possible accuracy scores on the *test* data set. However, for our purpose, which is the detection of the key passages, the description of the training data set is at least as crucial as the performance on the test data set. We stress that we are interested in understanding how the network accomplishes the classification task. We are now in a position to take a first step in that direction based on the following idea. The number of neurons in the last layer of a deep neural network generally coincides with the number of classes, previously denoted by K. Let us denote by[5] $x_i \in \mathbb{R}^K$ the value of that layer for the ith text. Thus, x_{ik} is the value of the kth neuron and it is a real number. A softmax activation function is usually applied to x_i in such a way to obtain K probabilities p_{ik} lying in the $K-1$ simplex

[5] In general, $x \in \mathbb{R}^N$ can be read as "a vector of N real components."

$$p_{ik} = \frac{\exp(x_{ik})}{\sum_{j=1}^{K} \exp(x_{ij})}. \tag{3}$$

The highest probability corresponds to the class the observation is assigned to by the network. However, if one entry of x_i is significantly higher than the others, it is mapped to 1 by the softmax transformation and all the other entries are mapped to zero. Assuming that two layers $x_i \neq x_j$ (corresponding to two different texts) both have the kth component sufficiently high, after applying the softmax transformation it will be impossible to know which of them had the highest activation rate toward cluster k. Thus, we make unconventional use of the trained deep neural network and observe the activation rate of neurons *before* applying the softmax transformation. Doing that allows us to sort the learning data (texts) based on their activation strengths. Eventually, only the texts which most strongly activate their classes are retained. This simple but efficient method provides us with the most relevant passages in the corpus for each class. Those passages will be referred to as "deep learning key passages" in what follows.

3.2 *Learning Linguistic Features*

Once the deep learning key passages have been collected, we still need to consider the reason why the neural network treated them as most relevant. What are the linguistic markers learned by the network that contribute to the classification? How to display them? Whereas in standard statistical analysis those markers are the word specificities and they are computed before the classification, in deep learning the textual features are learned during the training process. In order to visualize this type of linguistic material, we must analyze the deep neural network's hidden layers.

There are several types of neural networks that seem to be sensitive to different markers in the text. In this work, we mainly focus on convolutional neural networks (CNNs), usually employed for image classification. Convolution is a powerful tool, invariant with respect to translation and scale, allowing the CNN to detect key features in images. See [4] for more details. In order to analyze text data with CNNs, fixed-length input sequences are needed [2]. Therefore, the whole corpus is split into texts of 50 tokens each.[6] Then, an embedding on the word-level is adopted to convert words into real vectors of fixed size, being a network hyper-parameter. Several embedding techniques are available, such as Word2Vec [11], GloVe [12], or FastText [1, 3]. Note that working with textual features the size of words rather than the size of characters is more practical for linguists and leads to results which are easier to interpret. The word vectors are stacked by row in an embedding matrix whose entries are considered parameters to be learned. In our model, the convolutional layer is the most important one and it is expected to capture most of the linguistic features of the training data set. The next section describes this layer in detail.

[6] The number of tokens for each text is one of several hyper-parameters of the network.

3.2.1 Convolution

The convolution applied to the text uses filter mechanisms responsible for analyzing each word in its context to extract a salience score for each analyzed context. It is then the combination of all these features that pushes the network to make its decision. This combination can be as simple as a sum of contextual specificities (colocations or repeated segments) or as complex as detecting all the specific cooccurrences or even poly-occurrences of expressions [18].

As we can see with Fig. 2, the complete network is composed of three parts: embedding, convolution, and classification. The last one is a dense network whose last layer is composed of a number of neurons equal to the number of classes to be detected (8 for French and 22 for Latin).

3.2.2 Deconvolution

The idea of deconvolution as presented in [17] is to propose the opposite path. Taking the abstraction made by convolution (the feature map), and then returning to a vector space similar to the input embedding in order to be able to read the linguistic highlights learned by the network. It is a convolution transposed on all the filters that allow this operation to be performed (Fig. 3). Thus, we obtain a salience score for each word that represents the weight of each of the linguistic elements in the final deep learning decision. This score is the Text Deconvolution Saliency (TDS).

This is where the notion of passage takes on its full meaning. Each of the words is assigned a certain weight by convolution, but it is essential to understand that this weight is entirely dependent on the context of these words. The convolution filters all have a size that defines the window in which each word will be considered.

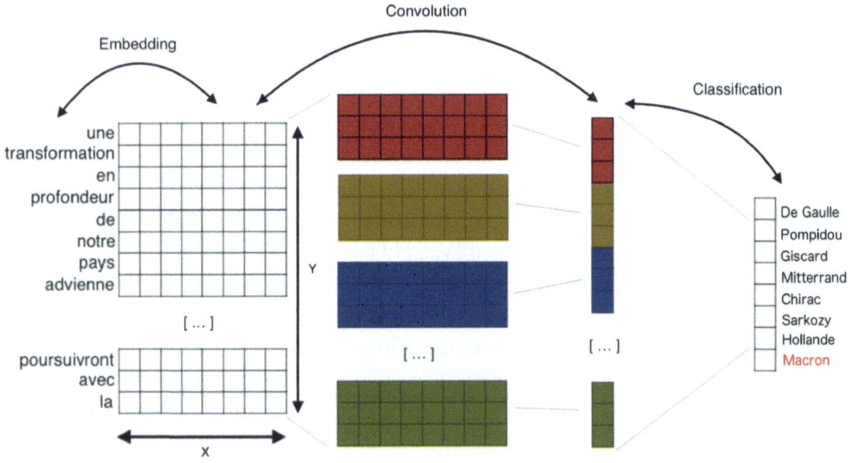

Fig. 2 Convolutional neural network (CNN)

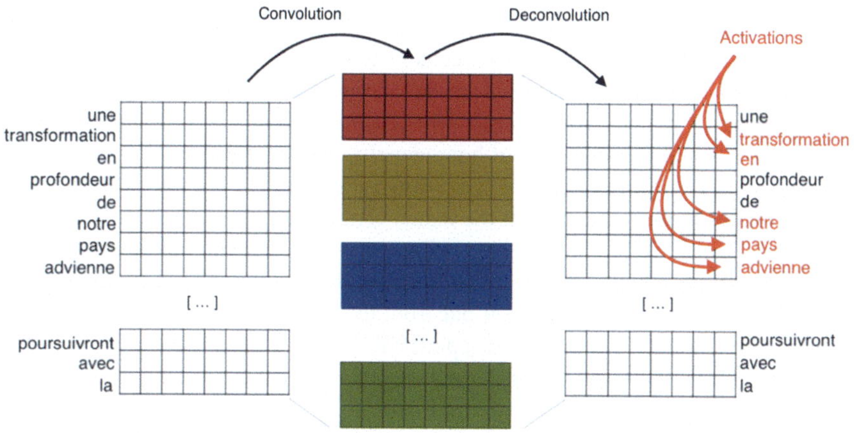

Fig. 3 Deconvolutional neural network (DNN)

This weight can therefore differ greatly from one context to another. We can well imagine here that the network might be sensitive to any type of micro-distributional markers (colocation, repeated segments, phrases, etc.). But in deep learning, the network must be considered as a whole and therefore the layers that follow convolution (MaxPooling and Dense) can also largely affect the network's decision. Thus, each linguistic marker is itself considered in the context of the entire word sequence (cooccurrence and poly-cooccurrence of expressions). One last notion that is important to emphasize is that the repetition of examples and the size of the training corpus necessarily pushes the network to also be sensitive to frequency type markers, such as specificities for example.

We can see here that deep learning potentially plays on several levels at the same time to learn and detect strong linguistic markers characteristic of each author. There is no a priori on the object being studied; the word or pattern, the micro-distribution or macro-distribution. These methods can therefore potentially detect all types of linguistic patterns described by [9]. Unfortunately, deconvolution does not take into account the rest of the network, so we still lack some elements for automatically observing such patterns. But what we do know now is that each strong activation in a sequence of words visibly points to a prominent passage in the text. These very Rastierian passages can be simple specific words or longer segments with cooccurrences or poly-cooccurrences. This method offers us new linguistic markers that scholars can use today to complement statistical analysis of textual data to provide new avenues to explore in the course of their research.

3.2.3 Filtered Processing

As with the statistical approach, our linguistic and statistics background allows us to add certain filters prior to the deep learning analysis to avoid all kinds of bias in the training corpus and thus refine the textual features detected by the model. For example in the corpus of French presidents, one-letter words and cardinals have been removed. Indeed, this corpus is strongly constrained by chronology. The dates and the changing punctuation strongly mark each president's speech and have to be removed to maintain a maximum of homogeneity in the corpus.

Some words or expressions are also strongly guided by chronology. For example it is totally implausible to find the name "Trump" in the C. De Gaulle speech. F. Hollande and E. Macron are the only French presidents who talk about the current American president. This information could be considered by deep learning as sufficient to characterize a president but from the point of view of linguistics, it is definitely not enough to characterize a political speech. The analysis of word distribution therefore allows us to remove those unique words or expressions which could mask the true key passages of a text.

The following examples therefore use this filter mechanism to observe the most relevant text features that characterize a text.

3.3 Examples

3.3.1 French Data Set

In the presidential corpus, the passage that the AI identifies as characteristic of Macron's speech gathers remarkable features of the current French president's language, and the deconvolution highlights observable linguistic elements with multiple interpretations:

> [...] entreprises de manière légitime, qui permettra aux acteurs europ éens d' émerger dans un march é loyal et qui permettra aussi de compenser les profondes d ésorganisations sur l' économie traditionnelle que cette transformation parfois cr ée. Les grandes plateformes num ériques, la protection des donn ées sont au cœur de notre souverainet é [...]

(Macron, the 26th of September 2017, speech about Europe at the Sorbonne)

One of the most interesting linguistic features is the relative pronoun *qui* which is a typical element in Macron's construction of phrases. With this relative pronoun, which drives the speech, the complexity of the phrase is brought to the forefront. The second characteristic revealed is the sequence of two verbs in the future (*permettra* [...] *permettra*) and one verb in the present tense (*sont*): here too it is a form of promise, most often combined with the future, and a pragmatism declined in the present tense that is brought to light. Last but not least, a characteristic lexicon is constructed to form an ambitious, dynamic and entrepreneurial discourse that distinguishes Macron's discourse from that of a De Gaulle, a Pompidou or a Mitterrand: *entreprises*, *acteurs*, *europ éens*, *March é*, *transformation*, *cr ée*, *num ériques*, *donn ées*.

3.3.2 Latin Data Set

Deep learning can also be easily adapted to the subtleties of each language. This is what we will try to demonstrate here by using Latin and the variation that is specific to it. In the next example, we focus on the following deep learning key passage from Julius Caesar (the one of most relevant according to the AI).

> [...] ut tum accidit . munitionem quam pertinere a castris ad flumen supra demonstrauimus dextri Caesaris cornus cohortes ignorantia loci sunt secutae cum portam quaererent castrorum que eam munitionem esse arbitrarentur . quod cum esset animaduersum coniunctam esse flumini prorutis munitionibus defendente nullo transcenderunt omnis que noster equitatus eas cohortes est [...]

(C ésar, Belum ciuile, 3, 68)

In this excerpt, deep learning recognized an occurrence of a "textual motif" whose pattern has been clearly identified before [conjonction ut or relative pronoun + supra/antea + 1ers person of the verbs demonstro/dico/memoro] [7, 8]. Simple variants of the motif as *ut supra demonstraui* ("as I have explained above") or *quem antea memorauimus* ("whom we mentionned before") were quite easily identified by using standard data mining methods. However, such methods are unable to retrieve a variant of the motif as complex as the one found in the excerpt, *munitionem, quam pertinere a castris ad flumen supra demonstrauimus* ("the entrenchment, which, as we have explained above, extended from the camp to the river") where the relative pronoun is not the complement of the verb as in *quem antea memorauimus*, but the subject of an infinitive clause depending on the verb *demonstrauimus*. Once again, deep learning has shown its efficiency in identifying texts, but also in mining multidimensional structures. In the same passage, deep learning detects another structure which can also be considered as textual motif, *Quod cum esset animaduersum coniunctam esse flumini, prorutis munitionibus defendente nullo* ("And as it had been observed that it was connected with the river, the defenses having been thrown down, no one defending it"). This syntactic motif consists in a *cum* clause followed by two absolute ablatives [9]. Moreover, this syntactic p includes an occurrence of the multidimensional motif *quod ubi cognitum est* ("when this has been known") made of the coordinating relative *quod* followed by a subordinative conjunction and a passive verb form [8]. As [7] defines the "textual motif" as a multidimensional unit fulfilling characterizing and structuring textual functions, it is not very surprising that deep learning confirms the saliency of such patterns for text identification.

4 Conclusion

4.1 Discussion

Standard statistical analysis and deep learning are not fully disconnected fields. Indeed, the word frequency-based approach, typical of the standard statistical techniques, is also used by deep learning approaches [6], sensitive to the repetition of certain phenomena during the learning phase. Moreover, the sequential approach typical of deep learning architectures is also used by statistical analyses sensitive to the contextualization of terms. See, for instance, the analysis of repeated segments [16] or motifs [7].

Combining purely statistical approaches with artificial neural networks, this contribution allows us to identify key passages (either of a text or of an author) and linguistic features bringing us toward a deeper understanding of texts. It introduces a novel approach to explore the hidden layers of a convolutional neural network, trying to explain which are the relevant linguistic features used by the network to perform the classification task. While the features which allow the standard statistical text analysis to detect key passages are well known (the word specificity), the areas of activation in deep learning seem to uncover new linguistic units that the traditional corpus semantics struggled to show.

The statistical text analysis can be qualified as paradigmatic: The task is to tally the discrete elements, identified as being representative, over a given text portion (a sentence, a paragraph, en extract of *n* words). Some filters, either linguistic (for instance, only taking nouns into account) or statistical (for instance, only considering words with a positive z-score) can be applied to the discrete elements, in order to optimize the analysis. Note that the pertinence of these filters changes with the different corpora (different languages, different genres, specific speakers).

As for the deep learning approach which may be qualified as syntagmatic: Through the use of sliding windows, the goal is to uncover the areas in the texts that, in a particular context, allowed the neural network to distinguish one author from another or one text from another in a given corpus.

In social sciences, these areas of activation seem to fulfill the essential hermeneutics of the corpus semantics carried out by the machine [14, 15] Based on the material substance of a text—since the machine needs tokens—for the linguist, the areas of activation are the key to a deeper interpretation of texts.

4.2 Perspectives

Given that areas of activation detected by deep neural networks bring a dramatic improvement to our interpretation capabilities, this contribution needs to be extended in two directions. First, from a linguistic perspective, lemmatization, and part-of-speech tagging should be implemented. Doing that would allow for a straightforward description of the text, not only accounting for a lexical level, but also for a grammatical and syntactical level.

From a methodological perspective, in order to extract not only discriminative spatial linguistic markers (using CNNs), recurrent layers (RNN) might be introduced in the neural network. Those layers take into account the temporal dimension of texts and produce a dynamic view. Future works might focus on the combination of both approaches to extract spatial linguistic patterns with CNNs and be able to analyze their sequential organization with RNNs.

References

1. Bojanowski P, Grave E, Joulin A, Mikolov T (2017) Enriching word vectors with subword information. Trans Assoc Comput Linguist 5:135–146
2. Ducoffe M, Precioso F, Arthur A, Mayaffre D, Lavigne F, Vanni L (2016) Machine learning under the light of phraseology expertise: use case of presidential speeches, de gaulle - hollande (1958–2016). Actes de JADT 2016:155–168
3. Joulin A, Grave E, Mikolov PBT (2017) Bag of tricks for efficient text classification. EACL, 427
4. Krizhevsky A, Sutskever I, Hinton GE (2012) Imagenet classification with deep convolutional neural networks. Adv Neural Info Proc Syst 1097–1105
5. Lafon P (1984) Dépouillements et statistiques en lexicométrie. In Genève-Paris, Slatkine-Champion
6. Lebart L (1997) Réseaux de neurones et analyse des correspondances. In: Modulad, (INRIA Paris), vol 18, pp 21–37
7. Longrée D, Mellet S (2013) Le motif : une unité phraséologique englobante ? étendre le champ de la phraséologie de la langue au discours. Langages 189:65–79
8. Longrée D, Mellet S, Lavigne F (2019) Construction cognitive d'un motif : cooccurrences textuelles et associations mémorielles. In: CogniTextes. http://journals.openedition.org/cognitextes/1202
9. Mellet S, Longrée D (2009) Syntactical motifs and textual structures. Belg J Linguist 23:161–173
10. Mellet S, Longrée D (2012) Légitimité d'une unité textométrique : le motif. Actes de JADT 2012:715–728
11. Mikolov T, Sutskever I, Chen K, Corrado GS, Dean J (2013) Distributed representations of words and phrases and their compositionality. Adv Neural Inf Process Syst 3111–3119
12. Pennington J, Socher R, Manning C (2014) Glove: global vectors for word representation. In: Proceedings of the 2014 conference on empirical methods in natural language processing (EMNLP), pp 1532–1543
13. Quiniou S, Cellier P, Charnois T, Legallois D (2012) Fouille de données pour la stylistique : cas des motifs séquentiels émergents. In: Actes de JADT 2012
14. Rastier F (2007) Passages. Corpus 6:25–54
15. Rastier F (2011) La mesure et le grain: sémantique de corpus. Champion; diff, Slatkine
16. Salem A (1987) Pratique des segments répétés. essai de statistique textuelle. Klincksieck, Paris
17. Vanni L, Ducoffe M, Precioso F, Mayaffre D, Longree D et al (2018) Text deconvolution saliency (tds) : a deep tool box for linguistic analysis. In: Proceedings of the 56th annual meeting of the association for computational linguistics (volume 1: Long Papers). Melbourne, Australia. Association for Computational Linguistics, pp 548–557
18. Vanni L, Mittmann A (2016) Cooccurrences spécifiques et représentations graphiques, le nouveau thème d'hyperbase. Actes de JADT 2016:295–305

Concentration Indices for Dialogue Dominance Phenomena in TV Series: The Case of the Big Bang Theory

Andrea Fronzetti Colladon and Maurizio Naldi

Abstract Dialogues in a TV series (especially in sitcoms) represent the main interaction among characters. Dialogues may exhibit concentration, with some characters dominating, or showing instead a choral action, where all characters contribute equally to the conversation. The degree of concentration represents a distinctive feature (a signature) of the TV series. In this paper, we advocate the use of a concentration index (the Hirschman–Herfindahl Index) to examine dominance phenomena in TV series and apply it to the Big Bang Theory TV series. The use of the concentration index allows us to reveal a declining trend in dialogue concentration as well as the decline of some characters and the emergence of others. We find the decline in dominance to be highly correlated with a decline in popularity. A stronger concentration is present for episodes (i.e. by analysing concentration of episodes rather than speaking lines), where the number of characters that dominate episodes is quite small.

1 Introduction

TV series are a steadily growing business, with the number of original scripted TV series in the USA exhibiting a CAGR (Compound Annual Growth Rate) of 11.1% from 2009 to 2017.[1] Writing their script is a delicate task, carried out with due care, with the final aim of making the TV series successful [1]. A major task is the balance between characters and the prevalence given to some of them.

[1] Data from the Statista website https://www.statista.com/statistics/444870/scripted-primetime-tv-series-number-usa/.

A. Fronzetti Colladon
University of Perugia, Department of Engineering, Via G. Duranti n. 93, 06125 Perugia, Italy
e-mail: andrea.fronzetticolladon@unipg.it

M. Naldi (✉)
Department of Law, Economics, Politics and Modern Languages, LUMSA University, Via Marcantonio Colonna 19, 00192 Rome, Italy
e-mail: m.naldi@lumsa.it

D. F. Iezzi et al. (eds.), *Text Analytics*, Studies in Classification, Data Analysis, and Knowledge Organization,
https://doi.org/10.1007/978-3-030-52680-1_5

Analysing the relationship between characters through the tools of graph theory is a relatively recent area of research. The resulting networks are typically called character networks; a survey of the tools employed to automatically extract the character network is reported in [9], and the difficulties of the task are highlighted in [16] and [6]. Examples of the application of social network analysis to TV series are the analysis of the *Game of Thrones* performed in [2], and the analysis of narration of *Breaking Bad*, again *Game of Thrones*, and *House of Cards* in [4]. A two-mode network has been employed in [5] to analyse the dynamics of character activities and plots in movies. Social networks in fiction have also been employed to validate literary theories, e.g. to examine the relationship between the number of characters and the settings [8]. An even more ambitious example of trying to obtain the signature of a novel's story through the topological analysis of its character network is shown in [17].

In this paper, we wish to further explore the relationship among characters through the use of social network analysis tools by focussing on dialogues. In particular, we wish to identify dominant characters, i.e. characters that dominate dialogues. For that purpose, we advocate the use of a concentration index borrowed from industrial economics, namely the Hirschman–Herfindahl index (HHI).

We report the results obtained for the Big Bang Theory (BBT) series. We show that:

- a steady declining trend is present for dominance in the number of speaking lines;
- the rank-size distribution of speaking lines among the major characters is roughly linear;
- a pattern is also present in the characters dominating the scene, with some declining while others emerge;
- episodes are dominated by a very small number of characters.

The paper is organized as follows. We describe our Big Bang Theory dataset in Sect. 2 and the associated graph in Sect. 3. In Sect. 4, we introduce the concentration (dominance) index and show the results of its application to the BBT series.

2 The Dataset

Our analysis is based on the set of scripts for the Big Bang Theory (BBT) TV series. In this section, we describe the main characteristics of that dataset, which has also been employed in [7].

The Big Bang Theory is an American television sitcom, premiered on CBS in 2007. It has now reached its twelfth season. After a slow start (ranking 68th in the first season and 40th in its second one), it ranked as CBS's highest-rated show in that evening on its first episode in the third season.

The dialogues have been retrieved from the BBT transcripts site.[2] The script reports the dialogues as a sequence of speaking lines, each line being formed of the speaking character name and the text of his/her speech (an uninterrupted text by a character counts as one speaking line, irrespective of the actual length of the text; a speaking line ends when another character takes turns). An excerpt of the dialogues is shown hereafter.

LEONARD : I'm sure she'll still love him.
SHELDON : I wouldn't.
LEONARD : Well, what do you want to do?
SHELDON : I want to leave.

We have examined all the episodes from the initial one of Season 1 to the 24th episode of Season 9, for a total of 207 episodes.

3 Representation of Dialogues

Our aim is to detect the presence of dominant characters in the series. We will focus on dialogues, since the series is essentially a sitcom (see [14] for an introduction to the genre, and [10] for the collocation of the Big Bang Theory within that genre) and is therefore based on dialogues (this is different, e.g. from what happens in an action movie, where the interaction is mainly physical). The notion of dominance is associated with those characters speaking most of the time. In this section, we see how we can extract the interaction structure associated with dialogues for the purpose of detecting dominance phenomena.

As reported in Sect. 2, the dialogues are shown in the script as sequences of speaking lines, each line representing a talkspurt by a single character. Each character interacts with the following character taking turns. We can represent that interaction through a graph, which describes the social network embedded in the dialogues. Our dialogue-based approach is different from what is done in [2], where any kind of interaction is included, or in [3], where two characters are connected if their names occur a certain number of words apart from each other. However, for the time being, we neglect the actual content of the dialogues or the sentiments conveyed by them.

The graph is built by connecting two characters if they talk to each other. Actually, we build a weighted directed graph. The nodes represent the characters; there is a link from character A to character B if A speaks to B (we infer that A speaks to B when a speaking line by A is followed by a speaking line by B); the weight of the link represents the number of times when A speaks to B.

As an example, the resulting dialogue graph is shown in Fig. 1 for Episode 1 of the first season. In order not to garble the graph, we have not reported the links concerning single instances, i.e. when character A speaks just once to character B.

[2]https://bigbangtrans.wordpress.com.

Fig. 1 Dialogue graph for Episode 1 of Season 1

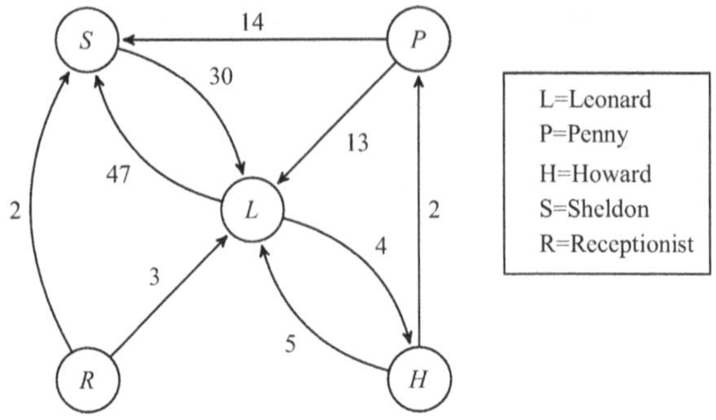

As we can see, just two links are two-way, which means that just two talking relationships are reciprocated (namely those between Sheldon and Leonard, and between Howard and Leonard). For this example, we can compute the density (the ratio of actual links to the maximum potential number of links for a graph of that size), which is just 0.45.

An alternative way to describe the interaction between the characters is the associated (weighted) adjacency matrix D, which is reported below, where we have employed a mapping between character names and indices as suggested by the order in the legend in Fig. 1 (i.e. 1 for Leonard, 2 for Penny, and so on). The element d_{ij} of D is the number of times A speaks to B.

$$\mathbf{D} = \begin{pmatrix} 0 & 0 & 4 & 47 & 0 \\ 13 & 0 & 0 & 14 & 0 \\ 5 & 2 & 0 & 0 & 0 \\ 30 & 0 & 0 & 0 & 0 \\ 3 & 0 & 0 & 2 & 0 \end{pmatrix}$$

In the graph, the out-degree of each node represents the number of times that the character speaks before somebody, i.e. its number of speaking lines in the script. Again for the first episode, we show the speaking lines for the group of 5 characters in the following vector, which are, otherwise stated, the out-degrees of each node, or the row sums of the adjacency matrix.

$$\begin{pmatrix} 51 \\ 27 \\ 7 \\ 30 \\ 5 \end{pmatrix}$$

4 Dominance in Dialogues

After building the dialogue graph and explained the notion of dominance in this context, we wish to analyse if such dominance phenomena are present, and to what extent, in the Big Bang Theory series. For that purpose, in this section we introduce a dominance index and apply it to the BBT scripts. In addition, we identify the dominant characters throughout the series.

As a dominance index, we borrow a concentration index from the field of industrial economics, named the Hirschman–Herfindahl Index (or HHI, for short) [11, 13]. In order to see if an industry is concentrated in the hands of few companies (i.e. if a few companies dominate the market), the HHI was defined as the sum of the squared market shares of all the companies in the market: the higher the HHI, the more concentrated the market (i.e. dominated by a few firms). In our case, we similarly define HHI as our dominance index by considering the number of times a character speaks to any other character (i.e. the number of speaking lines of that character, in the theatre jargon). The equivalent of the *market share* in this context is therefore the fraction of speaking lines of each character with respect to the overall number of speaking lines in the episode. On the graph, it is the out-degree of that character normalized by the sum of all the out-degrees.

By recalling the definition of the weighted adjacency matrix **D**, for an episode in which n characters appear, the HHI is

$$\mathrm{HHI} = \frac{\sum_{i=1}^{n}\left(\sum_{j=1}^{n} d_{ij}\right)^2}{\left(\sum_{i=1}^{n}\sum_{j=1}^{n} d_{ij}\right)^2} \tag{1}$$

The HHI takes values in the $[1/n, 1]$ range, with $1/n$ representing perfect equipartition of the market (in our case, perfect equidistribution of speaking lines among all the characters), and 1 representing a monopoly (in our case, a monologue by one character). As to intermediate cases, in order to determine whether an HHI value denotes a strong dominance by one or more characters, we can adopt the classification suggested in [15]: a value lower than 0.15 denotes unconcentrated markets, while a value larger than 0.25 denotes highly concentrated markets, and intermediate values represent moderately concentrated markets. The HHI has already been used to analyse dominance in dialogues in [7] for TV series and [12] for personal finance forums.

We report an example of computation of the HHI for the graph of Fig. 1 in Table 1.

We can now compute the HHI for each episode in the series. In Fig. 2, we report the average HHI across each season. We can now see that the concentration in dialogues has been steadily decreasing over the years. The dominance index fell into the moderate concentration region as early as Season 3 and bordered on the unconcentrated region starting from Season 7. We do not know whether this was a deliberate choice of the series authors, but we note that the series has moved from episodes with a few dominant characters to a more choral action. At the same time, we observe a moderate dispersion around the average value, accounting for roughly ±20% of the average.

Table 1 HHI computation for the graph of Fig. 1

Character	Out-degree	Dialogue share	Squared dialogue share
Leonard	51	0.425	0.1806
Sheldon	30	0.25	0.0625
Penny	27	0.225	0.0506
Howard	7	0.058	0.0034
Receptionist	5	0.042	0.0018
HHI			0.2989

Fig. 2 Time evolution of HHI

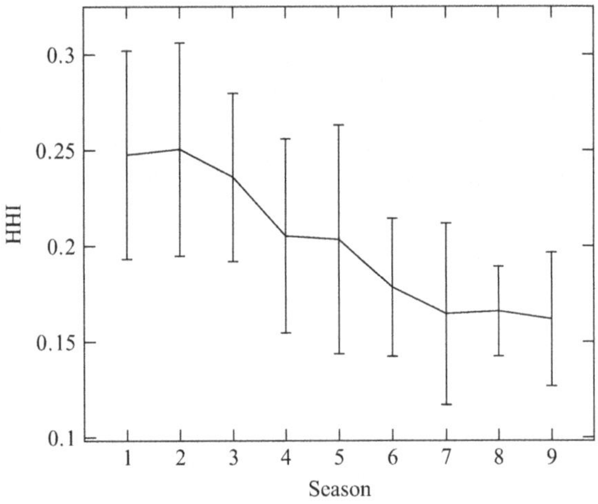

It is interesting to compare that trend in dominance with a similar trend observed for viewers' ratings collected on the IMDB (Internet Movie Database) platform, reported in Fig. 3 [7]. The resulting correlation coefficient is 0.847, showing that the two trends are quite similar. Though we cannot state a causal relationship, we observe that the declining popularity has been associated to the shift from a dominance situation to a more choral action.

So far, we have considered the presence of dominance phenomena without specifying who's dominating the dialogues.

We consider first the dominance throughout the series by analysing the overall number of speaking lines spoken by each character. In Fig. 4, we report that number for the major characters. We see that we are far from an equidistribution even among the major characters. The character speaking most over the series is Sheldon, followed by Leonard and Penny. The decay appears to be quite linear in the rank.

Delving deeper, we may wish to see if the dominant character was the same throughout the series, or an alteration of dominant characters was present, or some characters faded along the series to give room to emerging characters. We can assess

Fig. 3 Viewers' ratings

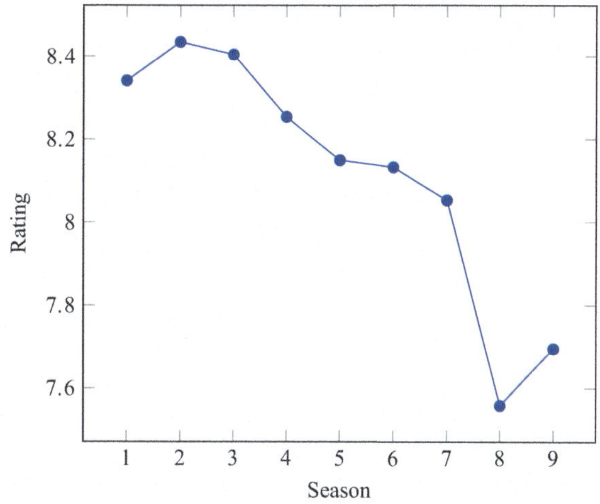

Fig. 4 Number of speaking lines per character (whole series)

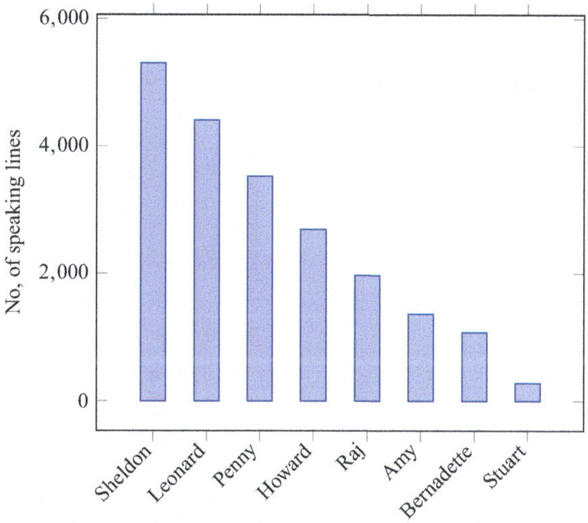

that by identifying the dominant character in each episode. In Fig. 5, we report the number of episodes where each character was dominant. Actually, we see several patterns here. After building up in Seasons 1 to 4, Sheldon lost ground in Season 5 to regain his leading role afterwards. Instead Leonard following a rather steady decline throughout the seasons. Penny gradually gained prominence, but reached her peak in Season 7 and fell somewhat during the last two seasons.

The cumulative plot of speaking lines in Fig. 6 shows the gradual takeover by Sheldon, who became the overall leading character as early as in Season 3.

Since we have now moved to considering episodes rather than speaking lines as the measure of dominance, it is interesting to examine if the concentration metric is different from what we saw for the number of speaking lines. If we apply the definition

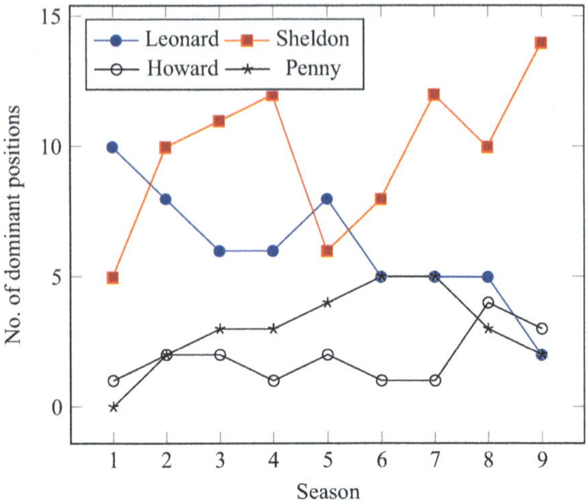

Fig. 5 Dynamics of dominating characters

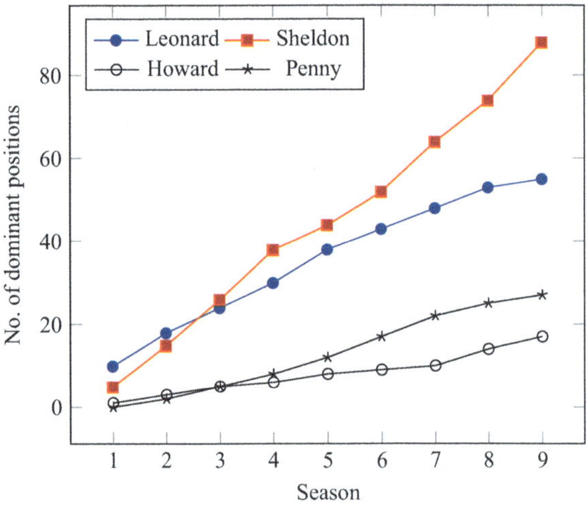

Fig. 6 Cumulative dynamics of dominating characters

of the HHI to the number of episodes where a character appears as dominant, we obtain the graph in Fig. 7. If we compare this graph with that of Fig. 2, we get a bit different story. First, the figures are quite higher: the concentration is much larger when we measure it over the number of dominated episodes rather than over the number of speaking lines, which means that episodes are assigned to a small number of dominating characters. Secondly, we do not observe the steady declining trend as in Fig. 2. Rather we have a stronger concentration in the first and the last season, but in between we see a concentration fluctuating around a central value (roughly 0.35).

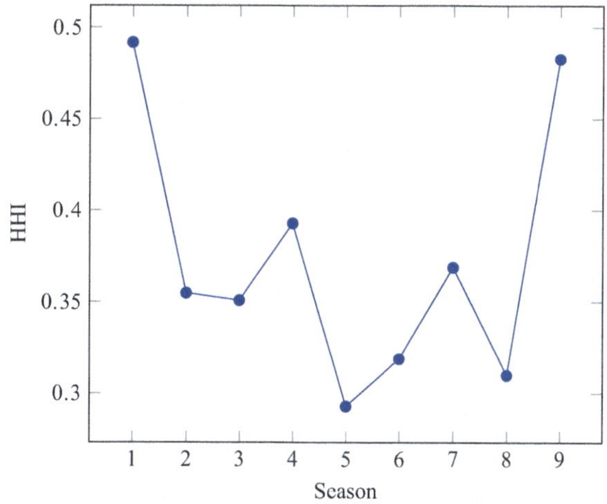

Fig. 7 HHI dynamics for the number of episodes

5 Conclusions

We have introduced the use of a concentration index, namely the Hirschman–Herfindahl Index, borrowed from industrial economics, to analyse dominance phenomena in the dialogues of TV series. We have performed a concentration analysis of the Big Bang Theory TV series. The analysis allows to reveal a declining trend in the concentration of dialogues over the years, i.e. the passage from a few dominating characters to a more choral action. However, the number of characters that dominate episodes is rather small. The distribution of speaking lines among characters over the whole series exhibits a linear rank–size relationship. However, the analysis of dominance by specific characters over seasons allows to detect a pattern where some characters decline in importance and others emerge.

The adoption of a dominance index is therefore a valuable aid to detect a relevant stylistic feature of a TV series such as the dominance of some characters in dialogues and to investigate the association between stylistic features of the series and its performance. A refinement of the analysis can be envisaged where the recipient of each speaking line is more accurately identified.

Acknowledgements Maurizio Naldi has been partially supported by the Italian Ministry of Education, University, and Research (MIUR) under the national research projects PRIN AHeAd #20174LF3T8.

References

1. Allrath G, Gymnich M, Surkamp C (2005) Introduction: towards a narratology of tv series. In: Narrative strategies in television series. Springer, pp 1–43
2. Beveridge A, Shan J (2016) Network of thrones. Math Horizons 23(4):18–22

3. Bonato A, D'Angelo DR, Elenberg ER, Gleich DF, Hou Y (2016) Mining and modeling character networks. In: Algorithms and models for the web graph: 13th international workshop, WAW 2016, Montreal, QC, Canada, December 14–15, Proceedings 13. Springer, pp 100–114
4. Bost X, Labatut V, Gueye S, Linarès G (2016) Narrative smoothing: dynamic conversational network for the analysis of tv series plots. In: Proceedings of the 2016 IEEE/ACM international conference on advances in social networks analysis and mining. IEEE Press, pp 1111–1118
5. Chao D, Kanno T, Furuta K, Lin C (2018) Representing stories as interdependent dynamics of character activities and plots: a two-mode network relational event model. Digital Scholarship in the Humanities
6. Edwards M, Mitchell L, Tuke J, Roughan M (2018) The one comparing narrative social network extraction techniques. arXiv preprint arXiv:1811.01467
7. Fronzetti Colladon A, Naldi M (2019) Predicting the performance of tv series through textual and network analysis: the case of big bang theory. PloS one 14(11)
8. Jayannavar P, Agarwal A, Ju M, Rambow O (2015) Validating literary theories using automatic social network extraction. In: Proceedings of the fourth workshop on computational linguistics for literature, pp 32–41
9. Labatut V, Bost X (2019) Extraction and analysis of fictional character networks: a survey. ACM Comput Surv (CSUR) 52(5):89
10. Ma Z, Jiang M (2013) Interpretation of verbal humor in the sitcom the big bang theory from the perspective of adaptation-relevance theory. Theor Pract Lang Stud 3(12)
11. Naldi M (2003) Concentration indices and Zipf's law. Econom Lett 78(3):329–334
12. Naldi M (2019) Interactions and sentiment in personal finance forums: an exploratory analysis. Information 10(7):237
13. Rhoades SA (1993) The Herfindahl-Hirschman index. Fed Res Bull 79:188
14. Smith ES (1999) Writing television sitcoms. Penguin
15. The U.S. Department of justice and the federal trade commission: horizontal merger guidelines (19 August 2010)
16. Vala H, Jurgens D, Piper A, Ruths D (2015) Mr. bennet, his coachman, and the archbishop walk into a bar but only one of them gets recognized: On the difficulty of detecting characters in literary texts. In: Proceedings of the 2015 conference on empirical methods in natural language processing, pp 769–774
17. Waumans MC, Nicodème T, Bersini H (2015) Topology analysis of social networks extracted from literature. PloS one 10(6):e0126470

A Conversation Analysis of Interactions in Personal Finance Forums

Maurizio Naldi

Abstract Online forums represent a widely used means of getting information and taking decision concerning personal finance. The dynamics of conversations taking place in the forum may be examined through a social network analysis (SNA). Two major personal finance forums are analysed here though an SNA approach, also borrowing metrics from Industrial Economics such as the Hirschman–Herfindahl Index (HHI) and the CR4 index. The major aim is to analyse the presence of dominance phenomena among the speakers. A social network is built out of the sequence of posts and replies. Though no heavy dominance is found on the basis of the HHI, a few speakers submit most posts and exhibit an aggressive behaviour by engaging in monologues (submitting a chain of posts) and tit-for-tats (immediate reply by a speaker to a post replying to that same speaker). Most replies occur within a short timeframe (within half an hour).

1 Introduction

Decisions concerning personal finance are often taken by individuals not just on the basis of factual information (e.g., company's official financial statements or information about past performance of funds), but also considering the opinions of other individuals. Nowadays, personal finance forums on the Internet have often replaced friends and professionals in that role. In those forums, the interaction occurs among people who typically do not know one another personally and know very few personal information (if any) about other participants. Anyway, they often create online communities that can bring value to all participants [1]. Examples of such forums are SavingAdvice (http://www.savingadvice.com/forums/) or Money Talk (http://www.money-talk.org/board.html).

M. Naldi (✉)
Department of Law, Economics, Politics and Modern Languages,
LUMSA University, Via Marcantonio Colonna 19, 00192 Rome, Italy
e-mail: m.naldi@lumsa.it

D. F. Iezzi et al. (eds.), *Text Analytics*, Studies in Classification, Data Analysis, and Knowledge Organization,
https://doi.org/10.1007/978-3-030-52680-1_6

The actual influence of such forums on individuals' decisions has been investigated in several papers, considering, e.g., how the level of activity on forums impacts on stock trading levels [17], how participation in such forums pushes towards a more risky-seeking behaviour [18], or introducing an agents-based model to determine how individual competences evolve due to the interaction [6].

It has been observed that such forums may be employed by more aggressive participants to manipulate more inexperienced ones [3], establishing a dominance over the forum. In addition to being undesirable for ethical reasons, such an influence is often contrary to the very same rules of the forum.

Here we investigate the subject by adopting a different approach from the semantic analysis of [3]. We consider the social network structure embedded in the online forum. Online forums have been examined by exploiting their intrinsic social network structure in several papers. The network structure has been employed to study opinion formation and influence in [5] with the aim of identifying the most influential speakers. Other authors have instead focussed on the learning experience in online forums [4, 14]. Underground forums have been analysed in [7]. However, so far no social network analysis has been conducted on personal finance forums.

In this paper, we wish to investigate the presence of imbalances in online forums devoted to personal finance issues and the dynamics of the interaction between participants, by extracting the social network structure of the forum. The rationale is that participants wishing to manipulate others would try to take control of the discussion by posting more frequently and being more reactive. Unlike what was reported in [9], we do not deal with the actual content of the posts (e.g., the sentiment expressed in them), but focus on the conversation structure, i.e., the social network made of the post submitters, whose connections are related to post replies.

For that purpose, we employ two datasets extracted from two major personal finance threads on the *SavingAdvice* website. For the purpose of the analysis, the thread is represented as the sequence of participants taking turns, with dates and times of each post attached.

We conduct a conversation analysis, wishing to assess if: (1) there are any dominant participants (in particular the thread starter); (2) repetitive patterns appear such as sustained monologues or sparring matches between two participants; (3) replies occur on a short-time scale.

The paper provides the following contributions:

- through the use of concentration indices we find out that, though no dominance exist, the top 4 speakers submit over 60% of the posts (Sect. 3);
- both recurring reply sequences and monologues appear, representing a significant portion of all posting activity (Sect. 4);
- reply times can be modelled by a lognormal distribution, with 50% of the posts being submitted no longer than 14 or 23 min (for the two datasets, respectively) after the last one (Sect. 4).

Table 1 Datasets

Thread	Creator	No. of speakers	No. of posts
Struggling	jpg7n16	25	155
Guru	james.hendrickson	18	104

2 Datasets

We consider the two most popular threads on the *SavingAdvice* website. The topics are the following, where we indicate an identifying short name between parentheses:

1. Should struggling families tithe? (Struggling)
2. African-American Personal Finance Gurus (Guru).

The main characteristics of those datasets are reported in Table 1.

For each thread, we identify the set of speakers $S = \{s_1, s_2, \dots, s_n\}$, i.e., the individuals who submit posts. We identify also the set of posts $P = \{p_1, p_2, \dots, p_m\}$ and a function $\mathcal{F} : P \to S$, that assigns each post to its submitter. For each speaker, we can therefore compute the number of posts submitted by him/her. If we use the indicator function $\mathbb{1}(\cdot)$, the number of posts submitted by the generic speaker s_i is

$$N_i = \sum_{j=1}^{m} \mathbb{1}_{\mathcal{F}(p_j)=s_i}. \tag{1}$$

The mapping from posts to speaker induces an order in the sequence of speakers. By defining the j-th speaker in the sequence as

$$s_{(j)} = \mathcal{F}(p_j), \tag{2}$$

we have the sequence of speakers $\mathcal{S} = \{s_{(1)}, s_{(2)}, \dots, s_{(m)}\}$.

3 Dominance in a Thread

As hinted in the Introduction, dominance may be a dangerous phenomenon in online forums. By acting more prominently than others, some speakers may exert an excessive influence on the general direction of the discussion and influence the silent observers on the forum. In this section, we wish to assess if some dominance emerges in the two threads we have chosen to examine. We adopt a set of indicators that may be employed on similar forums, so that a methodology to examine dominance may be drawn out of them.

We consider the following aspects of dominance:

- Concentration of posts;
- Distribution of posts;
- Most frequent speakers;
- Monologues.

As to the first aspect, we wish to see if the posts are submitted by one, or a few, speakers rather than having all the speakers contributing equally to the discussion. This is an issue not unlike what we may encounter in economics, when we examine whether a company, or a few companies, dominates a market. It is therefore natural to borrow pertaining concentration indices from Industrial Economics. Namely, we adopt the Hirschman–Herfindahl Index (HHI) [8, 11, 15], and the CR_4 index [10, 13]. For a market where n companies operate, whose market shares are $v_1, v_2, \ldots, v_n$, the HHI is defined as

$$\mathrm{HHI} = \sum_{i=1}^{n} v_i^2. \tag{3}$$

The HHI satisfies the inequality $1/n \leq \mathrm{HHI} \leq 1$, where the lowest value corresponds to the case of no concentration (perfect equidistribution of the market) and the highest value represents the case of monopoly. Therefore, the larger the HHI, the larger the concentration. Instead, the CR_4 measures the percentage of the whole market owned by the top four companies: Similarly, the higher the CR_4, the heavier the concentration.

In our case, the fraction of posts submitted by a speaker can be considered as his/her market share, so that the HHI can be redefined as

$$\mathrm{HHI} = \sum_{i=1}^{n} \left(\frac{N_i}{m} \right)^2 = \frac{\sum_{i=1}^{n} N_i^2}{m^2}. \tag{4}$$

Similarly, the CR_4 can be redefined as

$$\mathrm{CR}_4 = \frac{\sum_{i=1}^{4} N_i}{\sum_{i=1}^{n} N_i}. \tag{5}$$

For our datasets, we get the results reported in Table 2. In order to see if those values represent a significant degree of concentration, we can resort to the guidelines provided by the US Department of Justice, for which the point of demarcation between unconcentrated and moderately concentrated markets is set as $\mathrm{HHI} = 0.15$ [16]. Since the values in Table 2 are slightly below that threshold, we can conclude that we are still in the unconcentrated region, but close to having a significant concentration phenomenon.

However, the CR_4 index shows that the top 4 speakers submit more than 60% of all the posts.

Table 2 Concentration measures

Thread	HHI	CR_4 (%)
Struggling	0.1220	61.94
Guru	0.1396	67.31

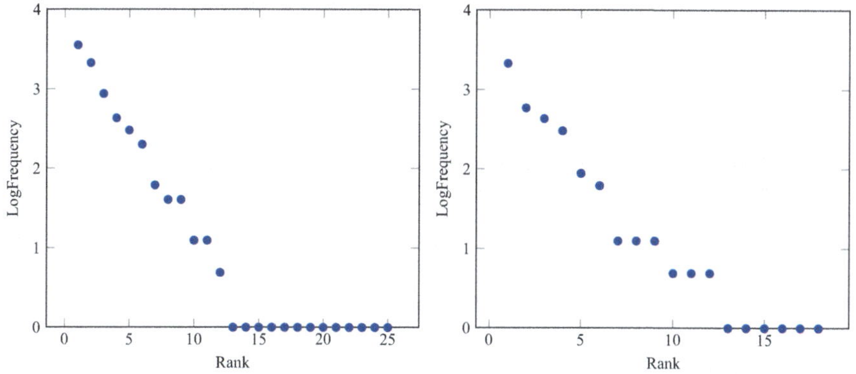

(a) Struggling dataset (b) Guru dataset

Fig. 1 Rank-size plot

We can delve deeper than the single figure provided by those concentration indices and see how the posts are distributed among the speakers. For that purpose, we take a look at the rank-size plot: after ranking speakers by the number of posts they submit (in decreasing order), the frequency of posts is plotted vs the rank of the speaker. In Fig. 1, we see that a linear relationship appears between $\log N_{(i)}$ and the rank i, so that a power law $N_{(i)} = k/i^{\alpha}$ may be assumed to apply roughly, where k is a normalizing constant and α is the Zipf exponent (see, e.g., [8, 12]). That exponent (whose name is due to the fact that the power law is also called a generalized Zipf law) measures the slope of the log-linear curve, hence the imbalances between the contributions of all the speakers. By performing a linear regression, we get a rough estimate of α, which is 0.25 for both datasets (actually 0.2545 for the Struggling dataset and 0.2501 for the Guru one). This cannot be considered a large value (the original Zipf law was formulated for $\alpha = 1$ [2]).

Delving deeper into the top 4, we also see that the most frequent speaker typically contributes around 1/4 of the overall number of posts, which can be considered as a major influence. In the Struggling dataset, the most frequent speaker is the thread originator itself (with 22.6% of posts), while that's not true in the Guru dataset, where the most frequent speaker contributes 26.9% of posts and the originator just 2.88%.

Finally, an additional measure of dominance is the presence of monologues. In this context, we define a monologue as a sequence of posts submitted by the same speaker, i.e., an AA or AAA or even longer identical subsequence. In a more formal setting, we define an AA subsequence is one such that $\mathcal{F}(p_{j+1}) = \mathcal{F}(p_{j+1})$,

$j = 1, 2, \ldots, m - 1$. More generally, a k-long subsequence of this kind which is one such that $\mathcal{F}(p_{j+k}) = \mathcal{F}(p_{j+k-1}) = \cdots = \mathcal{F}(p_{j+1})$, $j = 1, 2, \ldots, m - k$. Monologues represent an attempt to monopolize the discussion and are therefore an additional measure of dominance. The occurrence rate of such monologues in our datasets is, respectively, 8.3% for Struggling and 8.7% for Guru, which is far from being negligible. It is also interesting to compute how many speakers resort to monologues. In our datasets, the monologues are to be ascribed to just 6 speakers in the Struggling dataset (out of 25 speakers) and 4 speakers in the Guru dataset (out of 18 speakers).

4 Interactions Between Post Submitters

After examining dominance on its own, i.e., by examining the activity of each speaker, we turn to interactions, by examining how each speaker interacts with the others. In this section, we analyse the structure of replies, both from a static and a dynamic viewpoint. We use the dynamic and static qualifiers according to whether time is considered or not.

From the static point of view, we wish to see the amount of interactions among the speakers over the whole thread existence. A natural approach is to extract the social network (as represented by a graph) embedded in the thread.

In particular, we build a graph representing how speakers reply to each other, considering each post as a reply to the previous one. We build the replies graph by setting a link from a node X to a node Y if the speaker represented by node X has replied at least once in the thread to a post submitted by the speaker represented by node Y. The resulting graphs are shown in Fig. 2, which is ordered from the core to the periphery in order of decreasing degree of the nodes, laid out on concentric rings. Here the degree of a node represents the number of speakers to which it has replied. In both cases, an inner core of most connected nodes appear, which represent the speakers replying to most other speakers. Reply patterns emerge as bidirectional links (couples of speakers who reply to each other). Loops represent monologues instead, i.e., speakers submitting two or more posts in a row. It is to be noted that, though we built the social network this way, we can add more information by building the weighted graph where the link from X to Y has a weight representing the number of replies of X to Y.

We can gain another view of those interactions by building the adjacency matrix for that social network. In particular, we may be interested in examining how diffuse are the interactions. A measure of the degree of interaction is the sparsity of the adjacency matrix, i.e., the percentage of zero elements in that matrix. In our case, we get a sparsity of 87.4% for the Struggling dataset and 83% for the Guru dataset. Those are very large figures, meaning that just a small percentage of the possible couples of speakers interact.

If we turn to a dynamic view of replies, we have to consider the timing factor. We can therefore examine the patterns of replies. In particular, we wish to see if there are rejoinder patterns, i.e., any ABA pattern (speaker A immediately counteracts to a post

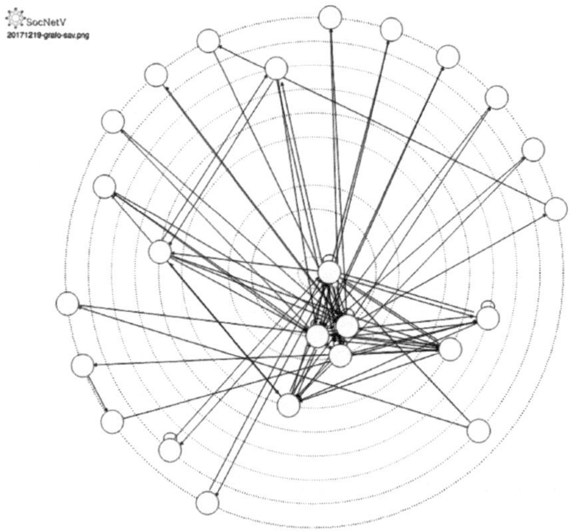

(a) Struggling dataset

(b) Guru dataset

Fig. 2 Replies graph

Table 3 Reply time statistics (in minutes)

Thread	Mean	Median	Standard deviation	95% percentile (%)
Struggling	70.5	23	156.2	254.7
Guru	58.9	14	112.7	406.7

by B replying to its previous post), or ABCA pattern (where the reply comes after two posts). These patterns can be called tit-for-tats, since they denote an immediate wish to take centre stage again. More formally, we can define an ABA pattern through the condition $\mathcal{F}(p_{j+2}) = \mathcal{F}(p_j)$ with $\mathcal{F}(p_{j+1}) \neq \mathcal{F}(p_j)$, $j = 1, 2, \ldots, m-2$, and the ABCA pattern through the condition $\mathcal{F}(p_{j+3}) = \mathcal{F}(p_j)$ with $\mathcal{F}(p_{j+1}), \mathcal{F}(p_{j+2}) \neq \mathcal{F}(p_j)$, $j = 1, 2, \ldots, m-2$. We find that ABA patterns represent 19.6% and 16.3% of posts, respectively, in the Struggling and Guru dataset. Instead, ABCA pattern occur less frequently, representing 4.4% and 12.5% of all posts, respectively. We see that tit-for-tats account for a significant portion of the overall number of posts and may be seen as a further appearance of dominance phenomena.

Further, we are interested in how fast the interactions are between contributors to the thread. We define the reply time as the time elapsing between a post and the subsequent one. The main statistics of the reply time are reported in Table 3. In both dataset, the mean reply time is around 1 hour, but 50% of the replies take place within either 14 min (Guru dataset) or 23 min (Struggling dataset), i.e., with a much smaller turnaround. There is therefore a significant skewness to the right.

A more complete view of the variety of reply times is obtained if we model the probability density function. In Fig. 3, we report the curves obtained through a Gaussian kernel estimator, an exponential model, and a lognormal model (whose parameters have been estimated by the method of moments). By applying the Anderson-Darling test, we find out that the exponential hypothesis is rejected at the 5% significance level, while the lognormal one is not rejected, with a p-value as high as 0.72 for the Struggling dataset and 0.076 for the Guru dataset.

5 Conclusions

We have analysed two major threads within a personal finance forum as a conversation between submitters acting as speakers. Personal finance forums, not unlike other online forums, exhibit an embedded social network structure, which may be used to analyse the behaviour of forum participants. We find that, though no significant concentration exists, the top four speakers submit over 60% of the posts. Patterns of interaction emerge as the presence of several couples of speakers who reply to each other (tit-for-tat), several monologues, and short reply times. These patterns represent some degree of dominant behaviour.

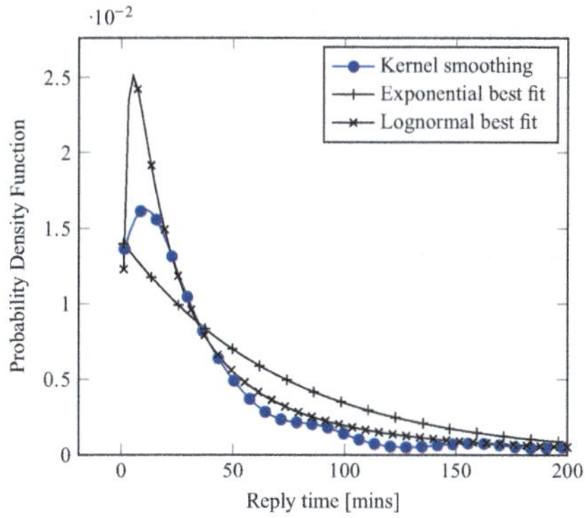

(a) Struggling dataset

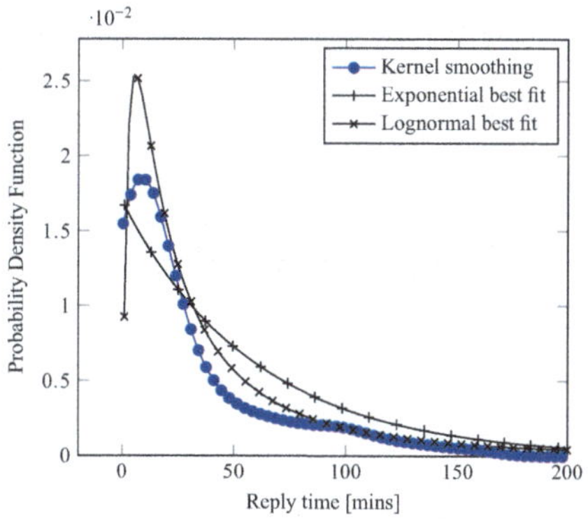

(b) Guru dataset

Fig. 3 Reply time

The set of metrics adopted to analyse and may adopted to conduct similar analyses over a wider set of personal finance forums, to see if such phenomena are widespread, or other online forums, to see if they are affected by the same kind of aggressive behaviour. This analysis may be complemented by a sentiment analysis of the posts.

Acknowledgements Maurizio Naldi has been partially supported by the Italian Ministry of Education, University, and Research (MIUR) under the national research projects PRIN AHeAd #20174LF3T8.

References

1. Armstrong A, Hagel J (2000) The real value of online communities. Knowl Communities 74(3):85–95
2. Baixeries J, Elvevåg B, Ferrer-i Cancho R (2013) The evolution of the exponent of zipf's law in language ontogeny. PloS one 8(3):e53227
3. Campbell J, Cecez-Kecmanovic D (2011) Communicative practices in an online financial forum during abnormal stock market behavior. Inf Manage 48(1):37–52
4. DeSanctis G, Fayard AL, Roach M, Jiang L (2003) Learning in online forums. Eur Manage J 21(5):565–577
5. Kaiser C, Bodendorf F (2012) Mining consumer dialog in online forums. Internet Res 22(3):275–297
6. Mastroeni L, Vellucci P, Naldi M (2017) Individual competence evolution under equality bias. In: 2017 European modelling symposium (EMS)
7. Motoyama M, McCoy D, Levchenko K, Savage S, Voelker GM (2011) An analysis of underground forums. In: Proceedings of the 2011 ACM SIGCOMM conference on internet measurement conference. ACM, pp 71–80
8. Naldi M (2003) Concentration indices and Zipf's law. Econ Lett 78(3):329–334
9. Naldi M (2019) Interactions and sentiment in personal finance forums: an exploratory analysis. Information 10(7):237
10. Naldi M, Flamini M (2014) Correlation and concordance between the CR_4 index and the Herfindahl-Hirschman index. SSRN Working paper series

11. Naldi M, Flamini M (2017) Censoring and distortion in the Hirschman–Herfindahl index computation. Economic papers: a journal of applied economics and policy
12. Naldi M, Salaris C (2006) Rank-size distribution of teletraffic and customers over a wide area network. Eur Trans Telecommun 17(4):415–421
13. Pavic I, Galetic F, Piplica D (2016) Similarities and differences between the CR and HHI as an indicator of market concentration and market power. Br J Econ Manage Trade 13(1):1–8
14. Rabbany R, ElAtia S, Takaffoli M, Zaïane OR (2014) Collaborative learning of students in online discussion forums: a social network analysis perspective. In: Educational data mining. Springer, pp 441–466
15. Rhoades SA (1993) The Herfindahl-Hirschman index. Fed Res Bull 79:188
16. The U.S. Department of justice and the federal trade commission: horizontal merger guidelines (19 August 2010)
17. Tumarkin R, Whitelaw RF (2001) News or noise? internet postings and stock prices. Fin Anal J 57(3):41–51
18. Zhu R, Dholakia UM, Chen X, Algesheimer R (2012) Does online community participation foster risky financial behavior? J Market Res 49(3):394–407

Dictionaries and Specific Languages

Big Corpora and Text Clustering: The Italian Accounting Jurisdiction Case

Domenica Fioredistella Iezzi and Rosamaria Berté

Abstract Currently, big corpora, coming from Open Government Data projects, in several research areas, allow quickly to available a large number of documents. In the legal field, many information are stored and therefore made it necessary to develop a specific dictionary to classify text data. The paper aims to present the workflow to visualize and classify Big Legal Corpora, identifying the main steps of a chain process. The analyzed corpus is composed of 123,989 judgments, in Italian language, published by the Court of Audit, from 2010 to 2018.

1 Introduction

In the last twenty years, the exponential growth of computing, combined with the development of statistical methods for Linguistics, allowed the achievement of important goals for the analysis of text data, also for big size one. Currently, it is possible to manipulate quickly big corpora, but in the preprocessing step, disambiguations of words remains an open issue [1, 2]. In many cases, a solution is to build a manual custom lists, that requires very long work times. Big corpora are characterized by **variety**, **velocity**, **variability**, and **veracity**, as well as volume [3]. Big corpora can be in various formats, so the preprocessing step is very complicated compared with structured data, e.g., tweets contain more noises with respect to opened questions of a survey. Moreover, information flow, came from various sources like the Internet or open data, is updated very frequently (velocity) and present many issues linked to different kinds of data (variability). In those corpora, there are very several noises,

D. F. Iezzi (✉)
Department of Enterprise Engineering, Tor Vergata University,
Via del Politecnico, 1, 00133 Rome, Italy
e-mail: stella.iezzi@uniroma2.it

R. Berté
Corte dei Conti, Direzione Generale dei Sistemi Informativi
Automatizzati della Corte dei Conti (DGSIA), 00195 Rome, Italy
e-mail: rosamaria.berte@corteconti.it

D. F. Iezzi et al. (eds.), *Text Analytics*, Studies in Classification, Data Analysis, and Knowledge Organization,
https://doi.org/10.1007/978-3-030-52680-1_7

then it takes too much time for cleaning them (variability). The reliable of this text information is not known (veracity). The treatment of these corpora, then, requires a very careful work especially in the pre-treatment phase, with a long process of disambiguation of the words. In the broadest definition, the meaning of "lexical disambiguation" is to determine the meaning of every word context. It is a fundamental characteristic of language because words can have more than one meaning. Words are assumed to have a finite and discrete set of senses from a dictionary, a lexical knowledge-based on an ontology [4]. In the legal field, the deluge of electronically stored information has therefore required developing a framework for intelligent extraction of useful knowledge for documents needed for categorizing information, retrieving relevant information as well as other useful tasks [5]. Automatic models were presented able to automatically detect the type of provisions contained in a regulatory text [15]. The linguistic profiling on corpora made up of various types of Italian legal texts, has studied the characterization of the different sub-varieties of the Italian legal language with the use of Natural Language Processing (NLP) tools and techniques that automatically manage to identify in the text the linguistic specificities [16]. Moreover, different clustering models have been developed that are able to group legal judgments into different clusters in effective manner [17]. Palmirani and Vitali [6, 7] have introduced a XML standard (Akoma Ntoso—Architecture for Knowledge-Oriented Management of African Normative Texts using Open Standards and Ontologies), which is an international technical standard for representing executive, legislative, and judiciary documents in a structured manner. Despite this enormous effort, a very large number of legal document are still unstructured [8]. Currently, *Open government* has allowed citizens to access a huge number of documents, but the consultation, through metadata, is still very complex. The *OECD Open Government Data project* also contributes to the OECD Going Digital project which supports governments and policymakers advance in the digital transformation of their public sectors, economic activities, and societies overall.[1]

In this paper, we analyze the judicial measures issued by the Court of Audit, from 2010 to 2018 in matters of responsibility and pension.[2] The paper aims at creating a vocabulary for jurisdictional functions of public accounting assigned by the Constitution to the Court of Auditors and classify the contents. The long-term objective is to support the accounting magistracy in drafting the final measures the proceedings: starting from a set of input data and the information that gradually accompany the various procedural steps, to develop a semiautomatic workflow that reflects the judge's decision-making process according to the subject matter and the reference standard.

This paper is structured as follows. In Sect. 2, the data and methods are introduced; in Sect. 3, the main results are presented; in Sect. 4, conclusions and future steps are discussed.

[1] See the website https://www.oecd.org/going-digital/.

[2] It can be consulted on the institutional database at the following link: https://banchedati.corteconti.it/.

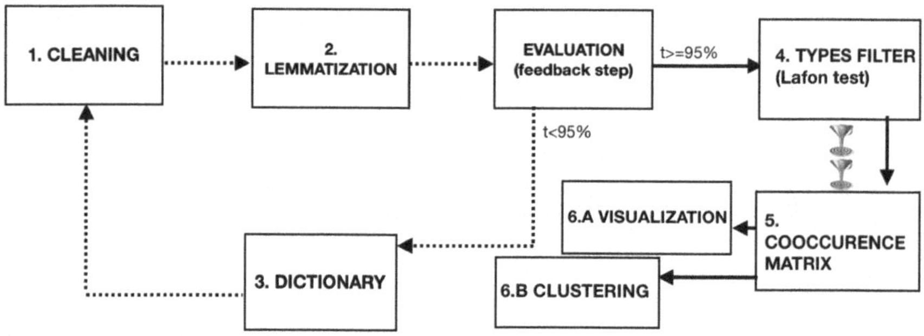

Fig. 1 Workflow

2 Data and Method

We collected a corpus of 123,989 judgments, in Italian language, published by the Court of Audit, from 2010 to 2018. The corpus is 1.8 GB in size; and before the preprocessing, it was composed of 22,5362,318 tokens, 513,527 types, and 223,610 hapax legomena. The legal texts, written in various digital formats (docx, doc, rtf, pdf), have been subjected to automatic acquisition and normalization procedures in the txt-UTF8 format, using different Python libraries, and the IraMuTeQ software.[3]

The method has six basic steps, plus one feedback step (Fig. 1):

1. CLEANING: We normalized the corpus with Python libraries. We used regular expressions and other automatic and semiautomatic data cleaning techniques. The presence of abbreviations important for the context and standardization was verified, just as the complex graphic forms constituting a unity of sense which were identified, special characters have been eliminated which would have led to errors of orthographic and elaborative type, checks were carried out on accents, apostasy, acronyms.
2. LEMMATIZATION: We adopted a morphological normalization to study the lemmas present in the Iramuteq dictionary. In the first phase of lemmatization, it was observed that for many inflected forms, although belonging to the dictionary of the Italian language, the algorithm did not identify the part of speech (POS). In particular, 73% of the forms were not recognized.
   0. EVALUATION (feedback step): We fixed a quality level of 95%, based on the percentage of POS recognized.
3. DICTIONARY: Manual disambiguation of approximately 22,300 unrecognized grammatical forms was required: abbreviations, e.g., dlg for law decree, proper names, nouns, and adjectives specific to the context (e.g., responsibility, case, judge, pension). To put constant emphasis of the legal lexicon considered particularly significant to textual analysis purposes, the set of words which discounts the concept of rule of law has been lexicalized. Specifically, the sequence: *art* –

[3]See the website http://www.iramuteq.org.

$number - comma - typelaw - numberlaw - yearlaw$ identifies a new multiword with unit occurrence. In this way, we created an external resource, a specific dictionary of 357.045 grammatical forms.

4. TYPES FILTER: Assuming that the corpus is a population and each part of it is a sample (subcorpora), we calculated the specificities of subcorpora on the Terms Documents Matrix (TDM) **X** of size (511.544 × 123.990). We cut **X** using only the terms with a frequency lower than 10 occurrences. To reduce further the sparsity of the **X** matrix, we worked on three Lexical Tables:
   (a) **Y**(words x year);
   (b) **S**(words x subject);
   (c) **O**(words x court order).
   Y, **M**, and **O** we selected the specificity words using Lafon Test [9] by years, matters, and court orders.
5. COOCCURRENCES MATRIX or adjacency matrix: We have built four co-occurrence matrices **A** (Year X Year), **B** (Subject X Subject), **C** (court order X court order), and **D** (Term X Term), that we used as input to visualize text data [10].
6. A.VISUALIZATION: We transformed the co-occurrence matrices (**A**, **B**, **C**, and **D**) in weighted undirected network graphs, where weights represent the co-occurrence count in the matrix. We created a layout using the Fruchterman and Reingold algorithm [11].

B. CLUSTERING: We applied two clustering methods:

(1) Ward2, also known as the minimum variance method, that is an agglomerative hierarchical clustering procedure. The object of Ward method is to minimize the increase of the within-class sum of the squared errors, [12]

$$E = \sum_{k=1}^{K} \sum_{x-i \in C-k} ||x_i - m_k||^2 \tag{1}$$

where K is the number of clusters and m_k is the centroid cluster C_k. At each step in the analysis, union of every possible pair of clusters is considered and two clusters whose fusion result in the minimum increase in the error sum of squares are combined. We applied intertextual distance [13] on **X** matrix, to measure proximities and oppositions in a corpus.

(2) Alceste method is a divisive hierarchical clustering algorithm whose aim is to maximize the inter-cluster Chi-squared distance [14]. This algorithm is applied to a special Document Term Matrix **U**, where documents are called uce (elementary context units). If an uce does not include enough terms, it is merged with the following one into an uc (context unit). This matrix is then weighted as binary, so only the presence/absence of terms are taken into account, not their frequencies (see Table 2).

The aim is to split the **U** matrix into two clusters by maximizing the Chi-squared distance between those. This method suggests to do a double clustering to get more robust clusters [14].

Once we have determined all the partitions, we calculate for each of them two criteria: the total size, i.e. the total number of documents belonging to the cross-classes of the partition the total Chi-square, or the sum of the Chi-square association of each crossed class of the partition. In the end, we keep for each value of k (therefore for partitions in classes 2, 3 …max_k), the partitions that have the maximum total size or the maximum total Chi-square (therefore, we have one or two partitions for each k).

Before undertaking any type of analysis, the corpus has been enriched with specific metadata associated with measures to better qualify automatic processing and improve knowledge extraction (Table 1; Fig. 2)

In the paper, we transformed the metadata "DATE" into "YEAR," and we obtained nine modalities, one for each year (from 2010 to 2018). "SUBJECT" is classified into 9 groups (1. conflicts of jurisdiction, 2. account, 3. request of a party, 4. civil pension, 5. war pensions, 6. military pensions, 7. issues, 8. realized, 9. responsibility). "COURT ORDER" into four groups (1. decree, 2. ordinance, 3. judgment, 4. judgment-ordinance).

Table 1 Input matrix for clustering

	w_1	w_2	…	w_j	…	w_p
uc_1	0	0	1	0	0	1
uc_2	1	0	1	0	0	0
…	..	…	…	…	…	…
uci	1	0	1	0	0	0
…	..	…	…	…	…	…
uc_k	1	0	1	0	0	0

META INFORMATION	
SEZ	code of jurisdictional section that issued the decision
DATE	date of pronunciation
NUMPROV	chronological number of decisions
SUBJECT	subject of public accounting
COURT_ORDER	Court order
OBJ.	judgment object
ID TEXT	progressive number of texts

Fig. 2 Metadata of the CORPUS

3 The Main Results

The work chain that accompanied the corpus analysis included all the steps illustrated in Fig. 1. In this section, we will focus attention on the main results.

3.1 Preprocessing and Building of a Dictionary

The corpus has been tokenized to detect the basic units of the analysis as lexical units and to calculate some lexicometric indexes which help the researcher to identify the existence of a peculiar language, i.e. a set of keywords that differentiate the corpus from other corpora.

After the lemmatization and lexicalization, the exclusion of stop words and words with a frequency lower than 10, the first statistical analysis returns quantitative measures relating to the corpus. The statistical analysis shows that the first 50 most frequent active forms are those referable to the subject of the corpus: the accounting jurisdiction. From a grammatical point of view, considering in the statistical analysis in addition to the active forms also the supplementary ones, a percentage prevalence of "unrecognized" forms is denoted, immediately followed by the presence of nouns, adjectives and verbs (See Table 2).

It has been observed that for many lexical forms, although belonging to the current dictionary of the Italian language, the automatic lemmatization algorithm did not correctly identify the so-called part of speech (POS), or the grammatical function of the single type within the sentence or fragment. In fact, the grammatical construction of the segments that make up the legal provisions holds particularities that are not easily recognized by the automatic system. The consistency of these "unrecognized" forms implies a linguistic deepening of the corpus and the need to carry out text disambiguation and grammatical tagging. The structure of a judicial order, for example, a sentence, follows specific forms also dictated by regulatory requirements. It is made up of a series of essential parts:

Table 2 Part of speech distribution

POS	Effectifs	Percentage (%)
nr	45,698	73.2
nom	8961	14.35
adj	4428	7.09
ver	2751	4.41
adv	541	0.87
versup	11	0.02
nomsup	39	0.06
Tot.	62429	100

- the header, the indication of the judge,
- the trial parties, the defenders,
- the conclusions of the parties,
- the description of the progress of the trial,
- the factual and legal reasons for the decision,
- the device of the measure and the signing of the judge.

It was therefore decided to intervene on the dictionary of forms by going to tag the part of the discourse of the active and additional forms not recognized, but referable to well-known grammatical forms. Furthermore, to give greater emphasis to the legal lexicon, considered particularly significant for the purposes of this textual analysis, the set of words that are believed to synthesize a juridical/legislative concept, has been suitably tagged. The corpus was again tokenized and lemmatized based on the new dictionary of forms built for the purpose. The analysis of the distribution of the frequency of words in the corpus has allowed us to study the so-called textual specificities, that is the set of terms that in our case typify a collection of legal checks on public accounting and pensions.

3.2 Visualization and Clustering

The first 100 most frequent active forms are strictly attributable to the accounting jurisdiction, both in terms of pensions and liability. It is no coincidence that the graphic form with the highest number of occurrences is the word "law" (LEGGE), followed by the terms "pension" (PENSIONE), "section" (SEZIONE), "court," (CORTE) "judgment" (GIUDIZIO), and "sentence" (SENTENZA). These results suggest the adoption of a specific language, from which legal/administrative technicalities emerge (see Fig. 3).

The legal provisions that best distinguish the corpus are those relating to tax liability and pension matters (See Fig. 4).

The variable "type of measure" indicates the type of judicial measure with which the legal treatment is defined; it can take the following modalities: sentence, ordinance, sentence/ordinance, decree. From the analysis of the specificities in relation to this variable, for each mode, the following situation is represented:

- Sentences (9%);
- Ordinances(28%);
- Ordinances/Sentences (29%);
- Decrees (8%).

For the variable "year" about 30% of the forms are under or overused respect the all corpus.

From the analysis of the lexical correspondences on the active forms concerning the variable "type of provision," most of the information, almost 70%, is represented by the first two factor axes. The first latent factor can represent 47% of the variance

Fig. 3 Wordcloud of the corpus

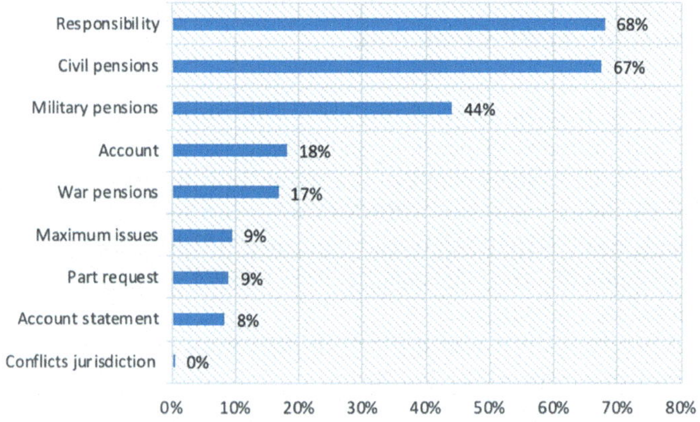

Fig. 4 Percentage of over/under terms used for the "Subject" variable

and the second 20.06%. Thus, by interpreting the position of words to the factorial axes, we can observe that terms referring to the type of action known as a decree, better characterize the corpus; these active forms are more distant from the origin of the factorial axes in which there are trivial forms, used on average throughout the corpus (Fig. 5). From the observation of the factorial representation (Fig. 5) 4 classes were found: 2 very separate (red and purple clusters) and 2 overlapping (blue and green clusters). The red cluster represents the bureaucratic work done by secretarial work and protocols; the purple cluster are the keywords of pension treatments; the

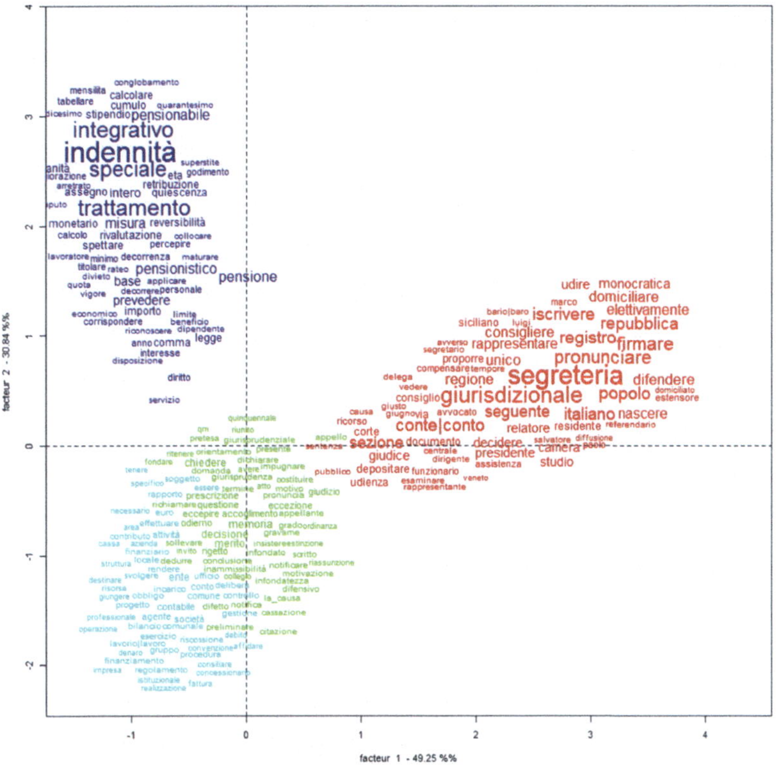

Fig. 5 Factorial representation

blue and green groups all the other categories. This processing was carried out with the help of the Iramuteq software. [4]

The first ten words that contribute most to the determination of the first two latent factors are mostly adjectives and verbal forms typical of language for drafting provisions on matters of accounting jurisdiction (Table 3).

The hierarchical clustering algorithm ward-2 has been applied to the corpus with reference to two notes: "Section" and "Object." Specifically, the clustering calculated in relation to the "Year" variable showed interesting results in terms of data information. As shown in the heatmap and tree graph below (Figs. 6 and 7), hierarchical clustering divides the corpus very well into two large groups: one in which provisions are drawn up between the years 2010 to 2013, the second in which the most recent ones are found from 2014 to 2018. This division certifies the regulatory changes occurred in accounting legislative framework that, in 2015, introduced the new accounting justice code. The first section for how to name the 26 jurisdictional sections of the Court of Auditors, a central and regional level; the second variable,

[4] See the website www.iramuteq.org.

Table 3 The first 10 words and their factorial contribution

	COR-Factor 1		COR-Factor 2
andare (TO GO)	0,994381797	Risarcitorie (COMPENSATION)	0,989894132
previsione (FORECAST)	0,994246389	Danaro (MONEY)	0,976470009
Corsista (STUDENT)	0,994160331	Ammissibilità (ADMISSIBILITY)	0,972772139
Ultimo (LAST)	0,993413387	Elusivo (ELUSIVE)	0,956157427
Motivo (REASON)	0,991566302	Quantificare (QUANTIFY)	0,955656406
Sez (SECTION)	0,991440731	Dissentire DISAGREE)	0,952501864
Purché (PROVIDED)	0,991170343	Pecuniario (PECUNIARY)	0,946510851
Inviare (TO SEND)	0,990859354	Irrogazione (INFLICTION)	0,944260734
Giurisprudenza (LAW)	0,988134126	Catalogare (CATOLOGUE)	0,943971861
Presidenziale (PRESIDENTIAL)	0,987186653	Arricchimento (ENRICHMENT)	0,941145478

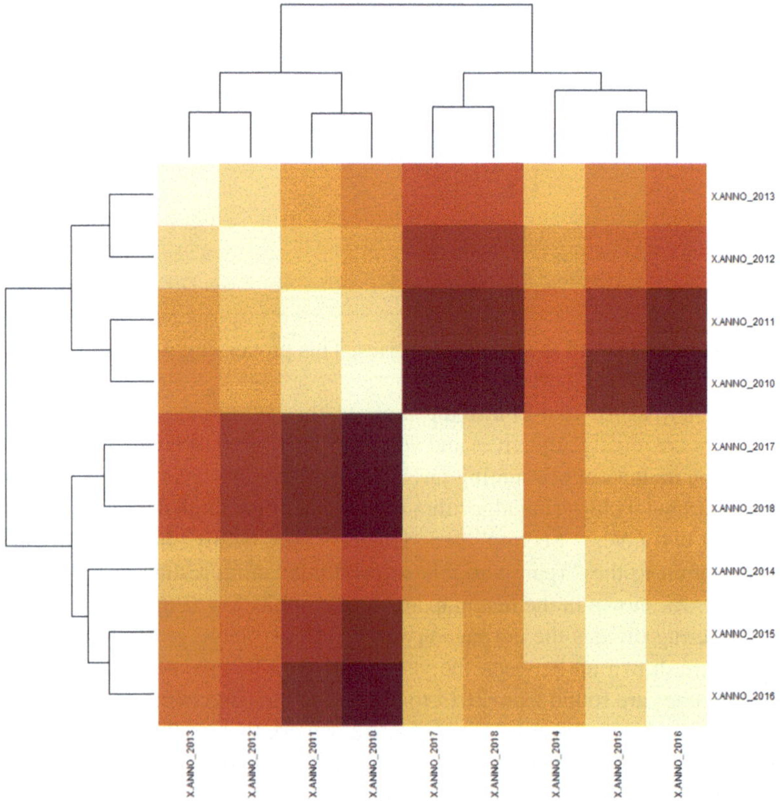

Fig. 6 Headmap obtained with hierarchical clustering (word2—intertexual distance)

Fig. 7 Tree graph

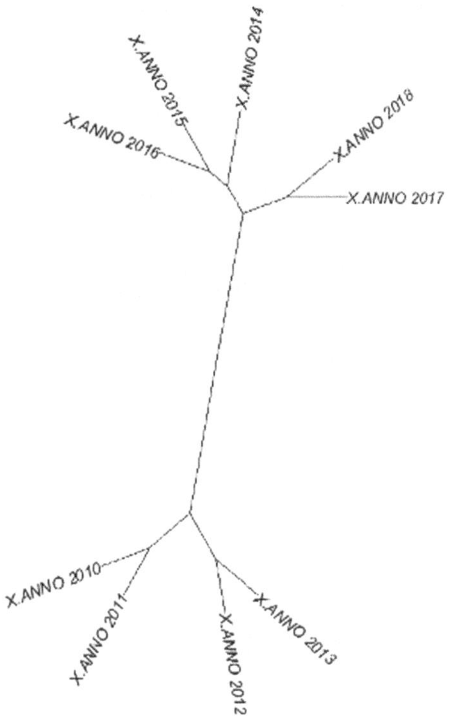

the methods of which represent the object of the cause: correction of material error, release from seizure, incidental, interpretation, application for facilitated definition, application for nullity, merit, compliance, provisional execution, complaint, revocation, seizure, suspension, other. The results of the clustering are clearly visible in the following figures, where the heat patterns and heat maps for both variables are listed. Co-occurrence statistics have proven very effective in discovering associations between concepts. In computational linguistics, recurrence statistics were applied for disambiguation of word sense, word grouping, and lexicon construction.

As can be seen from the image in Fig. 8 in the corpus, 13 clusters can be identified and each of them expresses a concept associated with the co-occurrence of words.[5]

4 Conclusions and the Future Steps

In recent years, the evolution of scientific research has made it possible to analyze big corpora. This paper describes the first results of an analysis on a thematic corpus consisting of measures issued by the Italian Court Audit in carrying out the jurisdictional

[5]For greater readability of the graph, the first 500 most frequent words were represented.

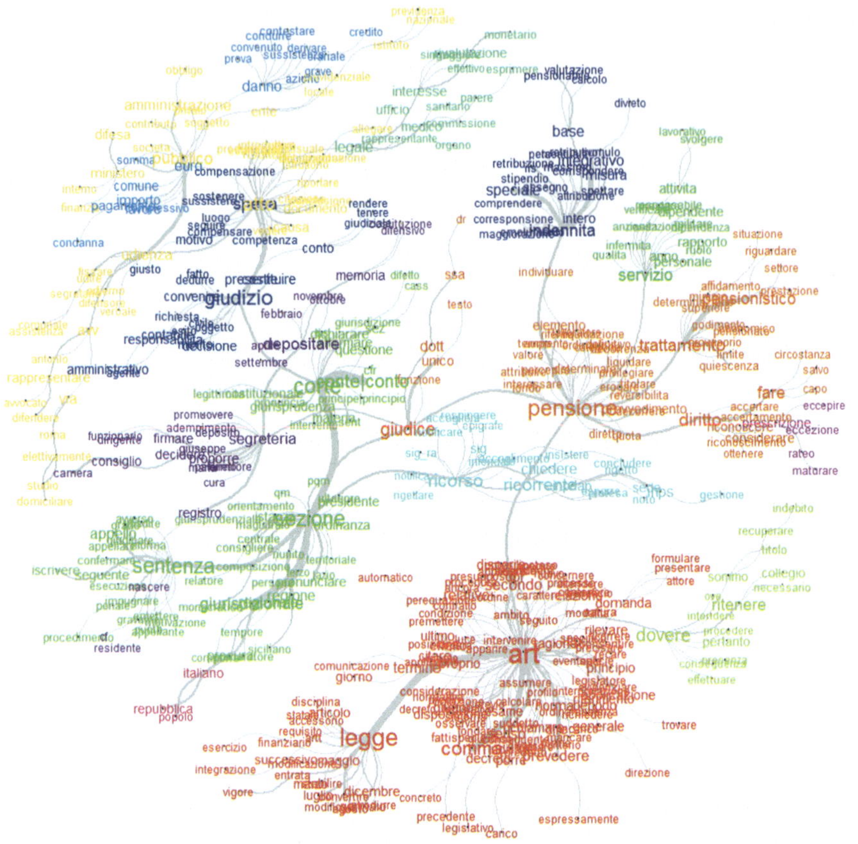

Fig. 8 Representation of the co-occurrence analysis

function in matters of liability and pension, trying to highlight the peculiarities of the legal-accounting language. The analyses required a very extensive work on data normalization and cleaning, which involved then a further work dictionary improvement to support analysis through semantic tags and lexical disambiguation, further contextualized to the context of the study.

On an open issue that has to be further investigated concerns the profiling of the corpus, where the help of supervised classification techniques may facilitate the preparation of the services by the accounting judges, offering them a focus on the motivating parts of the controls, considered strictly related to free conviction and legal appreciation.

References

1. Deriu F, Iezzi DF (2020) Text analytics in gender studies. Introduction., Int Rev Soc 30(1):1–5
2. Bolasco S (2013) L'analisi automatica dei testi. Carocci, Roma
3. Wei D (2014) Data mining with big data. IEE Trans Knowl Data Eng 26:97–107
4. Aguirre E, Edmonds P (2006) Word sense disambiguation. Text, speech and language technology. Springer, Hedemberg
5. Adebayo JK, Di Caro L, Boella RB (2017) A Legalbot: a deep learning based conversational agent in the legal domain. International conference on applications of natural language of information systems. Springer, pp 267–273
6. Palmirani M, Vitali F (2011) Akoma-Ntoso for legal documents. Legislative XML for the semantic web. Models, standards for document management. Springer Verlag, Principles, Berlin, pp. 75–100
7. Palmirani M, Pagallo U, Casanovas P, Sartor G (2012) AI approaches to the complexity of legal systems. Models and ethical challenges for legal systems, legal language and legal ontologies, argumentation and software agents. In: Lecture notes in computer science AI approaches to the complexity of legal systems. Models and ethical challenges for legal systems, legal language and legal ontologies, argumentation and software agents. Springer Berlin Heidelberg, Heidelberg
8. Zeleznikow J (2005) Using toulmin argumentation to develop an on line dispute resolution environment. To appear in proceedings of ontario society for the study of argumentation conference on the uses of argument, McMaster University, Hamilton, Ontario, Canada, May 18–21
9. Lafon P (1980) Sur la variabilité de la fréquence des formes dans un corpus. In **Mots**: 1 . Saussure, Zipf, Lagado, des méthodes, des calculs, des doutes et le vocabulaire de quelques textes politiques. pp 127–165
10. Iezzi DF(2012) Centrality measures for text clustering. Commun Stat 41(16–17):3179–3197
11. Fruchterman TMJ, Reingold EM (1991) Graph drawing by force-directed placement. Software: practice and experience 21(11)
12. Everitt BS (1980) Cluster analysis, 2nd edn. Heineman Educational Books Ltd., London
13. Labbé D, Labbé C (2001) Intertextual distance an authorship attribution. J Quant Linguist 8(3):213–228
14. Reinert M (1990) ALCESTE, une méthodologie d'analyse des données textuelles et une application: Aurélia de G. de Nerval. Bulletin de méthodologie sociologique 28:24–54
15. Biagioli C, Francesconi E, Spinosa P, Taddei M (2003) The NIR project. Standards and tools for legislative drafting and legal document Web publication. In: Proceeding of the international conference of artificial intelligence and Law. Edinburgh, June 24
16. Venturi G (2013) Investigating legal language peculiarities across different types of Italian legal texts: an NLP-based approach
17. Nanda R (2017) Legal information retrieval using topic clustering and neural networks. In: Coliee et al (2017) Fourth competition on legal information extraction/entailment. EPiC Ser Comput 47(2017):68–78

Lexicometric Paradoxes of *Frequency*: Comparing VoBIS and NVdB

Luisa Revelli

Abstract An emblematic "anti-spoken" canon, the linguistic variety offered as the standard model in Italian schools following national unification, while showing some signs of evolution over time, has remained relatively artificial and resistant to change. The lexical frequencies documented for scholastic Italian can therefore be intrinsically unaligned with those of the *base vocabulary* as well as with data for apparently similar varieties of Italian. This implies a need for interpretive models that assess quantitative data in the light of the complex paradigmatic relations among potential competing usages as well as the multi-layered connections between the number and type of meanings observed across different contexts of use. In this paper, I review the scholastic Italian modelled by teachers in the first 150 years after unification, with a view to assessing the strengths and weaknesses of applying lexicometric parameters to a linguistic variety that was targeted at an inexpert audience but specialist in nature, informed by lofty ideals but conditioned by practical educational needs, and constantly evolving yet resistant to the pull of contemporary living varieties.

1 Introduction

The first 150 years since the unification of Italy have seen a drastic shift in the status of the national language within Italians' sociolinguistic repertoire. In practice, Italian has gone from being a foreign or second language for most speakers to becoming their mother tongue. For as long as varieties other than Italian continued to be those most widely spoken and speakers had little opportunity to use Italian outside of school, teachers' consistent modelling of Italian forms and lexemes exerted a particularly strong influence on their students' linguistic practices and skills.

By increasing Italians' level of familiarity with their national language and in particular with certain lexemes and structures, the school system played a decisive part in the later fortunes of these same linguistic forms. In other words, the language

L. Revelli (✉)
Dipartimento di Scienze Umane e Sociali, Università della Valle d'Aosta, Aosta, Italy
e-mail: l.revelli@univda.it

D. F. Iezzi et al. (eds.), *Text Analytics*, Studies in Classification, Data Analysis, and Knowledge Organization,
https://doi.org/10.1007/978-3-030-52680-1_8

model presented by teachers contributed to the establishment of a shared variety of reference among literate speakers, and this variety in turn influenced the school variety, leading it to change more rapidly and in specific ways.

Nevertheless, within the overall architecture of Italian [3], *scholastic Italian* (henceforth IS)[1] persists as a distinct variety to *everyday Italian.* IS displays specific, divergent characteristics because it was intentionally devised for educational purposes, as is particularly evident from its lexicon: while on the one hand, an ideal of expressive richness induced teachers to discourage all forms of repetition or vagueness and to prefer lexical variety to the point of affectedness, on the other hand, the urgent need to teach children literacy skills imposed reliance on particular synonyms perceived as more correct, appropriate or ornamental than others, thereby reducing the available range of expressive possibilities. At the same time, the need to convey learning contents through language demanded the use of metalanguages, technical terms and peculiar semantic meanings that widened the gap between the lexical reservoir of IS and everyday Italian usages. To what extent and in what terms is this gap reflected in the relative *base vocabularies* of the two varieties? This is the question I now address based on a qualitative interpretation of lexicometric data.

2 Scholastic Italian

While the label "scholastic Italian" was coined in the mid-1970s [2], De Mauro [8]—in his *Storia linguistica dell'Italia unita*—had already observed that the specific variety of Italian used in educational settings was distinct and distant from the spontaneously spoken everyday variety. Scholars went on to identify typical or even exclusive features of IS [12]. Cortelazzo [6] provided a systematic description of the main phenomena characterizing what he defined as an "artificial variety of Italian that was (is) offered by the school system to young Italians setting about acquiring the national language". The features reviewed by this author included the use of the pronoun *ciò* instead of *questo* or *quello*, *il quale* in place of *che* and *in cui* instead of *dove*; a tendency to position adjectives before nouns; frequent use of the absolute past in place of the present perfect; use of the third-person singular pronoun in the anaphoric form *egli*; a preference for periphrasis; the substitution of direct speech with indirect speech; and frequent use of exclamations. With regard to lexicon, Cortelazzo noted a universal hostility towards the use of generic nouns and repetition, the rejection of regional idioms, a preference for synonyms from a higher register in place of the forms spontaneously used by the children, and extensive use of diminutives. Cautioning that the phenomena he reported did not all bear the same weight or occur with the same frequency and that his list of key features was not necessarily exhaustive, Cortelazzo dated the use of scholastic Italian to the period spanning the late 1920s and the 1970s.

[1] For general background on the stable and evolving characteristics of IS, see De Blasi [7], Cortelazzo [6], Benedetti and Serianni [1], Revelli [14].

In contrast, when observing that a preference for linguistic forms that were distant from the living language had represented a constant feature of how Italian was taught in schools, De Blasi ([7]: 19, note 42) dated the appearance of a specific scholastic variety to an even earlier period, theorizing that scholastic Italian had taken form "immediately after unification (and probably even beforehand)" and that while other scholars had associated it with the dominant rhetoric of the fascist period (1922–1943), the latter phenomenon had "perhaps only contributed to perpetuating a trend that was already flourishing".

The hypothesis that exclusively scholastic canons date back to the earliest beginnings of the Italian educational tradition and, albeit with somewhat modified characteristics, are still in use today has been confirmed by research conducted as part of the Progetto CoDiSV (*Corpus Digitale delle Scritture scolastiche d'ambito valdostano*)[2] and based on a collection of one thousand exercise books produced by school children in the Aosta Valley region from the late nineteenth to the early twenty-first centuries. The data analysed in this paper is drawn from this corpus, which has been more fully presented and explored in Revelli [14].

3 Methodological Aspects

With a view to investigating how the IS lexicon has evolved over time and which of its features have proven to be diachronically stable, we selected and computationally analysed 150 elementary school exercise books evenly divided into six chronological subcorpora corresponding to consecutive twenty-year periods between 1881 and 2000.

The computerized transcripts of the linguistic models presented by teachers[3] yielded a vocabulary of 152,151 tokens, based on 18,898 types and 11,751 lemmas.

The transcripts were initially analysed using the software program *Concordance*,[4] which supports *Key Word in Context* (KWIC) searches and generates frequency lists with the number of occurrences of each graphic form in the text and its *relative frequency* or the ratio—in percentage form—between the occurrences of this form and the total number or words in the corpus. This enabled us to conduct basic quantitative and concordance analyses for the six subcorpora.

We then conducted a more in-depth statistical-textual assessment using the program T-LAB 7.3.,[5] which segments the text into elementary units and supports

[2]The CoDiSV Project, initiated in 2004 at Aosta Valley University with the aim of collecting and studying samples of childhood writing [4]; [13]; [15], has yielded an archive with digital reproductions of over 1200 documents (mainly books, but also school registers, diaries, report cards, etc.) which is freely accessible at www.codisv.it.

[3]The types of text included in the present corpus were: instructions for the completion of exercises, composition titles, dictations, corrections, assessments and feedback, all sourced in the students' exercise books.

[4]www.concordancesoftware.co.uk.

[5]www.tlab.it.

flexible lemmatization in that the software's automatic choices may be edited to customize the *dictionary*. The manual operations required to disambiguate homographs and correctly lemmatize antiquated, polysemic and higher forms proved to be onerous, as did the effort required to work around the program's in-built constraints[6] and automatic settings, which are designed for *content analysis* as opposed to the formal linguistic assessment of texts. However, once these steps had been taken, we could avail of T-LAB's suite of tools to conduct valuable co-word analysis, textual analysis, and comparative analysis of the subcorpora.[7]

The specific nature and the size of the sample under study preclude meaningful direct comparison with other varieties of Italian based on exclusively quantitative parameters applied to the entire Italian lexicon. The IS vocabulary is inextricably bound up with contents and goals that are peculiar to educational settings: for example, the number of hapaxes in our sample was very high, especially in the context of certain kinds of exercise; the measures of terminological sophistication suggested an overall high rate of lexical richness, which however was extremely variable and discontinuous; the relationship between *full words* and *empty words*[8] was strongly influenced by the specific types of text selected for analysis, which included a large number of vocabulary exercises, list structures, syntactically simple sentences devised for the purposes of grammatical analysis, etc.

The IS lexicon may however be usefully compared with that of other school-related varieties (e.g. the language used in textbooks or the interlanguages productively used by learners) or varieties in other fields of usage, via qualitative–quantitative analysis of their relative *base vocabularies*. To this end, from the 18.898 types in our sample, we selected all the panchronic nouns, adjectives, and verbs (classified as such because they occurred more than five times in at least four of the six chronological subcorpora or in three non-consecutive ones) and tagged them grammatically and semantically: the 2.022 lexemes obtained following this procedure were taken to constitute the VoBIS—*Basic vocabulary of scholastic Italian* [14, 15].

The lexical repertoire adopted as a basis for comparison with the VoBIS was the 2016 edition of the *Nuovo Vocabolario di base della lingua italiana* (henceforth NVdB) curated by Tullio De Mauro and Isabella Chiari (2012),[9] which divides the approximately 7000 words that are statistically most frequent and accessible to twenty-first century Italian speakers into three reservoirs, respectively, labelled as *lessico fondamentale* (FO, around 2000 high-frequency words that are used in 86% of discourse and texts), *lessico ad alto uso* (AU, around 3000 frequently used words that cover 6% of occurrences) and *lessico di alta disponibilità* (AD, around 2000

[6]For example, Version 7.3 of T-LAB can only process a maximum of 1500 lexical units at a time when identifying word associations, comparing pairs of key words, conducting co-word analysis, and modelling emerging themes.

[7]For a detailed description of the guiding principles and methodologies adopted, see Revelli [14].

[8]*Full words* are those with meaningful semantic content (nouns, adjectives, verbs and adverbs), while "empty words" fulfil grammatical functions such as completing, connecting, and supporting *lexical words*.

[9]https://www.internazionale.it/opinione/tullio-de-mauro/2016/12/23/il-nuovo-vocabolario-di-base-della-lingua-italiana.

words that are "only used in some contexts but are understandable to all speakers and are perceived as equally available to, or even more available than, the most frequently used words"). We chose the NVdB, which encompasses the relative frequencies of spoken varieties of Italian and is based on a later time period than that of the VoBIS, as our standard of comparison so as to verify whether and to what extent the written model offered by IS has influenced subsequent practical usages.

4 Base Vocabularies *Compared*: *Frequencies* in the NVdB and VoBIS

Initial comparison of the two lexical repertoires under study yielded the following general outcome: of the 2022 lemmas contained in the VoBIS, 1784 were also present in the NVdB, with 53% featuring in the FO, 26% in the AU and 9% in the AD lexicons.

While we cannot offer an in-depth treatment of the convergences between the two vocabularies here, it should be pointed out that many apparently similar frequency distributions actually conceal significant differences, mainly due to IS's tendency to offer more restricted or even alternative semantic definitions: among the many words that take on specifically educational meanings in IS (e.g. *diario*, *interrogazione*, *nota*, *pensierino*, *voto*), some retain no connection whatsoever with their everyday meanings, as in the case of *tema*, which rather than a subject or topic to be dealt with, refers, in IS, to a specific type of text.

Turning to the 238 VoBIS words that do not feature in the NVdB (12%), grouping these into categories helps to point up the critical issues that can arise when the comparative frequency parameter is applied.

A first, large category that is exclusive to IS consists of metalinguistic terms which are highly characteristic of the school setting, such as *alfabetico*, *apostrofo*, *coniugazione*, *preposizione*. Despite their polysemic potential, many of these words—such as *coniugare*, *derivato*, *imperfetto*, *possessivo*, *primitivo*—are frequently used in IS but exclusively as metalinguistic labels.[10] Their strong quantitative incidence does not therefore imply that students were exposed to the different meanings that these terms can bear, but rather that—for purely educational purposes—they were insistently presented with their specialized meanings.

A second group comprises terms that are typically used in the context of teaching children to read and write. These are mainly nouns whose referents, though concrete, scarcely ever feature in everyday life. They are used at school because their written form both provides and demands knowledge of key but counterintuitive conventions for the spelling and decoding of written words. Examples include *acquaio*, *acquavite* and *acqueo*, which are clearly not introduced to students for pressing thematic reasons, but rather with the aim of consolidating correct graphemic representations.

[10]For example, *dimostrativo*—which is always preceded by *aggettivo* or *pronome*—never occurs in combination with *atto*, *gesto*, etc.

The high frequencies of other terms are explained by their pertinence to subject-specific or broad educational themes. Examples include historical–geographical terms (*legione*, *vetta*), words for describing the world of nature (*arto*, *astro*) and rural life (*semina*, *vendemmia*) and sets of verbs (*castigare*, *disobbedire*), adjectives (*diligente*, *ordinato*) and abstract nouns (*umiltà*, *penitenza*) conventionally drawn on in the context of civic or moral education and student assessment.

Unlike the NVdB, the diachronic nature of the VoBIS means that it features numerous outdated terms: these may be formal variants that have fallen out of use (e.g. *annunziare* in place of *annunciare*) or are currently less preferred (*ubbidire* in place of *obbedire*); terms whose referents have been rendered irrelevant or anachronistic by the social changes of the past fifty years (*manto*, *ricamatrice*); old-fashioned or formal members of pairs/sets of synonyms that only in the school setting have continued to be preferred over competing terms perceived by speakers as more contemporary (*persuadere* in place of *convincere*).[11]

Again in relation to the differences between the two repertoires, while the absence from the NVdB of somewhat affected terms in the school lexicon such as *diletto*, *garbato*, *vezzo* and *soave* is only to be expected, it is surprising to find that terms which had remained stably in use over time and all over Italy are also missing from it. This is the case of zoonyms such as *bue*, *elefante*, *formica*; names of fruits regularly consumed by Italians such as *fragola*, *noce* and *uva*; and concrete nouns in common use such as *carezza*, *martello*, *ombrello*. The fact that terms from these categories are not to be found in the NVdB can only partly be explained by sociolinguistic factors. On the one hand, it is true that—given its target audience—IS makes more frequent reference to themes and referents from the material and experiential culture than do varieties targeted at and produced by adult speakers; but on the other hand, the fact that certain words are emphasized and learnt at school should theoretically make them even more likely candidates for inclusion among those terms which—though only used sporadically in everyday life—are universally known by virtue of their association with commonplace objects, events and experiences. We would thus expect to find them at least in the AD repertoire. However, De Mauro himself has on several occasions pointed out the elusive nature of these repertoires, which should not be viewed as statistical but as conjectural.[12] Ultimately, even the most powerful and meticulous quantitative analyses would not suffice to eliminate the unpredictable and random factors that inescapably affect the measurement of frequency. In our own

[11] *Bambagia*, *cagionare*, *figliolo*, *focolare*, *garzone*, *uscio* are further instances of this last-mentioned category. Overall, the IS repertoire is markedly different to that of the contemporary lexicon, systematically preferring words that are missing from the NVdB, even in place of terms classified as FO within the latter (e.g., *appetito* in place of *fame*, *ardere* in place of *bruciare*, *sciupare* in place of *rovinare*, etc.).

[12] "It is challenging, it is extremely difficult to systematically identify following objective criteria those everyday words which we all know but may not utter for months and years on end" [11]. In the preface to the NVdB, it is specified that the AD words "were obtained based on the old VdB's list of 2300 highly available words, from which groups of university students were asked to delete the words they no longer perceived as among the most widely used and to approve new words they perceived as currently highly available".

data, given that the IS corpus is of limited size, we also observe unexpected gaps in the other direction: certain keywords in the NVdB—especially the FO repertoire—which might reasonably be expected to feature in the VoBIS are not included in it. While we may surmise that these terms were known to the children who wrote the texts in the sample, they do not feature in the resulting base vocabulary for purely accidental reasons.[13] Other types of omission may certainly be viewed as intentional and as specific traits of IS. Such deliberately omitted categories include: neologisms and loanwords, which teachers—possibly in some cases to avoid creating spelling difficulties for their students—tended to avoid, even after they had become well established in standard Italian (*jeans*, *quiz*, *smog*); terms deemed inappropriate for a young audience (*aborto*, *droga*, *sesso*); slang, vulgar expressions, insults, swear words (*coglione*, *culo*, *ruttare*); derogatory labels (*ebreo*, *nano*, *negro*); and even words that teachers likely avoided on prudential grounds because they perceived them as potentially divisive, propagandistic or at least as bearing ideological and political connotations. It is not possible to overgeneralize in relation to this last category, which embodies the intimate relations between lexicon, schooling and social and cultural climate. Indeed, the available evidence as to how the phenomenon may have varied over time and even up to the present day is based on low or null frequency rather than high-frequency units of the base vocabulary.

5 Conclusions and Future Prospects

As I set out to show, our qualitative–quantitative analysis of IS confirmed that, despite displaying signs of modernization over time, the language model presented by Italian teachers to their pupils has stably featured words that are not part of the standard base vocabulary, while simultaneously excluding commonly used terms deemed unsuitable, inappropriate or simply overused. The data mining process implemented revealed the presence of numerous logonyms and labels that are typical of, or exclusive to, the metalanguage of teaching and grammar, hapaxes only occasionally used in the context of particular types of exercises but indispensable for teaching purposes, and formulas in which common words take on specialized semantic meanings—different to their everyday meanings—to refer to tasks and routine communications that are typical of the school setting. The lexical frequencies documented for varieties of scholastic Italian thus prove to be intrinsically unaligned with those of the standard base vocabulary, as well as with data for apparently similar varieties of Italian. This implies a need for interpretive models that assess quantitative data in the light of the complex paradigmatic relations among potential competing usages as well as the multi-layered connections between the number and type of meanings observed across different contexts of use. This line of inquiry, which has already been partly

[13]One example of the extent to which random factors can come into play is the case of ethnonyms, of which some are missing the CoDiSV (e.g., *cinese*, *iugoslavo*) while others that would appear to have been equally widely used are included in it (e.g., *giapponese*, *inglese*).

explored in psycholinguistic and language learning studies on text comprehension and readability, may be further pursued by systematically comparing chronologically similar corpora of IS and VdB in relation to at least two specific areas.

The first of these areas, which is more specifically learning related, concerns the actual outcomes of protracted exposure during the school years to IS words that do not feature in the base vocabulary of standard Italian. In the light of the incremental and adaptive nature of vocabulary acquisition, but also the ease with which lexical knowledge can be lost if it is not put into practice, the following research questions might usefully be investigated. To what extent does repeatedly encountering a given term in the IS context affect its subsequent use in other life domains? To what extent does being taught a specific meaning of a term in the school context influence (positively or negatively) the subsequent learning of additional and different meanings of the same word? How do the preferred models and paradigmatic options offered by IS fare out, at least in terms of receptive competence, in the face of competition from other varieties encountered by speakers in potentially more meaningful domains? And indeed, how authoritative, significant and communicatively salient can the scholastic lexical model actually be in a country where Italian has become the mother tongue of the majority of citizens, and competing input—lexical and otherwise—from alternative sources appears to be quantitatively overwhelming?

A second area, which is related to the first but with a more lexicographic focus, regards the hypothesis that part of the *scholastic base vocabulary* is shared by all literate adult speakers and may contribute to shaping their *highly available lexical repertoire*. The objective difficulties that arise when we attempt to identify the "words we view as the most common, both mentally and culturally, but which are actually of minimum—almost zero—frequency, especially in written usage" ([10], 142) might be partly overcome by investigating that portion of speakers' shared lexical patrimony that they acquired, potentially also through other channels, but certainly via the IS. Although statistically insignificant within their adult production, the terms that are familiar to all speakers because they occur highly frequently and perform key functions in the Italian that are directed at school children—for example terms typically featured on alphabet charts (*oca*), words used to teach unusual spellings (*chamois*), the names of popular educational games and exercises (*cruciverba*), the lexicon of fairy tales and children's stories (*carrozza*), school subjects (*geografia*) or school routines (*giustificazione*)—could likely be readily elicited from speakers, and therefore, albeit barely identifiable in the adult lexicon, be included in the AD repertoire of the standard Italian *base vocabulary*.

Of course, in this case too, to avoid any semantic ambiguity or misinterpretations, appropriate instruments should be identified and used to map out the phenomenology of the effectively active meanings associated with these words, as well as to assess and critically interpret the relationships between frequency of lexical (and semantic) input and that of both actual and potential lexical (and semantic) output. In short, the descriptive paradigm deployed in future research should bring a dialectic, dynamic and comparative approach to bear on the dimensions of *receptivity*, *productivity* and *availability* in relation to the different meanings of given lexemes. For only by attributing the appropriate relative weights to *quantity* and *quality*—the *indispensable duo* flagged by De Mauro [9]—can we correctly distinguish between true and merely apparent paradoxes in the domain of statistical frequencies.

References

1. Benedetti G, Serianni L (2009) Scritti sui banchi. L'italiano a scuola fra alunni e insegnanti. Carocci, Roma
2. Benincà P, Ferraboschi G, Gaspari G, Vanelli L (1974) Italiano standard o italiano scolastico? In: Dal dialetto alla lingua, Atti del IX Convegno per gli Studi Dialettali Italiani. Pacini, Pisa, pp 19–39
3. Berruto G (1987) Sociolinguistica dell'italiano contemporaneo. La Nuova Italia Scientifica, Roma
4. Bertolino F (ed) (2014) Stili di vita, stili di scuola. Aracne, Roma
5. Chiari I, De Mauro T (2012) The new basic vocabulary of Italian: problems and methods. Rivista di statistica applicata/Italian J Appl Stat 22(1):21–35
6. Cortelazzo M (1995) Un'ipotesi per la storia dell'italiano scolastico. In: Antonelli Q, Becchi E (eds) Scritture bambine. Laterza, Roma-Bari, pp 237–252
7. De Blasi N (1993) L'italiano nella scuola. In: Luca Serianni, and Pietro Trifone (eds) Storia della lingua italiana, vol I "I luoghi della codificazione", Einaudi, Torino, pp 383–423
8. De Mauro T (1963) Storia linguistica dell'Italia unita. Laterza, Roma-Bari
9. De Mauro T (1995) Quantità-qualità: un binomio indispensabile per comprendere il linguaggio. In: Sergio Bolasco, and Roberto Cipriani (eds) Ricerca qualitativa e computer. Teorie, metodi e applicazioni. Franco Angeli, Milano, pp 21–30
10. De Mauro T (2004) La cultura degli italiani. In: Erbani F (ed) Laterza, Roma-Bari
11. De Mauro T (2014) Storia linguistica dell'Italia repubblicana. Laterza, Roma-Bari
12. Moneglia M (1982) Sul cambiamento dello stile della lingua scritta: scrivono i bambini. In: Nencioni G et al (eds) La lingua italiana in movimento. Accademia della Crusca, Firenze, pp 230–276
13. Revelli L (ed) (2012) Scritture scolastiche dall'Unità d'Italia ai giorni nostri. Aracne, Roma
14. Revelli L (2013) Diacronia dell'italiano scolastico. Aracne, Roma
15. Revelli L (2015) Dal micro- al macro-: storie del mutamento linguistico attraverso il CoDiSV. In: Gianmario Raimondi, and Hélène Champvillair (eds) Il CoDiSV in classe. Proposte metodologiche e didattiche di ricerca applicata. Aracne, Roma, pp 93–202

Emotions and Dense Words in Emotional Text Analysis: An Invariant or a Contextual Relationship?

Nadia Battisti and Francesca Romana Dolcetti

Abstract Emotional textual analysis (ETA) analyses the symbolic level of texts as a part of applied research and interventions. It is based on the study of the association of dense words, words that convey most of the emotional components of texts. In this approach, language is thought of as an organizer of the relationship between the individual contributor of the text and his or her context, rather than as a detector of the individual's emotions. This paper presents the findings of a research project whose goal is to advance the reflection on the theoretical construct of dense words based on its use in psychosocial interventions. Seventy-nine dictionaries of 79 ETAs, which were conducted over a period of 11 years by two different groups of researchers, were analysed in order to: (1) measure the concordance between the two groups of researchers in identifying dense and non-dense words in the analysed texts; and (2) explore, by correspondence factorial analysis, the relationship between the range of consistency of the words and their presence as dense or non-dense in the groups of dictionaries that were aggregated whenever their original corpuses were thematically similar. The results of these analyses show the level of agreement among researchers in identifying emotional density, confirm the hypothesis that there are affective invariants in words and suggest that the relationship between words and emotions can be represented along a continuum in which the context assumes a discriminating function in identifying density or non-density. Thus, it would seem that the construct of density has fuzzy boundaries and is not dichotomously polarized as in sentiment analysis, which classifies words as positive or negative.

N. Battisti (✉) · F. R. Dolcetti
Studio RisorseObiettiviStrumenti, Rome, Italy
e-mail: studio@studio-ros.it

D. F. Iezzi et al. (eds.), *Text Analytics*, Studies in Classification, Data Analysis, and Knowledge Organization,
https://doi.org/10.1007/978-3-030-52680-1_9

1 Emotional Text Analysis: Specific Characteristics and Issues in Research

Classifying the graphic forms of a text into new categories and using them as active variables is a long-standing practice in text analysis [1]. As in other methodologies, emotional text analysis (ETA) conceived by Carli and Paniccia at the beginning of the 2000s [2, 3] seeks to respond to the need for research to identify invariants within language.

ETA is characterised by the fact that it seeks invariants in the language process rather than only in the content of the word and that only the emotionally dense words of the text are its active variables. In order to carry out an ETA, words are classified on the basis of two references: (1) psychological, where words are identified as having or not an emotional density according to models rooted at the crossroads of where psychoanalysis, sociology, anthropology and statistics meet and integrate; and (2) linguistic and semiological, where graphic forms are reduced to a "quasi-lexeme", which in this paper is called "stem". The etymology allows to recover the polysemic richness of the word as it evolved in its use from its origins to the present day. Since the 1960s, the philosophy of language has been historically and culturally decisive when, with authors such as Foucault [4], it focused on the problem of language no longer only as the transmission of content but also as an action on the real.

The dense word construct is at the centre of the ETA model and is the basis of a methodology that traces emotions in discourses and texts with the aim of capturing the affective dynamics of psychosocial intervention [2, 3, 5–7].

Selecting emotional words has always been an integral part of studies on sentiment analysis. These studies, which have adopted various research approaches and methods, have a long-standing tradition, starting from the pioneer studies by Osgood et al. that began in the 1950s [8].

The construct of emotion used in ETA is distinct from others in that it is not based on an a priori categorization of the emotions present in the language, which would definitively link an emotion to its expression. Indeed, this is what occurs in the construct of primary emotions whose expression would be innate [9, 10] and, moreover, based on facial expressions rather than on language. It is also what is suggested in the theoretical approach of secondary emotions, that is, those acquired and influenced by cultures [11, 10].[1] In the psychological models underlying ETA, the presence of a relationship between words and emotions is inferred due to the extensive and constantly changing process of affective meaning with which people, through language, organize the reality in which they live. Words, as traces of the affective and symbolic process, are both subjective and social expression. As Reinert [12] said: "Lieux d'énonciation entre les représentations personnelles et les pre-construits sociaux, donc partagés" [Places of utterances between personal representations and social pre-constructions, therefore shared].

[1] Indeed, the process begins with identifying some of them and ends with detecting hundreds. The primary ones are typically: joy, sadness, anger, surprise, interest and disgust. To these, many others are added from time to time, such as shame, guilt, embarrassment, contempt and shyness.

In this theoretical and operational framework, an emotion expressed through words is an act [13]. More precisely, it is a collusive proposal [14–17], which is organised within a relationship, confirming its nature as an individual phenomenon with a social value. Operationally, the emotional sphere, which is a way for the mind to be infinite, is channelled into words and is expressed both through individual words [18] and through interwoven co-occurring words [12].

Hence, ETA continues in the tradition of distinguishing between empty, ambiguous and instrumental words and words that bear full meaning, marking, in their occurrence, the meaning of discourse and also of the affective one [12, 19]. Empty, ambiguous and instrumental words have the only function of establishing links with full and dense words. While dense words are organized by association, such as in interweaving words where a given word leads towards an emotional direction that will then be compiled, confirmed or challenged by the next word and so on. The almost endless combinations of words and the possibility of inventing new ones, which can resemble each other without ever being identical, see emotionality as an unending construction of new directions that drive coexistence.

If we assume that the construction of linguistic connections serves the operational function of language—a typical function of the dividing and "asymmetrizing" mode of the mind [18]—ETA captures the emotional process by deconstructing those very connections in order to revert to the associative links between words, those with a high emotional value. It also captures the local configuration with the statistical support of co-occurrence calculation. Thus, ETA is intended as a process that is isomorphic to that of the symbolic emotional functioning of the mind, operationalizing the knowledge that Sigmund Freud discovered he could attain through the method of free association. Moreover, this method has inspired various schools of thoughts regarding psychosociological interventions in France and Italy since the 1970s [20].

ETA is carried out with the aid of various software such as Alceste, developed by Reinert [12, 21] and T-Lab by Lancia [22]. In order to carry out an emotional analysis of a textual corpus, the internal dictionaries of the software are modified to edit a specific dictionary for that particular corpus; just like for other types of textual analysis, in ETA the active variables are not the types and a preliminary stemming must be carried out.[2] As for other analyses, the stemming process peculiar to ETA begins with excluding from the active variables the grammatical functors, also called function, denotative or empty words. It is followed by "lexematization".[3] [23], and finally, only the words that are a trace of the emotional nature of the text are identified and used for the purposes of analysis.

[2]We used the term stemming and not tagging to indicate that in ETA the labelling is distinct from, albeit with some elements of, the best-known linguistic or semantic models and in other aspects, also from semiometry.

[3]According to Torres-Moreno, an alternative to lemmatization and stemming is lexematization, that is, clustering words in the same family according to spelling. In principle, lexematization enables the words (chanter, chantaient, chanté, chanteront, chanteuse, chanteuses, chanteurs, chant) to be clustered in the same graphic form: "chant".

This selection process relies on the psychosocial competence of the researcher, who, in turn, combines psychological and sociological knowledge with that of the etymology of words. Etymology can explain the historical and cultural background that the word introduces into the context in which it is used. For this reason, as already mentioned in previous works [5], stemming can also be limited to only some of the graphic forms of the same lexeme when an emotional value is recognized only to some of its forms, or it can involve multi-words. We will discuss it in more detail later in this paper.

2 The Research Project

This paper presents the findings of a research project whose goal is to advance the reflection on the theoretical construct of dense words based on its use in psychosocial interventions. The research project consisted of two stages:

(1) Testing, through standardized deviation, the concordance between two groups of researchers who use ETA, both skilled in this method.
(2) Identifying, through correspondence factorial analysis, the best possible representations of the relationship between the levels of agreement on the words and their presence as dense or non-dense and the groups of dictionaries that were aggregated whenever their original corpuses were thematically similar.

2.1 Testing the Consistency of Dense and Non-dense Words: Data Construction and Methodology

In order to create basic lists of dense and non-dense words, the research project followed two intertwined directions: (a) built a database consisting of emotionally dense words chosen by two groups of researchers, with different experience, but with the same training; (b) verified the variability and coherence of the choices of these researchers through different statistical methods.[4]

This research project consisted of a meta-analysis of the concordance among researchers, which was based on choices that were already made during the interventions, each having their specific characteristics.

The 79 dictionaries that were collected came from ETAs carried out over a period of 11 years (1998–2009), by two groups of researchers that we named "Senior" and "Junior".[5] They differ primarily in terms of experience: the Senior group having

[4]Remember that ETA users proceed by manually selecting dense words as active variables in the dictionary of the text of their analyses.

[5]Since an ETA is carried out within research–interventions, they are the result of the work carried out by several researchers, and the choice of the words on which the analysis is based on is usually arrived at through discussion among several researchers. In every case that we examined, the final choices were made by three researchers who took on this task.

developed this methodology, also used it first, while the Junior group was trained in it later. Specifically, 38 dictionaries from ETAs that were carried out between 1998 and 2002 were collected from the Senior group and 41 dictionaries from ETAs that were carried out between 2002 and 2009 were collected from the Junior group. In addition, the Senior group consisted of researchers who shared a steadier working relationship with respect to those in the Junior group. There are some researchers who participated in the latter group on a more continuous basis, while others participated occasionally in order to carry out specific ETAs, which are performed not only for external clients, but also for training less-experienced researchers.

The dictionaries were created from a corpus consisting of texts that were homogenous by type: 41 corpora were created from interviews, 23 from essays, 6 from reports, 4 from focus groups, 2 from newspaper articles, 1 from clinical interviews, 1 from diaries and 1 from the text of a book.

Each dictionary varied in size from a minimum of 1182 word types to a maximum of 25,816 (with an average of 6602). The type–token ratio (TTR), the ratio between types and tokens [24], was below 20% for 92% of the dictionaries of the Senior group and for 81% of those of the Junior group, indicating, on the whole, a good level of lexical richness of the corpora.

The first stage of the project consisted in building the necessary data to identify and compare each researcher's choices regarding the emotional density of words in the different dictionaries. A relative normalized deviation was applied, and a concordance analysis was carried out on the extent that the Senior and Junior groups agreed on the density or non-density of the words. We then developed a correspondence analysis and some specificity analyses based on these data.

Some preliminary steps were needed before carrying out the analysis on concordance among researchers due to some problems, which were not always easy to solve. The first problem regarded the labels used for stemming the word types in each dictionary. They were different from case to case because the researchers had carried out the labelling manually for each analysis. Hence, the original files were revised to make them homogenous in order to be able to compare them between the different dictionaries.

Indeed, originally we had files with a.dat extension and an A2_DICO name, which is the name of the dictionaries of the corpora assigned by the Alceste software that was used by both groups of researchers for the 79 ETAs. This type of file contains various information in line form: the lists of the word types of the corpus and any word written in upper case, both in alphabetical order; lists of any number present in the corpus, the modality of the illustrative variables used in each analysis and the punctuation of that corpus; and in column form: the digital identification codes for each graphic element constituting the corpus; the occurrence of the graphic forms present in the corpus together with a ("a", "r" o "s") code to indicate the specific element as an active (accepté), non-active (rejeté) or supplementary variable. There is also a column for the possible merging of the different word types under one common stem. Only these three columns and the lines regarding the words were used in this research. Specifically, from the last column, we imported only the stems above the occurrence threshold used for each analysis—this threshold, which is usually 3, ranged up to a maximum of 6 in the 79 dictionaries.

For all 79 dictionaries, only the data on the dictionaries' graphic forms, their stems and the information—"a" = accepté or "r" = rejeté—indicating the researcher's choice for that specific word as dense or non-dense, were collected. At the end of this process, across the 79 dictionaries as a whole, there were 6,109,401 original occurrences.

These data were then inserted into a single file and modified so as to allow the TaLTaC software [25] combined with TreeTagger software[6] to import them as a cross-tabulation table, that is, dense or non-dense words x dictionaries.

Using Excel and through meticulous manual work to take into account similarities and differences in the behaviour of the researchers as regards to stemming, all the different labels used for the same stem were reduced to only one label.[7]

Since the resulting table was very large, we only provided the contents of its columns as follows: (1) the graphic form of the words in the original texts; (2) the stem used in each case by the researcher; (3) the reduction to a common stem for the purposes of this research; (4) indication of density or non-density of the word; (5) the successive 79 columns indicate how the stem was dichotomously processed, whether dense or non-dense, by the researcher.

After having removed the first two columns of the abovementioned table, the new stems field that was created for this research was merged with TaLTaC. Hence, in the resulting table, we have one or two rows for each stem: one row when, in the various analyses, it was always processed as dense, and two rows when, in different analyses, it was processed either as dense or as non-dense. In order to obtain homogenous information on how each stem was processed by the researchers in the various analyses and have it fit in a single row, we turned to a data processing expert. He produced a Boolean table, in full disjunctive normal form, with stems in rows and the dictionaries in columns, containing a total of 12,571 stems identified in the 79 dictionaries. In this table, for each dictionary, each stem was classified as "dense", "non-dense" or "not present in the dictionary".

We then proceeded in creating a new Excel table (an excerpt is shown in Table 1), where for each new stem we can see: (1) the number of dictionaries in which it was present; (2) the number of dictionaries in which it was classified as non-dense; (3) the number of dictionaries in which it was classified as dense; (4) how many times it was classified as non-dense by the Senior and Junior groups of researchers, respectively, and (5) how many times it was classified as dense by the Senior and Junior groups, respectively.

---

[6]The TreeTagger is a tool developed by Helmut Schmid at the Institute for Computational Linguistics of the University of Stuttgart, which can also be used for annotating text with lemma information.

[7]For example, the stem "to_abandon", which, as a rule, includes both the verb and adjective forms, and was labelled in the different dictionaries as abandon <, aband <, abandona < and abando < has been reduced to the single "to_abandon" label. However, if the lexeme "to_compete" was traced to at times as distinct "to_compete" and "competence" and others not, in the latter case, unification was not forced.

Table 1 Classification of the stems in the dictionaries, by the Senior and Junior groups

Stem	Present	Non-dense	Dense	Dense (Senior group)	Non-dense (Senior group)	Dense (Junior group)	Non-dense (Junior group)
Ancient	16	5	11	6	2	5	3
Academy	21	2	19	7	0	12	2
Explore	20	1	19	13	0	6	1
Compulsory	18	1	17	11	0	6	1
Alarm	14	1	13	7	0	6	1
Become ill	10	1	9	3	0	6	1
Observe	38	19	19	10	8	9	11
Manager	20	3	17	10	1	7	2
Actor	17	6	11	5	2	6	4
News	31	15	16	8	6	8	9
Deny	26	6	20	10	2	10	4
Shadow	13	3	10	5	1	5	2
Exclude	35	8	27	11	2	16	6
Nest	15	7	8	3	2	5	5
Ranking	6	1	5	1	0	4	1
Neutral	6	1	5	1	0	4	1

In order to establish a link between the observations on the agreement between the Senior and Junior groups [26] within the various dictionaries, we developed an index through the simple technique of combining two values. It is a standardized deviation formulated as follows:

$$\frac{\text{Relative frequency Junior dense} - \text{Relative frequency Senior dense}}{\sqrt{\text{Relative frequency Senior dense}}}$$

Two conditions were compared using this formula: the relative choice by the Senior group and the relative choice by the Junior group regarding the density of each stem in the dictionary. Through this index, the Senior group's choices were used as model to examine agreement. The root in the denominator is used to offset the effects of the comparison. The index ranges from −1.00 to 1.83, where the maximum agreement between the Senior and Junior groups is around 0.

There is maximum agreement between the groups when they converge on the density or on the non-density, as well as when they converge on the relative frequencies that each group attributes to the stem. The positive sign or negative sign of the index indicates which of the two groups, Senior or Junior, deviates more in choosing a word as being emotionally dense or non-dense. The closer the number of observations made by the two groups of researchers, the more reliable the index.

However, it does not yield a result under certain conditions: for the stems that have been observed only by one of the two groups (7079 out of 12,571), and for the stems that the Senior group has always deemed non-dense (indeed, it should be noted that the above equation is set to the relative frequency of dense words). Given this limit, for the latter cases (1112 out of 12,571), only the normalized relative deviation between the Senior and Junior groups was calculated on the basis of the relative frequency of the non-dense words, identifying, inter alia, many other agreements on the non-density of words. At the end of this process, there were 5492 cases out of a total of 12,571, that is, 43.69% of the stems present in the 79 dictionaries, for which we were able to calculate the concordance between the two research groups.

Before proceeding with the evaluation of these data, we had to address the frequency issue because, overall, many words were observed in a small number of dictionaries. Hence, in the subsequent processing we decided to use only the stems that were present in at least 10 of the 79 dictionaries, deeming this to be the minimum frequency on which we could make reliable considerations on the results. In order to overcome the problem of low frequencies in the future, perhaps a unit for monitoring the dictionaries used in new ETAs could be set up.

Looking back at the results of the analysis on concordance between the two research groups, the database that was used contained 2758 stems, that is, 21.94% of the initial 12,571 stems. Table 2 and Fig. 1 show some of the results on the trend of concordance between the Senior and Junior groups in classifying the words as dense and non-dense.

Figure 1 shows that agreements are more numerous in the area of negative scores, which is due to the Senior group's greater trust in its choices with respect to that of the Junior group. This could be in line with the elements with which we initially differentiated the two groups: attributing a possible greater expertise to the Senior group and observing less variability of this group with respect to that of the Junior group.

Table 2 Data on agreement per range

Agreement range	Number of stems	Percentage of stems (%)
From −1 to −0.76	54	1.96
From −0.75 to −0.51	212	7.69
From −0.50 to −0.26	526	19.07
From −0.25 to −0.01	810	29.37
0	699	25.34
From +0.01 to +0.25	304	11.02
From +0.26 to +0.50	91	3.30
From +0.51 to +0.75	37	1.34
From +0.76 to +1.83	25	0.91
Total	2758	100.00

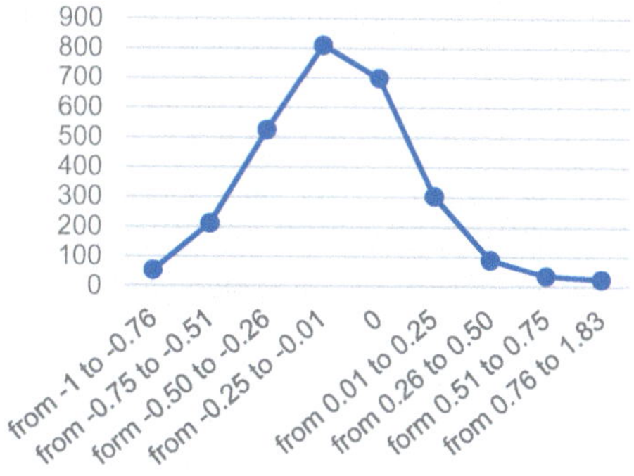

Fig. 1 Data on agreement per range

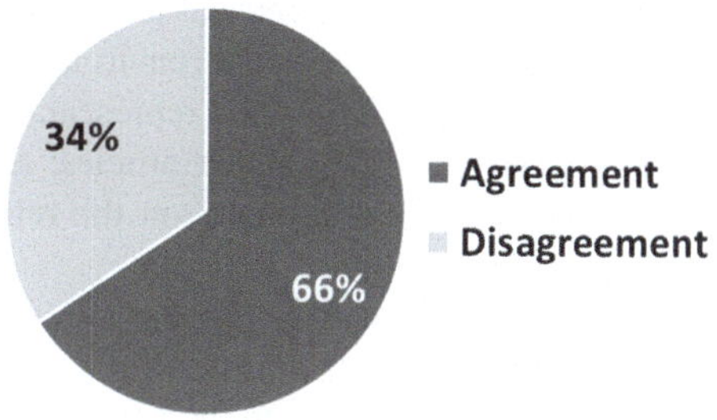

Fig. 2 Agreement and disagreement distribution

Table 3 Data on agreement

Agreement ranges	Number of stems
Agreement	1813
Disagreement	945
Total	2758

Figure 2 and Table 3 show that there is a significant share of agreement between the two groups that is 65.74% based on the agreement values in the −0.25 to +0.25 range.

This high percentage of agreement led us to further questions. Indeed, this agreement tells us very little about how it is distributed among the different dictionaries or their clusters or about how it is distributed based on density and non-density. A correspondence analysis can provide an answer to these questions.

2.2 *Study of the Relationship Between the Range of Word Consistency and the Groups of Dictionaries*

In this second stage of the research, we wanted to obtain, through correspondence analysis, the best possible representation of the relationship between the dense and non-dense stem agreement and eight different groups of dictionaries (that were aggregated whenever their original corpuses were thematically similar). We named each thematic group and indicated in parentheses the number of dictionaries that composed it as follows: "social coexistence crisis" (12), "psychological profession" (14), "experiences" (6), "new jobs" (9), "health care" (9), "training systems" (12), "media" (3) and "production systems" (14). In detail, the theme, "social coexistence crisis" includes ETAs on September 11, on terrorism, on Islam, on the representations of safety, occupational risks and road traffic; "psychological profession" includes analyses on the way to use specific theoretical constructs, on work expectations and on objectives; "experiences" includes analyses on experiments on experiencing suffering, guilt, shame, psychotherapy experiences, and finally, with respect to an event such as Christmas dinner; the theme, "new jobs", includes analyses on atypical contracts and on starting unusual jobs; "health" includes analyses for the organizational development of healthcare environments; "training systems" includes analyses on training demand and on the representation of the future after training; "media" includes analyses of newspaper articles; and "production systems" includes analyses on organizational climate, on the reputation of the enterprises and on local development.

In order to build the matrix for the correspondence analysis, we began by gathering the available data in a table. In the rows, we inserted the stems (each annotated by range of consistency), and in the columns, the number of dictionaries in which the specific word was present as dense, non-dense or absent in each of the eight groups of dictionaries (Table 4).

Table 4 was then converted into a contingency table, processing density/non-density as an active variable and the absence as a descriptive variable. These data were processed with SPAD [27] by correspondence analysis whose results are reported in the Fig. 3.

In Fig. 3, since there were numerous cases of stems with the annotations on the agreement, we limited the labels so as to reflect the variation with which they were distributed in space. The points with the labels represent the following variables: "present in the group of dictionaries as dense", "present in the group of dictionaries as non-dense" and "present in the group of dictionaries as absent", repeating them for each group of the eight dictionaries for a total of 24. The points without labels represent the stems with the annotations on the agreement; their values are essentially the same towards the positive polarity as well as towards the negative polarity, that is, between -0.25 and $+0.25$, where 0 is the maximum.

As it can be observed, the first two factors show that 61.24% of inertia is thus distributed: 51.89% along the first factor and 9.35 factors on the second factor. We did not include the factors following the first two because they did not provide any further significant information.

In Factor 1, the words are basically polarised according to their emotional density (on the right) and non-density (on the left). In Factor 2, the extent of the lexical richness of the various dictionary clusters, that is, their greater or lesser specific characteristics can be observed.

Table 4 Stem data with the range of consistency and their presence in the groups of dictionaries

	Group 1 "social coexistence crisis"			Group …	Group 8 "production systems"		
Stem with the range of consistency	Dense	Non-dense	Absent	…	Dense	Non-dense	Absent
High_from −0.01 to −0.25	0	10	2	…	2	11	1
Doubt_from −0.01 to −0.25	10	1	1	…	7	2	5
Competence_from −0.01 to −0.25	4	0	8	…	13	0	1
Research_from 0.01 to 0.25	2	8	2	…	6	6	2
Money_from 0.01 to 0.25	11	1	0	…	12	1	1
Share_from 0.01 to 0.25	5	4	3	…	7	4	3
In_front_from −0.01 to −0.25	0	12	0		1	11	2
Complain_acc0	6	0	6	…	8	0	6
Area_from −0.51 to −0.75	0	11	1	…	2	11	1
Definition_from −0.51 to −0.75	2	4	6	…	0	7	7
Construction_da −0.76a−1	1	2	9	…	2	3	9

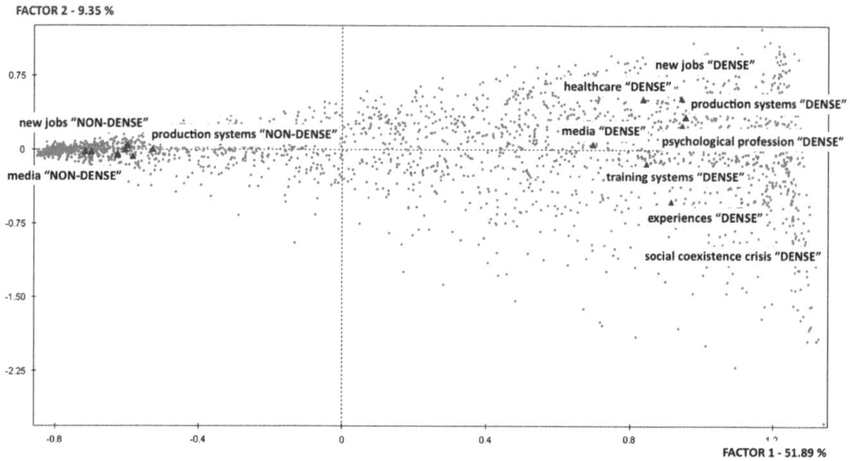

Fig. 3 Correspondence analysis

When observing the trend of the data along Factor 1, a figure shaped like a cone that widens gradually from left to right can be seen. The characteristic of this polarization is that there is less variability between the words as well as between the thematic clusters in the area of the non-dense words. Since the data tends to be concentrated, overlapping, it can be assumed that there is a strong commonality according to a given criterion. However, it does not seem to be due to the degree of agreement between the two groups of researchers since the values are essentially the same as those of the positive polarity on the right, that is, between −0.25 and +0.25, where 0 is the maximum. Extracting the words in order of significance along the factor we have: *it, would, have been, apart, everywhere, so much so that, finally, despite* and *a bit*—all with maximum agreement—words for which there is a strong consensus not only in ETA, but also in the literature that they belong to the realm of empty words or form words, as opposed to full or content words. This non-density area concerns the use of these words in support of others, according to the grammatical and syntactical models of Italian. To summarize, on the left side of the first factor we have non-dense words, little variability in words and categories and a high level of agreement.

As we move towards the central part of the figure, which corresponds to the intersection of the factors, always referring to the first factor, there is an area where the convergence of agreement between the Senior and Junior groups is less pronounced, as we can see from the annotations on the agreement for these words, including several words with low agreement (values ranging between −0.26 and +0.50 and between +0.26 and +0.50 of the index used here).

If the variability of the data along Factor 1 can be mainly explained by the distinction between density and non-density and with few variations in agreement, then a greater presence of disagreement between researchers as well as greater lexical variability can be observed in Factor 2.

Regarding the cases of disagreement along the second factor, groups of words with common aspects can be observed. For example, there is disagreement with proper nouns such as names of places as in the words *Germany, Switzerland, USA, Paris* and *Naples. China* is the sole exception, for which there is the maximum agreement, a sign of the times we could say. Other words for which there is disagreement are those related to professions and/or production areas (*chemistry, journalism, tourism, nurse, architecture, sociology* and *psychology*) or indicate means of transport (*train, bus, automobile* and *airplane*) and places that characterize daily life or the carrying out of an activity (*conference, hotel, house, hallway, department, classroom, bar, swimming pool, cinema* and *square*). In the future, we intend to discuss other groups of words for which there is no agreement: in the first group, there are words such as *order, against, right, left, civil, state* and *saint*; and in the second group, there are also verbs such as *to live, to retreat, to dress, to light, to aim* and *to trace*.

When observing the words with high disagreement along the second factor, there are criteria that appear to deny that this disagreement is only due to the individual idiosyncrasies of the researchers, but rather some words become emotionally dense in some contexts and not in others.

It can also be observed that the variability expressed on the right side of the second factor is in line with the differentiation of the various thematic clusters. This element was confirmed by a specificity analysis, but due to lack of space we will not discuss it here.

Finally, we would like to highlight how the second factor, which, on closer examination, consists only of dense words in the portion of high agreements, generally expresses the contextual differentiation of language. In particular, the clusters with greater significance are "coexistence crisis" and "experiences", in the lower right quarter of Fig. 3, while on the upper right there are "new jobs" and "production systems".

2.3 Definition of a Basic List of Dense and Non-dense Words

In the light of the results of the analysis of the concordance between researchers and the correspondence analysis that have been reported and discussed here, we believe that it is possible to identify some criteria in order to define the first list of dense and non-dense words to test and monitor over time.

Using the results of the correspondence analysis, we collected in different lists the words belonging to the four quadrants of the two factors. We thus obtained four lists containing 689 stems each, which is the number of words drawn from the analysis output. From each of these, we removed the words with low agreement, thus obtaining: (1) 618 stems with high agreement in the negative quadrant of the first factor, where, as we already mentioned, the non-dense words were concentrated; (2) 638 stems with high agreement in the positive quadrant of the first factor, where part of the emotional density was concentrated; (3) 840 stems with high agreement, in the second factor as a whole, where an additional and substantial share of emotionally dense words was concentrated, together with a really small share of non-dense words (5).

Further processing, by merging and intersecting these lists, led to the construction of a basic list of 948 dense words and a list of 618 non-dense words, for which, in the 79 dictionaries examined, a high agreement was found between the two research groups (ranging between -0.25 and $+0.25$ in the index). It is worth remembering that this result emerges from an initial list of words already selected, which contained stems that were present in at least 10 of the 79 dictionaries.

There are also these typically dense words (in addition to those that were observed in the factorial plane): *urgent, to entrust, to heal, to assault, experimental, to destroy, exchange, to accumulate, poor, to contradict, to hate, to force, luck, to frighten, to reform, violence, misfortune, excellence, duty, anger, bureaucracy, to esteem, to console, to succeed, conflict, blame, to credit, to facilitate, well-being, supervision, to defend, career, to betray*, etc.

This research provides a useful database to compare with other lists such as the "positive" and "negative" adjective list, which can be found in TaLTaC or others that can be drawn from the General Inquirer [28]. Here, for example, we can already report on a comparison that was made with the 210 full words identified in semiometry studies [19]: 202 of these words were defined as dense in at least one of the 79 dictionaries of the ETAs considered here; the remaining eight words are simply absent from these dictionaries.

3 Conclusion

This research project focused on two main areas:

(a) the operationalization of the distinction between dense and non-dense words, from the emotional point of view; and
(b) the agreement between researchers as a condition for measuring how much this distinction was shared, not only from a theoretical point of view, but also within working practice.

In this research, we have seen that the index used for observing agreement among researchers revealed a substantially high agreement between them in attributing the quality of density or non-density to a word within a specific context. However, it has also led us to disprove the hypothesis that words belong exclusively to a class, dense or non-dense. Although it is true that being aware of the historicity of language and its dynamism, it may seem naive to envisage such mutual exclusivity, we still deal with areas of work such as sentiment analysis in which the notion of clearly distinguishing words appears to be consolidated, such as when their affective meaning is divided into positive and negative [28, 29].

The correspondence analysis enabled us, on the one hand, to distinguish between agreement on density and that on non-density; on the other hand, it enabled us to observe an area that could be defined as the bad agreement area, which indicates the presence of words whose non-unanimity of choice seems to be linked in part to the thematic nature of the words and in part to realms of ambiguities already known in linguistic and textual statistics studies that lead us to envisage an area of density with more nuanced boundaries.

In conclusion, the results of this research do not confirm the construct of density in dichotomous terms but suggest a fuzzy structure. Hence, regarding the methodological aspect, we believe that the element that ought to disprove the consistency of researchers, that is, unanimity under specific circumstances; on the contrary, it contributes towards a more complex construct of dense words.

We have reached an important goal with respect to the initial objectives of the research project. We now have a basic list of 948 dense words and 618 non-dense words (in the form of stems) to be shared with the scientific community as a common ground for discussion.

We believe that we can use the lessons learnt in the data construction process to develop an ad hoc IT tool, which would make it possible to add to the data already collected, data from other ETAs produced by various research groups. This would gradually increase the number of observations made and could provide greater confidence to researchers in making assumptions about a given trend of the classification phenomenon, that is, on the value of the density of words.

The results of this research project indicate that further analyses should be carried out to study the trend among the three phenomena that were observed: the cluster of dense words with high agreement, the cluster of non-dense words with high agreement and the cluster of words for which there is disagreement. In addition, a more in-depth comparison could be carried out between the lists used in the area of sentiment analysis and connect them with the factorial studies that have been carried out regarding this area. With respect to the latter, from an analysis of the literature, it emerged that there are several interesting similarities between semiometric axes, the factors reported in the studies by Osgood et al. [8] and the results that have gradually emerged over the years in ETA.

References

1. Bardin L (1977) L'analyse de contenu [Content analysis]. Presses universitaires de France, Paris, France, p 233
2. Carli R, Paniccia RM (2000) Il colloquio come testo: l'analisi emozionale del testo [The interview as text: Emotional text analysis]. In: Trentini G (ed) Oltre l'intervista. Il colloquio nei contesti sociali. ISEDI, Turin, Italy, pp 125–158
3. Carli R, Paniccia RM (2002) L'analisi emozionale del testo. Uno strumento psicologico per leggere testi e discorsi [Emotional text analysis: A psychological tool for reading texts and discourses]. FrancoAngeli, Milan, Italy
4. Foucault M (1967) Le parole e le cose. Un'archeologia delle scienze umane [Words and things: an archeology of the human sciences] (trans: Panaitescu E). Rizzoli, Milan, Italy
5. Carli R, Dolcetti F, Battisti N (2004) L'Analisi Emozionale del Testo (AET): un caso di verifica nella formazione professionale [Emotional text analysis (ETA): A verification case in professional training]. In: Purnelle G, Fairon C, Dister A (eds) Le Poids des mots: Proceedings of the 7th international conference on textual data statistical analysis, vol 1. Presses Universitaires de Louvain, Louvain-la-Neuve, Belgium, pp 250–261
6. Carli R, Paniccia RM (1999) Psicologia della formazione [Psychology of training]. Il Mulino, Bologna, Italy
7. Carli R, Paniccia RM, Giovagnoli F, Carbone A, Bucci F (2015) Emotional textual analysis. In: Jason LA, Glenwick DS (eds) Hand-book of methodological approaches to community-based research: qualitative, quantitative, and mixed methods. Oxford University Press, New York, NY, pp 111–117
8. Osgood CE, May WH, Miron MS (1975) Cross-cultural universals of affective meaning. University of Illinois Press, Urbana, IL
9. Ekman P (ed) (1973) Darwin and facial expression: a century of research in review. Academic Press, New York, NY
10. Izard CE (1977) Human emotions. Springer, New York, NY
11. Damasio AR (1994) Descartes' error: emotion, reason, and the human brain. G.P. Putnam, London, England
12. Reinert M (1993) Les "mondes lexicaux" et leur "logique" à travers l'analyse statistique d'un corpus de récits de cauchemars ["Lexical worlds" and their "logic" through the statistical analysis of nightmare stories]. Langage et société 66:5–39
13. Austin JL (1975) How to do things with words. In: Urmson JO, Sbisà M (eds), 2nd edn. Harvard University Press, Cambridge, MA
14. Carli R (1987) Psicologia clinica. Introduzione alla teoria e alla tecnica [Clinical psychology: Introduction to theory and technique]. UTET, Turin, Italy

15. Carli R (1990) Il processo di collusione nelle rappresentazioni sociali [The collusion process in social representations]. Rivista di Psicologia Clinica 3:282–296
16. Carli R (1995) Il rapporto individuo–contesto [The individual–context relationship]. Rivista di Psicologia Clinica 2:5–20
17. Carli R (2006) La collusione e le sue basi sperimentali [Collusion and its experimental foundations]. Rivista di Psicologia Clinica 2(3):179–189
18. Matte Blanco I (1975) The unconscious as infinite sets: an essay in bi-logic. Duckworth, London, UK
19. Lebart L, Piron M, Steiner J-F (2003) La sémiométrie. Essai de statistique structurale [Semiometry: an essay on structural statistics]. Dunod, Paris, France
20. Barus-Michel J, Enriquez E, Levy A (2002) Vocabulaire de psychosociologie [Vocabulary of psychosociology]. Éditions érès, Toulouse, France
21. Reinert M (1996) Un logiciel d'analyse lexicale: ALCESTE [A lexical analysis software: ALCESTE]. Les Cahiers de l'analyse des données, XI, pp 471–484
22. Lancia F (2004) Strumenti per l'analisi dei testi. Introduzione all'uso di T-LAB [Tools for text analysis: an introduction on how to use T-Lab]. FrancoAngeli, Milan, Italy
23. Torres-Moreno J-M (2014) Automatic text summarization. Wiley, Hoboken, NJ
24. Tuzzi A (2003) L'analisi del contenuto [Content analysis]. Carocci, Rome, Italy
25. Bolasco S (2010) TaLTaC$^{2.10}$. Sviluppi, esperienze ed elementi essenziali di analisi automatica dei testi [Developments, experiences, and essential elements of automatic text analysis]. LED, Rome, Italy
26. Nobile S (2008) La chiusura del cerchio. La costruzione degli indici nella ricerca sociale [Closing the circle: the construction of indices in social research]. Bonanno, Rome, Italy
27. Lebart L, Salem A (1994) Statistique textuelle [Textual statistics]. Dunod, Paris, France
28. Stone PJ, Bales RF, Namenwirth JZ, Ogilvie DM (1962) The general inquirer: a computer system for content analysis and retrieval based on the sentence as a unit of information. Behav Sci 7(4):484–498
29. Bolasco S, Della Ratta Rinaldi F (2004) Experiments on semantic categorisation of texts: analysis of positive and negative dimension. In: Purnelle G, Fairon C, Dister A (eds.) Le Poids des mots: Proceedings of the 7th international conference on textual data statistical analysis, vol 1. Louvain-la-Neuve, Belgium: Presses Universitaires de Louvain, pp 202–210

Text Mining of Public Administration Documents: Preliminary Results on Judgments

Maria Francesca Romano, Antonella Baldassarini, and Pasquale Pavone

Abstract In the public administration (PA), most of the information (both qualitative and quantitative) is contained in texts in natural language and often the texts "hide" within them many numerical or quantifiable data. Among the documents of the PA, the judgments, which can contain information on criminal events that have an economic dimension, are particularly important: there is a flow of money generated by corruption or bribery and public procurement can be subject to disruption auction, just to mention some topics often brought to public opinion by the media. A possible source of data is that of the sentences issued by the Court of Cassation that can be obtained on the Web site http://www.italgiure.giustizia.it/sncass/. In this study, we present the preliminary results on 308 sentences of the Court of Cassation: information obtained with text mining—the names of companies, their legal form and the company name—is linked with additional information from business registers, creating the conditions for future economic analysis of criminal events.

1 Introduction and Objectives

In the public administration (PA), most of the information is contained in natural language texts and often the texts "conceal" many numerical data within them.

Among the various topics that can be identified within the documents of the PA, events that have an economic dimension assume particular importance. In particular, for criminal events, an important source of data is judgments.

M. F. Romano (✉) · P. Pavone
Istituto di Economia & EMBeDS-Scuola Superiore Sant'Anna, Pisa, Italy
e-mail: mariafrancesca.romano@santannapisa.it

A. Baldassarini
ISTAT, Rome, Italy

D. F. Iezzi et al. (eds.), *Text Analytics*, Studies in Classification, Data Analysis, and Knowledge Organization,
https://doi.org/10.1007/978-3-030-52680-1_10

Previous analysis [8] shows that this source of data may permit to verify:

- the criminal events;
- the nature of the companies involved, public and/or private;
- the role of each company in the criminal event (injured party, civil party, corruption agents);
- the economic amount of criminal actions;
- the places where the crimes took place.

The real cognitive advantage that can derive from the automatic collection of this information through text mining techniques consists of the possibility of linkage of the data obtained from this new source of data with the information contained in administrative or statistical databases on companies, such as the business register, managed by the National Statistical Office of Italy (Istat). A further integration between several archives—such as, for example, the police investigations, the public procurement database and the Court of Auditors' archive (already analyzed by [9]) —would lead, at least ideally, to big data structures [6].

In this paper, we present first results achievable with an automatic methodology for extracting the companies mentioned in the Court of Cassation's sentences. This exercise was carried out on a reduced number of judgments in order to assess the feasibility of an automatic integration between data from official statistical registers, tracing possible cognitive scenarios for an economic analysis of criminal events.

2 Methods

The automatic analysis of the texts allows to structure the textual information in a data warehouse [3], divided into two different databases: vocabulary and documents.

For the automatic lexical analysis, the reference database is the vocabulary, and the unit of analysis is the types,[1] to which grammatical and semantic notes are associated. These annotations are an example of meta-information attributed to the type, recoverable through the interrogation of the corresponding fields of the vocabulary database. The extraction/selection of the parts of the vocabulary is used to "narrate" the lexical characteristics of the corpus, highlighting the significant elements of each "part of the discourse," or to "illustrate" certain subsets of types and to measure the relations between them.

Through the regular expressions (RE)[2] we search for strings of text that—from time to time—can be the occurrences both of single types, of their classes and of relationships between classes or between single types and classes. The result of this processing consists in recovering the fragments in order to: verify the text query; invent the list of extracted strings; and annotate the fragments, if necessary. Through the integration in the automatic processing of texts between the level of lexical and

[1] Types can be: words or graphic forms, lemmas, stems, and multiwords.

[2] Algebraic notation that define string models in a formal and rigorous way [5].

textual analysis, it is possible to define a hybrid analysis system that uses dictionaries and rules in a repeated way, for the search of complex entities with variable elements [4]. In the field of information extraction, a structure with variable elements is equivalent to an entity of conceptual type articulated in various components to be searched for and extracted from a text.

In our analysis, the search for these expressions is aimed at identifying, on the one hand, the nominal multiword expressions and on the other the named entities, which in the specific case concern the quotations of companies in their various formulations.

3 Data and Results

3.1 A Document Warehouse of the Judgments

The corpus, analyzed with the TaLTac2.10 software,[3] consists of the texts of 308 judgments issued by the Corte di Cassazione in 2014 containing at least one of the following words: corruption, extortion, disruption, and public contract. In each of these 308 judgments, there is also a citation of subjects with public roles.[4]

The first step consisted of importing the textual information in TaLTaC2.10, transforming the text (which is not structured by its nature) into a document warehouse, including both the vocabulary database (lexical units) and the database of documents (text units).

The text units are the judgment, characterized by several variables: the year of issue (obtained in the input phase); other variables deriving from the results of subsequent queries, both qualitative (e.g., presence/absence of specific terms within the judgment) and quantitative (e.g., number of words, number of geographical locations, number of companies, and so on).

For each text unit (judgment), there are therefore statistical variables, the choice of which depends on the researcher's ability to use the linguistic resources provided by the software used. For example, the possibility of identifying single words (both as graphic form and as lemma), but also multiword expressions (MWEs), to assign them to semantic categories, to recognize the numbers (written as words or in Arabic numerals or in mixed mode), and to identify the presence of semantic categories within a defined sequence of k words or MWEs that allows to "tag" the presence of relevant content for the analysis.

[3]http://www.taltac.it/it/index.shtml, Bolasco [2].

[4]The presence of subjects with public roles was identified on 4800 sentences analyzed in a previous study [8].

Table 1 Distribution of occurrences between sentences partitions

Descrizione	Numero FG	Occorrenze
Sezione–Introduzione	9.089	59.318
Sezione–Motivi	9.608	58.098
Sezione–PQM	2.214	12.371
Sezione–Fatto	28.240	487.954
Sezione–Diritto	35.431	632.358
Sezione–Fatto Diritto	10.297	76.347
Totale vocabolario	52.738	1.326.497

The corpus obtained is divided into sections, which correspond according to the structure with which the magistrates' judgments are drafted.[5] Table 1 shows the number of words (FG) and their frequency (occurrences) for the corpus in the study.

In the creation of the corpus, it was also decided not to normalize the graphic forms with respect to upper/lower case, a functional choice for identifying the names of companies.[6]

3.2 *Companies' Text Mining*

The judgment is a text document, which only lawyers or legal experts can completely comprehend, so we had to imagine, construct, and test a more complex methodology to treat these texts in order to unambiguously identify the companies mentioned inside and export them for a linkage with the ASIA archive. This strategy is obliged to repair for the absence of a unique identifier of the cited companies, such as the tax code or VAT number.[7]

Our methodology therefore exploits the potential of text mining, allowing us to identify companies mentioned in a different way, or to identify companies whose names are not yet known [1, 3].

Firstly, the terms (FG) that indicated a legal form were selected from the vocabulary, obtaining the results reported in Table 2.

[5]Judgments generally respect a structure in sections: *Introduction* (which lists the defendants, the defenders, and the composition of the court), *Fatto* (describing the criminal events), *Diritto* and *Motivi* (description of legal sources), *per questi motivi (PQM)*—for these reasons—final decision). In some cases, the judge unified the Fatto and Diritto sections, a choice respected in the construction of the corpus.

[6]It was possible to differentiate, for example, the generic term "società" from that of "Società" with obvious different meaning.

[7]If, for example, was given a VAT number of companies, it would be possible to immediately obtain a list by selecting the FG with NUM grammatical tag and length 11.

Table 2 Type of identified legal form

Type of legal form	Number of FG	Occurrences
soc	5	946
impresa	23	564
srl	4	345
ditta	5	327
spa	9	231
azienda	4	182
consorzio	3	94
coop	10	63
ati	4	47
sas	4	42
snc	2	10
inc	2	2
Totale	75	2.853

Table 3 Distribution of the tagged legal form by section of the judgment

Section	Tagged legal form	%
Introd	173	6.06
Fatto	1595	55.91
Fatto Diritto	142	4.98
Diritto	782	27.41
Motivi	152	5.33
PQM	9	0.32
Total	2853	100.00

Alongside the "standard" legal forms (srl, spa, etc.), the less formal terms (e.g., *ditta*, *azienda*, *impresa*, and *società*) were also considered and all the different spellings with which the word is written in the judgments.[8]

A proof of the effectiveness of this tagging derives from the finding that the greatest number of graphic forms attributable to the legal form of the company can be found in the sections "*Fatto*," "*Diritto*," and "*Fatto e Diritto*" (see Table 3).

Subsequently, each of these graphic forms was assigned the same semantic tag ("ruolo_IMP"), functional for the extraction of names of companies in contiguous FGs each grammatically tagged as MAJ.

[8]For example, "ati" (temporary business association) is attributed to graphic forms that are different for upper/lower case characters: in our case, we found four different words: ati, Ati, ATI, ATi. The software is able to automatically reconstruct the lemma for verbs, the nouns of the Italian language, and the semantic tagging can be considered an analogue of lemmatization to FG considered semantically equivalent in the same semantic category.

Table 4 Distribution of sentences for the NumImp variable (frequency of segments with semantic tag nome_IMP

NumImp	num judgements	%
0	93	30.2
1 or more	215	69.8
Total	308	100.0

We then proceeded to identify the strings composed of the words classified as "ruolo_IMP" followed or preceded by one or more words with a capital letter. A crucial point was precisely to define the maximum number of words around the segment tagged as the ruolo_IMP trusting to include in the string the specific name of the company. The value chosen was set to 7, taking into account the possible fragmentation of the company name in several parts also separated by punctuation marks (dot) or abbreviations. The contiguity has been identified in 7 words,[9] progressively decreasing the value up to 2. At the recognized segments, we assigned the semantic tag impresa and then we proceeded to their extraction in a list. A list cleaning phase followed to eliminate possible repetitions.

The list of these segments (keeping the reference to the judgment of origin) has been exported to allow the subsequent merge with the ASIA register of the ISTAT, which will be discussed in the next paragraph.

The data warehouse was then updated by calculating for each document (judgment) a new variable (NumImp), a discrete quantitative one, containing the number of tags impresa in each document (judgment).

This variable takes values between 0 (no identified company) and 84 (the maximum value of identified companies).[10] The number of identified companies does not take into account the identity of the segments, but only counts the assigned semantic tag: the value 3 is assigned, for example, both in the case in which the list is "srl XYZ; XYZ srl; company XYZ" or "ZWS spa; ZWS spa; ZWS spa." Therefore, a value greater than 1 does not mean the identification of an equal number of companies within the sentence, but can derive both from the writing characteristics of the extensor and from the real presence of several companies.

We report in Table 4 the distribution of the Num_Imp variable by aggregating the values greater than unity: the result obtained by the automatic procedure described indicates an absence of companies mentioned in 30% of the corpus documents. We therefore wanted to verify the correctness of the automatic procedure by testing the presence of false negatives.

[9]The number k has been progressively increased to try to extract the names of companies composed of several words in queries of the type "CATGR (MAJ) LAG5 CATSEM (RuoloIMP)" which identifies for example "XYZ titolare della omonima ditta individuale" or "CATGR (PREP) CATGR (MAJ) LAG3 CATSEM (RuoloIMP)," to obtain "della ZWX & Company SPA."

[10]Obviously, the repetition does not take into account the identity of the segments, but only the assigned semantic tag: the value 3 is assigned, for example, both in the case in which the list is "srl XYZ; XYZ srl; company XYZ" or "ZWS spa; ZWS spa; ZWS spa."

After randomly extracting 20 of the 93 sentences,[11] the human reader has not identified any names of companies in any of them, dealing with cases in which illicit events only involved natural persons. The procedure used is therefore reliable at least in the correct recognition of the absence of companies' citations.

3.3 The Record Linkage Phase

The record linkage was carried out starting from the original list of types tagged as impresa. It was possible to link the information of the sentence archive with the Istat business register (ASIA) and others archives of public and private institutions with an automatic procedure. The linkage key is represented by the "company name" which does not always allow to reach univocal information.

Once the original file of the sentence types had been cleaned by the different duplicates of the same company, a starting list of 127 companies was defined and with the automatic procedure it was possible to obtain unique information on 32 companies/institutions.

In the cases in which it was not possible to univocally connect the information (95 companies), a direct verification of the information was carried out using the Infocamere database (Chamber of Commerce archive). In this way, it was possible to retrieve the fiscal code of another 45 companies, and then proceed with the automatic coupling with Istat's registers.

By also using the information deriving from the codification by region of the geographical locations mentioned in the sentences, it was possible to identify a further 14 companies; for 8 other companies, it was not possible to reach a univocal attribution.

The availability of the fiscal code has allowed us to reconstruct the information relating to the company/institution of the judgment archive with the information from the official statistical registers for the period 2011–2015.

For the univocally recognized companies, it was possible to integrate the information coming from the sentences with other variables present in the Istat databases: the fiscal code, the company name as declared to the Chambers of Commerce, the location (region/province), the sector of economic activity, the legal form, the number of employees, and the value added. Furthermore, other information of a personal nature relating to the company/institution was considered (start and end date of the activity), as well as other data on procedures in place (e.g., controlled administration, bankruptcy, liquidation, and other).

We do not consider it correct here to list the companies identified nor to provide the statistical structural variables that characterize them, as they do not constitute a statistically definable set of production units (sample or universe). We would only to underline that the linkage at the microlevel of the above information take it possible

[11] The list of 20 sampled judgments is available (Table 5) in the Appendix.

to trace the economic/legal profile of companies/institutions over time, allowing an economic–statistical analysis of criminal events.

4 Conclusions and Further Steps

We recall how the research path started by the mining of quantitative information associated with the phenomena of fiscal and/or legal illegality [7]. The ultimate goal of our research was to define and validate a methodology for text mining of PA documents, in order to be able to consider these documents as a source of statistical data.

The results presented here are only a first part of the construction of this methodology, limited only to the extraction of the names of companies mentioned in the judgments of the Corte di Cassazione and it also limited to a reduced set of sentences. However, we believe we have succeeded in setting up a feasible research path.

Transforming information—even if they are not collected for statistical purposes—in statistical data can lead to advantages both for public administrations but also for researchers. Referring only to judgments, it is possible to improve the proposed methodology: in fact, the contiguity of words semantically tagged consent to flag judgments about money exchanges (for corruption or as an electoral exchange), to count illegal events by time and space (region, cities) and to identify the role (corrupt or corrupting), the number of actors involved (professional or political role of people, without public or private companies, etc.); in short, all useful elements for a better knowledge of the criminal economy.

The results obtained so far demonstrate how it is possible to extract from the judgments the names of companies and institutions, public and private, involved in crimes and, above all, to be able to obtain more information on companies through a linkage with an official database. The next objective is to better define the proposed methodology (text mining and linkage procedure) in order to validate on a larger number of documents.

In summary, we have begun to validate a methodology of integration of qualitative data coming from judgments with other statistical databases. This road appears promising even with the limits already expressed before: the key for linking the archives is represented by the company name, which does not always allow to reach univocal information, often due to involuntary typos in the texts of the judgments. So, further improvements can be obtained both from a "direct" acquisition of the texts, and from the application of the methodology to a much greater number of documents. We recognize that many problems are still present: for example, we extract information of a non-sampling nature, often partial and incomplete; moreover, it is difficult to obtain a complete list of illegal events by year—due to the distance between year of the event and year of final judgment.

Acknowledgements We thank Guido Rey who led the research group on the criminal economy from 2010 to 2017 and which continues to follow the work of the research group.

Annex

See Table 5.

Table 5 List of the 20 random sampled judgments from the 93 judgments without companies' names

COD_document	Document number
182	sp01253
305	sp01254
220	sp03286
181	sp14838
99	sp15761
226	sp16157
308	sp19505
145	sp20483
273	sp26835
23	sp29505
288	sp32384
255	sp38083
120	sp45061
251	sp46000
24	sp46281
199	sp48038
78	sp48370
203	sp48649
93	sp48787

References

1. Baiocchi F, Bolasco S, Canzonetti A, Capo FM (2005) Estrazione di informazione da testi per la classificazione automatica di una base documentale: la soluzione di Text Mining per l'Authority della Concorrenza, in S. Bolasco, et al. Text mining: uno strumento strategico per imprese e istituzioni, CISU, Roma
2. Bolasco S (2010) Taltac2.10 Sviluppi, esperienze ed elementi essenziali di analisi automatica dei testi, LED, Milano. http://www.ledonline.it/Taltac/allegati/459-7-TALTAC-COL.pdf
3. Bolasco S (2013) L'analisi automatica dei testi. Fare ricerca con il text mining. Carocci Editore, Roma
4. Bolasco S, Pavone P (2010) Automatic dictionary and rule-based systems for extracting information from text. In: Palumbo F, Lauro CN, Greenacre M (eds) Data analysis and classification proceedings of the 6th conference of the classification and data analysis group of the Società Italiana di Statistica. Springer, Berlin-Heidelberg, pp 189–198
5. Lenci A, Montenagno S, Pirrelli V (2005) Testo e computer. Introduzione alla linguistica computazionale. Carocci Editore, Roma
6. Pramanik MI, Lau RYK, Youe WT, Ye Y, Li C (2017) Big data analytics for security and criminal investigation. WIREs Data Min Knowl Discov 7:e1208. https://doi.org/10.1002/widm.1208
7. Rey GM (a cura di) (2017) La mafia come impresa. Analisi del sistema economico criminale e delle politiche di contrasto. FrancoAngeli
8. Romano MF (2017) Dalle parole ai numeri: estrarre dati dalle sentenze della magistratura. In: Rey GM (a cura di) La mafia come impresa. Analisi del sistema economico criminale e delle politiche di contrasto. FrancoAngeli, pp 121–154
9. Zuliani A, Canzonetti A, Rispoli L, Aurisicchio G, Liverani A, Menichino P, Salvi S, De Benedetto M (2009) Un'analisi statistica delle sentenze della Corte dei conti: prime evidenze. Rivista trimestrale di diritto pubblico 3:673–706

Using the First Axis of a Correspondence Analysis as an Analytic Tool

Application to Establish and Define an Orality Gradient for Genres of Medieval French Texts

Bénédicte Pincemin, Alexei Lavrentiev, and Céline Guillot-Barbance

Abstract Our corpus of medieval French texts is divided into 59 discourse units (DUs) which cross text genres and spoken versus non-spoken text chunks (as tagged with q and sp TEI tags). A correspondence analysis (CA) performed on selected POS tags indicates orality as the main dimension of variation across DUs. Orality prevails over textual features which could fit in a one-dimensional model as well, such as text form (verse vs. prose) or time (composition century). We then design several methodological paths to investigate this gradient as computed by the CA first axis. Bootstrap is used to check the stability of observations; gradient-ordered barplots provide both a synthetic and analytic view of the correlation of any variable with the gradient; a way is also found to characterize the gradient poles (here, more-oral or less-oral poles) not only with the POS used for the CA analysis, but also with word forms, in order to get a more accurate and lexical description. This methodology could be transposed to other data with a potential gradient structure.

1 Research Question and Scientific Background

The way one can describe written texts with respect to oral linguistic uses is of primary importance for diachronic linguistics and for studying ancient stages of languages. In order to investigate this issue on Medieval French (ninth to fifteenth centuries), we propose a new approach, combining corpus linguistics [1], digital philology, textometric analysis [2], and open-source tools and data. Our research is based on the assumption that a continuum between orality and scripturality must substitute for

B. Pincemin (✉) · A. Lavrentiev · C. Guillot-Barbance
University of Lyon, CNRS & ENS de Lyon, IHRIM UMR5317, Lyon, France
e-mail: benedicte.pincemin@ens-lyon.fr

A. Lavrentiev
e-mail: alexei.lavrentev@ens-lyon.fr

C. Guillot-Barbance
e-mail: celine.guillot@ens-lyon.fr

D. F. Iezzi et al. (eds.), *Text Analytics*, Studies in Classification, Data Analysis, and Knowledge Organization,
https://doi.org/10.1007/978-3-030-52680-1_11

the old dichotomy between oral and written uses [3]. From that perspective, some written discourse may give us an indirect access to orality and to a hidden side of dead languages [4].

Most of the previous scholarship dealing with orality in Medieval French and using corpora have focussed on quoted speech. Since quoted speech is explicitly given as spoken and is linguistically or graphically separated from the rest of the text, it can be analyzed in a contrastive way with non-spoken discourse. This is also the starting point of our study, following previous research [5–7]. Thanks to updated and increased data, and thanks to new analytic tools, we would like to check if the opposition between quoted speech and plain text still constitutes the first dimension of variation in medieval written texts. In 2015 [7], we ascertained this within a 2.5-million-token corpus: orality was the main variation dimension over time variation (Old French vs. Middle French) and over five discourse domains. Since 2013 [6], we have also observed that orality was associated with verbs (and related parts of speech: personal pronouns, adverbs…), as opposed to nouns (and related parts of speech: prepositions, determiners…). In this research, we investigate if this is confirmed with improved and augmented data.

We relate this orality dimension to generic features. Koch and Oesterreicher's study [3] enables us to predict some correlations between text genres having affinities with orality and other genres, more written oriented. Here, we can distinguish 32 text genres into our 4-million-token corpus (69% of the genres are represented by at least 2 texts, 44% by 3 texts or more). We can also add another criterion, based on Koch [8], distinguishing text genres that were destined to oral performance in the early stages of French (for instance, the *chanson de geste* or hagiographic texts). We thus make the hypothesis that features of orality may be related to quoted speech, and also to text genres, such as those which are intended for oral performance.

We detail in [9] the linguistic results of this new experiment, concerning orality features and text genres in Medieval French. The present paper puts the emphasis on methodological outputs: how correspondence analysis may be used in a more analytic way rather than synthetically, and how results can be better controlled with new and complementary tools.

Resorting to a statistical multidimensional approach in order to investigate orality and text genres draws our research closer to Biber's work. Indeed, Biber initially developed his method to study speech and writing variation among registers in a corpus of modern English language [10]. Texts are characterized by linguistic features (such as word length, first person pronoun, or *that* deletion). 67 features were used in the seminal study, and this set was slightly augmented later [11]. Those features are selected because they are pointed out as relevant in linguistic studies about text genres. Moreover, they have to be automatically identifiable in digital corpora with a tagging program. Then, a principal factor analysis (PFA) is computed on the normalized relative frequency table crossing texts and tagged features. The first dimensions revealed by the factorial analysis are then interpreted, thanks to both the identification of the most contributive features for the dimension as well as the scores of genres on the dimension (defined as the mean scores of the texts), especially genres that the dimension contrasts. This method was subsequently applied to various corpora in order to determine if any dimensions could be stable across languages [12] and discourse domains [11], and then be considered as linguistic universals. It was also applied to identify patterns of diachronic register variation [12–14]. Thus, Biber's results concord with our first ones, as he points out the prominence of an oral versus literate dimension (which emerges as the first dimension in almost every study) [15]. Our study offers new insight not only because it deals with another language (Medieval French), but also because several technical choices are different, especially the three following ones: firstly, isolating quoted speech and processing it separately from the rest of the text systematically (as Biber and Finegan [14] did for one genre, dealing with fiction dialog and fiction prose as two separate genres); secondly, describing texts through their part-of-speech frequencies instead of elaborated linguistic features which thirdly implies a correspondence analysis processing rather than a principal factor analysis, since our data compose a two-way table instead of a table of measurements [2]. We have chosen to base the analysis on part-of-speech tags for several reasons: the overall quality of this tagging is satisfying in our corpus, and we manage its strengths and weaknesses; it is more stable and appropriate to identify linguistic units than word forms, due to the spelling variants and evolution in the medieval period; parts of speech account for some morphosyntactic information; this representation is very close to text expression and does not introduce much subjectivity; it is less partial and less sparse than bigrams [16].

Correspondence analysis [17, 18] is a key tool to corpus characterization in the textual data analysis community [2, 19, 20, 21]. It is also commonly applied to part-of-speech text characterizations since morphosyntactic taggers availability (see [22] for instance). Nevertheless, in this field, correspondence analysis is almost entirely devoted to a "synthetic" use, through 2D visualizations (and experimentation of 3D ones, cf. [23, 24]). The synthetic use is also based on information compression (selection of the first dimensions) and noise reduction (elimination of the last dimensions). This synthetic approach is even required in case of Guttman effect on serially structured corpora [25]. However, we would like to adopt here a methodological point of view and to insist on how an analytic use of factorial tools can contribute. Biber's approach [10, 11] already illustrates such an analytic use, but this paper aims to demonstrate this kind of use in the textual data analysis context (with a correspondence analysis computed on a part-of-speech text characterization, in Sect. 3), we evaluate and discuss the limits of this use (Sect. 4), and we implement new tools to guide interpretation (Sect. 5).

2 Methodology and Preparation of Textual Data

Our corpus is composed of 137 texts (4 million tokens),[1] taken from the *Base de français médiéval* (http://txm.bfm-corpus.org). This corpus is annotated with part-of-speech (POS) tags at the word level; speech quotation chunks and speech turns are marked up using TEI XML tags at an intermediate level between sentences and paragraphs. Every text is characterized with 28 metadata fields, including a 32-genre typology [9]. In order to perform a textometric analysis [2] on our XML-TEI annotated data, we used the TXM open-source corpus analysis platform [26] (http://textometrie.org) and DtmVic software [27] (http://www.dtmvic.com).

We have divided our corpus into 59 discourse units (DUs) obtained by splitting every genre into parts which represent spoken words on the one hand and the remaining parts on the other hand (some text genres have no spoken passages). Discourse unit labels, such as *q_rbrefLn*, combine four pieces of information: (i) the first letter is either *q* for quoted speech chunks, *sp* for speech turns, or *z* for the remaining (non-spoken) chunks; (ii) then there is the short name of the text genre (here, *rbref* means "récit bref," i.e., short narrative); (iii) the uppercase letter stands for the domain: literary (L), educational (D for "didactique"), religious (R), historical (H), legal (J for "juridique"), practical documents (P); and (iv) the last character indicates whether this DU is represented in our corpus by one (1), two (2), or more (*n*) texts.

We linguistically represented our texts with the Cattex POS tags[2] assigned by the TreeTagger tool [28], excluding punctuation, editorial markup, foreign words, and abbreviations.[3] The reliability of POS tags had been measured in a previous study [7] for a subset of 7 texts in which the tags had been manually checked. For the present analysis, we have eliminated low-frequency POS tags (freq. < 1700) that include many high error rate tags and do not carry much weight in the quantitative analysis. For the remaining high error rate tags (with more than 25% wrong assignments), we measured their influence on the correspondence analysis (CA) by checking their contribution to the first axis. Then, we removed the proper noun category (NOMpro) which showed both high error rate and high contribution to the first axis (14.67%).

A new correspondence analysis enabled two additional improvements from a linguistic perspective. We decided to remove compound determiners (DETcom, PRE.DETcom, like *ledit*). Indeed, as they emerged late thirteenth century, they introduced a singular and substantial diachronic effect (high contributions to the first axis). Moreover, the second axis described mainly the association between psalms (z_psautierRn) and possessive adjectives (ADJpos). This corresponds to very specific phrases with some distinctive nouns *(la meie aneme, li miens Deus, la tue misericorde*), and where the adjective is equivalent to a possessive determiner in other

[1]Since the first presentation of this work at the JADT 2018 conference in Roma, the corpus has been updated (better morphosyntactic tagging, new version for some texts). All experiments and results have been revised consequently.

[2]http://bfm.ens-lyon.fr/spip.php?article176.

[3]CQL query: [fropos! = "PON.*|ETR|OUT|RED|ABR"].

contexts. Therefore, we have merged the two categories (DETADJpos) and we finally obtained a contingency table crossing 59 DUs with 33 POS tags to explore with a CA.

3 Linguistic and Methodological Results from Correspondence Analysis

Our study reveals that the first axis can in fact be interpreted as an orality gradient. The factorial map (Fig. 1) shows z_ DUs on the left-hand side of the first axis, opposed to q_ and sp_ DUs on the right-hand side. Some genres intended for oral performance go to the right with speech chunks (especially plays – dramatiqueL, dramatiqueR), whereas genres related to written processing (especially practical or archive documents (P): charters, etc.) go to the left with out-of-speech chunks. As this opposition matches the first axis, orality appears as the first contrastive dimension for Old French (as regards POS frequencies), as it is in Biber's experiments with English [10], then numerous other languages [12, 15], with the same kind of linguistic features opposing verbs and clausal style with nouns and phrasal style (Table 1).

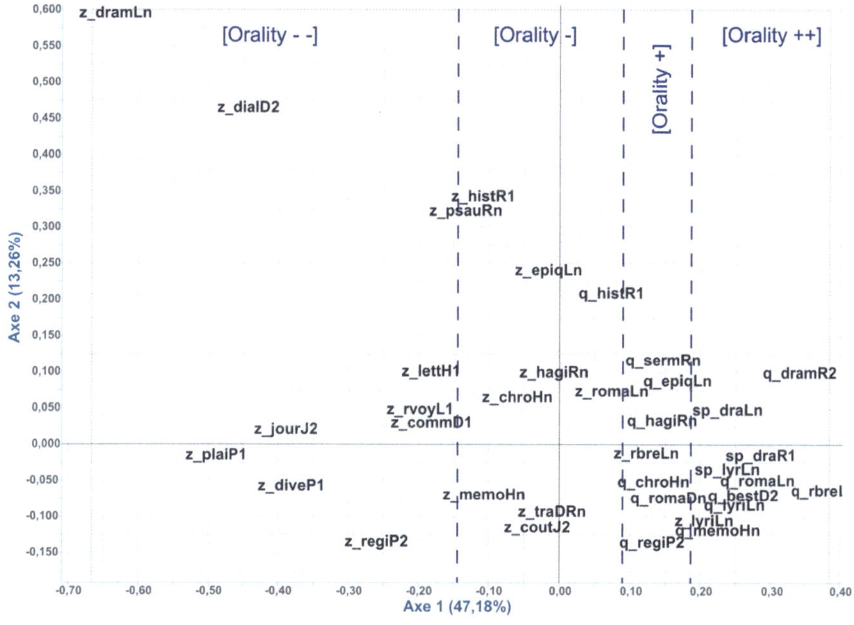

Fig. 1 CA map of the 59 DUs (TXM). 21 DUs with low representation quality (cosine squared to 1 × 2 plane < 0.3) and no significant contribution to this plane (ctrb1 < 2% & ctrb2 < 2%) have been filtered out (AFCWithStylesMacro.groovy), so that the figure is clearer. A few overlapping labels have been slightly shifted upwards or downwards, with no impact on axis 1 coordinate

Table 1 Six POS with the highest contributions on the first axis (ctrb1 > 2%), for both sides

	Less-oral pole			More-oral pole	
Label	Part of speech	Ctrb1	Label	Part of speech	Ctrb1
PRE	Preposition	15.22	PROper	Personal pronoun	18.08
NOMcom	Common noun	9.31	ADVgen	General adverb	7.91
VERppe	Past participle	5.59	ADVneg	Negative adverb	7.17
PRE.DETdef	Preposition + definite determiner	4.91	VERcjg	Finite verb	6.63
DETdef	Definite determiner	4.28	PROadv	Adverbial pronoun (*en*, *y*)	2.89
DETcar	Cardinal determiner	2.84	DETADJpos	Possessive determiner or adjective	2.26

Ctrb1 contribution to the first axis

Then, as a second result, DUs can be sorted according to their degree of orality, from "less-oral" to "more-oral" (Table 2). POS tags prove to be as efficient as more specialized linguistic features to arrange DUs (and genres) in a sharp order. Peculiar positions (for didactic dialogs or psalms for instance) can be explained by a formal use of language given by the rules of the genre. The linguistic analysis of the DU gradient is detailed in [9]. Improvements made to the statistical processing in 2018 and to the data in 2019 strengthen the linguistic interpretation published in 2017 [9]. No significant change is observed on the gradient given by the first axis, according to the four zones defined by the analysis, except for the spoken chunks of *Quatre livre des Rois* (q_histoireR1), which moves to the "less-oral" zone, because many unknown word forms (due to the anglo-normand dialect and to the peculiar use of diacritics in the edition) were erroneously tagged as NOMcom by the TreeTagger.

A bootstrap validation [29, 30] is applied to evaluate the stability of DU positions on the first axis (Fig. 2). Sizes of ellipses in the 1×2 map are correlated to sizes of DUs: the fewer the words there are in the DU, the less data the statistics process, and the greater is the confidence ellipse (Table 1). Only five DUs are ascribed a big ellipse which shows their uncertain position (Fig. 2): all of them are DUs from about ten words to about a hundred words, which are DUs for very singular linguistic usages, and are neither representative nor relevant for this overall linguistic analysis. The orality gradient is then confirmed throughout a statistic validation on our data.

The 2D factorial map provides a synthetic and efficient visualization. The second axis display reveals that the more-oral pole is more compact, more consistent, than the less-oral one, which is more heterogeneous (the cosine squared values corroborate this). Yet, what we want to stress in this methodological paper is that the main linguistic result is uniquely provided by the interpretation of the first axis. Benzécri illustrated the same kind of approach by using a 1D CA to reveal the hierarchy of characters in Racine's *Phèdre* [31, p. 68]. Biber's multidimensional analysis [10, 11] also adopted this kind of analytic use of factorial analysis, interpreting one dimension after the other. Our experiment emphasizes the analytical power of CA, which

Table 2 DU gradient on CA first axis

c1 rank	DU	c1	Ctrb1	$Cos^2 1$	Q12	Mass	Word Nb	Ellipse size
1	z_dramatiqueLn	−0.65	0.14	**0.30**	0.54	0.01	331	Medium+
2	z_plaidP1	−0.50	**2.48**	**0.72**	0.72	0.28	10,030	Small
3	z_dialogueD2	−0.45	0.05	0.20	0.40	0.01	266	Medium+
4	z_journalJ2	−0.40	**19.81**	**0.86**	0.86	**3.48**	124,042	Small
5	z_diversP1	−0.39	**1.38**	**0.70**	0.71	0.25	9002	Small
6	z_registreP2	−0.27	**22.38**	**0.69**	0.83	**8.60**	306,586	Small
7	z_rvoyageL1	−0.21	**1.03**	**0.34**	0.34	0.66	23,385	Small
8	z_commentaireD1	−0.20	0.28	**0.54**	0.54	0.21	7511	Small
9	z_lettreH1	−0.19	0.08	**0.35**	0.43	0.06	2245	Medium
10	z_psautierRn	−0.15	0.99	0.09	0.48	**1.29**	46,062	Small
11	sp_hagiographieR1	−0.14	0.30	0.14	0.15	0.44	15,760	Small
12	z_manuelDn	−0.12	0.10	0.07	0.18	0.19	6887	Small
13	z_memoiresHn	−0.12	**2.77**	**0.43**	0.61	**5.40**	192,524	Small
14	z_histoireR1	−0.12	0.77	0.08	0.17	**1.49**	53,154	Small
15	sp_dialogueD2	−0.10	0.25	0.17	0.43	0.66	23,424	Small
16	z_chroniqueHn	−0.07	**1.68**	0.29	0.04	**9.23**	329,059	Small
17	z_computD1	−0.07	0.06	0.04	0.11	0.39	14,082	Small
18	z_lettreR1	−0.06	0.12	0.10	0.06	0.84	29,992	Small
19	q_journalJ2	−0.06	0.00	0.01	0.24	0.03	1120	Medium
20	z_coutumierJ2	−0.04	0.31	0.04	0.61	**4.38**	156,306	Small
21	z_epiqueLn	−0.03	0.03	0.01	0.16	**1.23**	43,785	Small
22	z_traiteDRn	−0.02	0.08	0.01	0.53	**4.94**	175,972	Small
23	z_hagiographieRn	−0.02	0.05	0.03	0.00	**3.45**	122,923	Small
24	z_sermonRn	−0.01	0.00	0.00	0.00	**3.28**	116,895	Small
25	z_lapidaireD2	0.01	0.00	0.00	0.03	0.39	13,785	Small
26	z_romanDn	0.01	0.02	0.02	0.16	**2.66**	94,934	Small
27	z_rbrefsRn	0.03	0.06	0.03	0.03	**2.30**	81,894	Small
28	q_commentaireD1	0.04	0.00	0.00	0.28	0.00	123	**Large**
29	q_histoireR1	0.06	0.11	0.03	0.45	0.87	30,917	Small
30	z_romanLn	0.06	**2.03**	0.23	0.14	**15.27**	544,536	Small
31	q_traiteDRn	0.06	0.07	0.08	0.29	0.51	18,253	Small
32	z_bestiaireD2	0.08	0.21	0.29	0.14	0.96	34,164	Small
33	z_commentaireR1	0.09	0.33	0.13	0.15	**1.25**	44,671	Small
34	q_dialogueD2	0.09	0.00	0.02	0.26	0.00	92	**Large**
35	z_dramatiqueR2	0.09	0.02	0.10		0.07	2318	Medium

(continued)

separates the data (by the mathematical means of Singular Value Decomposition) into “deep” components (factors), just as a prism breaks light up into its constituent spectral colors. Despite its main use as a 2D illustration of a corpus structure in the textual data analysis field, CA is much more than a suggestive visualization or a quick sketch.

With that in mind, could any kind of feature be captured by this decomposition? If we look into our data, which cross genres and speech, one could question if orality could be the only available feature fitting a one-dimensional contrast: this would contest the prevalence of orality. Our next section is dedicated to this issue and tests

Table 2 (continued)

c1 rank	DU	c1	Ctrb1	Cos²1	Q12	Mass	Word Nb	Ellipse size
36	q_sermentJ1	0.10	0.00	0.05	0.20	0.00	98	**Large**
37	z_rbrefsLn	0.11	0.37	**0.38**	0.38	0.83	29,718	Small
38	q_registreP2	0.12	0.05	0.22	0.38	0.09	3308	Medium
39	q_chroniqueHn	0.12	0.89	**0.52**	0.58	**1.73**	61,524	Small
40	q_sermonRn	0.13	0.22	0.21	0.34	0.35	12,495	Small
41	q_hagiographieRn	0.13	0.60	**0.53**	0.54	0.96	34,152	Small
42	q_romanDn	0.15	**3.18**	**0.65**	0.84	**4.11**	146,631	Small
43	q_epiqueLn	0.15	0.85	**0.38**	0.46	**1.00**	35,791	Small
44	q_coutumierJ2	0.16	0.06	0.14	0.25	0.07	2368	Medium
45	q_rbrefsRn	0.16	0.50	0.27	0.28	0.57	20,257	Small
46	z_preceptesD2	0.17	0.21	0.20	0.22	0.20	7188	Small
47	q_manuelDn	0.18	0.51	0.29	0.29	0.44	15,786	Small
48	z_lyriqueLn	0.19	**3.09**	**0.56**	0.65	**2.31**	82,224	Small
49	q_preceptesD2	0.20	0.00	0.06	0.29	0.00	25	**Large+**
50	q_memoiresHn	0.21	0.09	**0.44**	0.49	0.06	2151	Medium
52	sp_lyriqueLn	0.22	0.04	**0.48**	0.66	0.02	885	Medium
51	sp_dramatiqueLn	0.22	**1.05**	**0.65**	0.74	0.59	20,970	Small
53	q_lyriqueLn	0.24	0.55	**0.67**	0.69	0.28	9945	Small
54	q_bestiaireD2	0.25	0.10	**0.65**	0.92	0.05	1667	Medium
55	q_romanLn	0.27	**26.58**	**0.89**	0.79	**10.47**	373,214	Small
56	sp_dramatiqueR1	0.27	0.84	**0.78**	0.34	0.32	11,411	Small
57	q_dramatiqueR2	0.33	0.04	**0.32**	0.05	0.01	405	Medium+
58	q_lapidaireD2	0.33	0.00	0.05	0.77	0.00	11	**Large+**
59	q_rbrefsLn	0.36	**2.13**	**0.76**		0.45	16,121	Small

c1 coordinate on axis #1, *Ctrb1* contribution to axis #1, *Cos²1* cosine squared to the axis #1, *Q12* representation quality in 1 × 2 plane (defined by cosine squared to 1 × 2 plane).Values that are of most relevance for current interpretation are bolded: *Ctrb1* ≥ 1%, *Cos²1* ≥ 0.3, *Mass* ≥ 1%

if more fine-grained data, with more "degrees of freedom" and internal variations along bipolar or linear variables, could overtake orality as the main contrast.

4 Discussion: Can the First Axis Match Any Data Feature, Is Orality Feature Really the Main One?

4.1 Challenging with Another Bipolar Textual Feature: The Verse/Prose Opposition

Every DU was divided according to the text form of the chunks: either prose (*p*), verse (*v*), or mixed form (*m*). The 4 smallest DUs, ranging from 18 to 27 tokens, were excluded. The CA was performed on these 76 new DUs characterized by the same 33 POS, and it showed the same overall opposition pattern between q_ and z_ DUs. To a

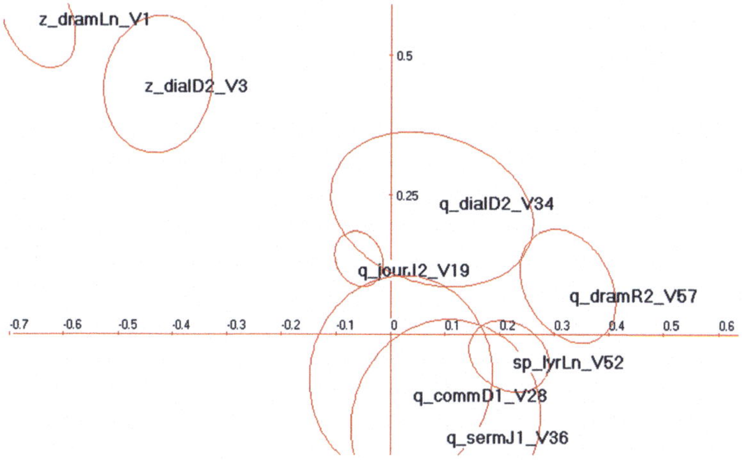

Fig. 2 CA map of the 10 DUs with the largest confidence ellipses (DtmVic). The two largest ones (q_proverbesD2, q_lapidaireD2) could not be drawn; the following three largest ones (q_commentaireD1, q_dialogueD2, q_sermentJ1) show that these DU positions cannot be interpreted; then other smaller ellipses indicate that the 54 remaining DU positions on axes 1 and 2 are stable

lesser extent, the verse displays some affinity with orality. New results were obtained from the nine genres that were divided into different forms (chroniqueH, hagiographieR, lapidaireD, preceptesD, rbrefsL, romanD, romanL, sermonR, traiteDR). The typical configuration is illustrated by historical chronicle (Fig. 3a). The only significant z/q inversion occurs in didactic novel (Fig. 3b) but it can be accounted for by the heterogeneity of this genre: the didactic novel in verse, which is clearly marked with orality, is represented by different parts of the *Roman de la rose*. This text is quite different from the other didactic novels and can be considered as a genre on its own.

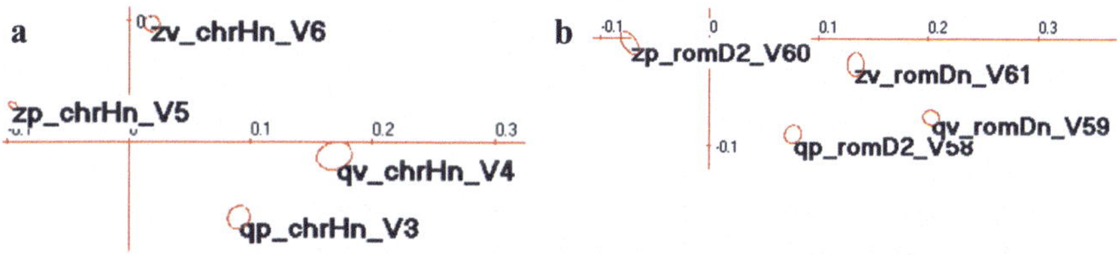

Fig. 3 Configuration of DUs for two genres: **a** Historical Chronicle, **b** Didactic Novel (the CA is computed on the full 76 × 33 table but only the selected DUs are displayed) (DtmVic)

4.2 Challenging with the Most Fine-Grained Units: Individual Texts

The 247 DUs represent up to 3 parts for a text: quoted speech chunks (q), speech turns (sp), and remaining (non oral) chunks (z). The 9 smallest DUs, ranging from 10 to 27 tokens, were excluded. Once again, orality emerged from the first axis: we calculated that for the DUs of at least 1000 words (which ensures that their position is stable), there were no more than 4 z/q inversions in 75 texts. These four negative differences between q_ and z_ first coordinates for a text are: −0.12 (rosem1), −0.11 (monstre), −0.03 (rosem3), and −0.01 (tdechamp). So, only two texts behave substantially in a peculiar way: the first volume of *Roman de la rose* by Jean de Meun (rosem1), strongly marked with orality, including non-spoken parts, and Monstrelet's *Chronicle* (monstre), a historical text where spoken parts display very clearly the features of the less-oral pole.

4.3 Challenging with a Linear Feature: Time

We still considered the 247 text-level DUs, but for balance and homogeneity considerations, we focussed on the period from the twelfth to the fifteenth centuries and took out the ninth, tenth, and eleventh centuries which were just represented by nine DUs (0.3% of words). In order to visualize the distribution of centuries on the factorial map, we used a grayscale coloring of the DU positions according to their century. It appears that orality prevails over diachrony (Fig. 4). Indeed, diachrony is essentially represented by the second dimension, whereas orality still remains the first dimension, as observed in § 4.2 on quite similar data. It should be noted that the diachrony dimension here does not account for the emergence of the compound determiners in the thirteenth century (*ledit*, etc.), since this part of speech was left out from the analysis (§ 2).

In their diachronic study on English from seventeenth to twentieth century, Biber and Finegan [14] noted that contrast between written and spoken registers increases over time. Such pattern does not appear clearly on our data. However, one could wonder if, in Biber and Finegan's analysis, orality contrast could be partially represented for past periods, given the fact that the orality dimension they used was computed to maximize the contrast in the modern language, it may be less efficient to represent all of the orality variation in a different time. In our study, all centuries contribute to the axis composition.

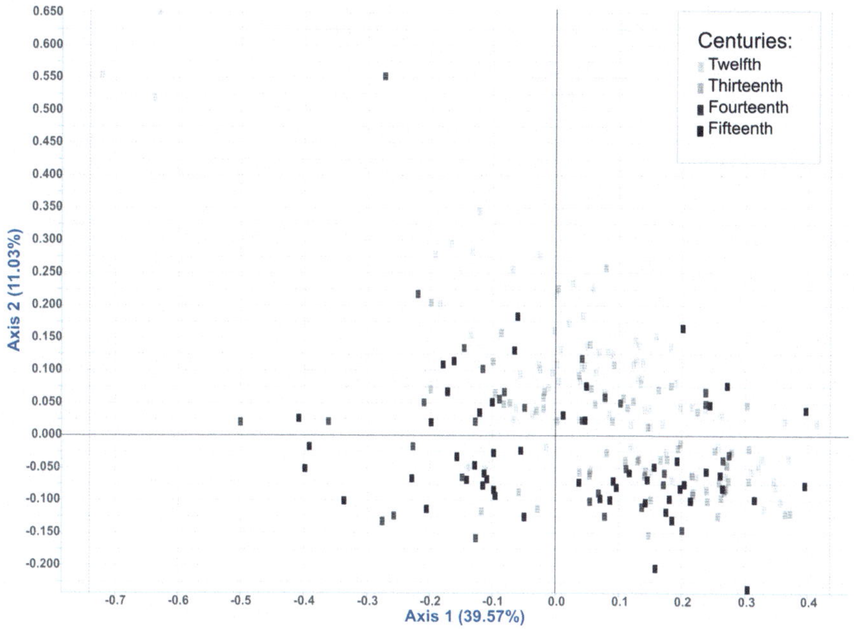

Fig. 4 Correspondence analysis on text-level DUs (238 DU × 33 POS) colorized according to centuries (twelfth to fifteenth) (TXM with AFCWithStyles macro)

4.4 Corpus Scale Effect

If the corpus is reduced to a few texts, the first axis may no longer represent orality. Our experiment is based on a previous research conducted by Mazziotta and Glickman [32] on three texts with mixed genres.

The main contrast we notice within the six DUs is the opposition between texts (*Chanson de Roland* vs. the two other texts) (Fig. 5), which concurs with the conclusions reached by Mazziotta and Glickman [32] from an analysis of these texts based on their syntactic properties, whereas in the entire corpus, the opposition highlighted between spoken DUs and other DUs remains predominant (Fig. 6). Nevertheless, from the POS perspective, in both cases, the main contrast still opposed common nouns (and determiners) to personal pronouns (and adverbs; finite verbs standing at an intermediate position in the 6-DU analysis).

In brief, orality stands as the first structural dimension when the corpus is large and diversified enough. In a small corpus composed of a few texts, the peculiarities of individual texts may overtake other general linguistic properties and achieve an overall structuring effect.

Our experiments lead us to conclude that the degree of orality proves to be a fundamental linguistic dimension which can summarize the main contrast arising from a large variety of texts. It should be related to a deep grammatical competition between common nouns, on the one hand, and verbs and personal pronouns, on

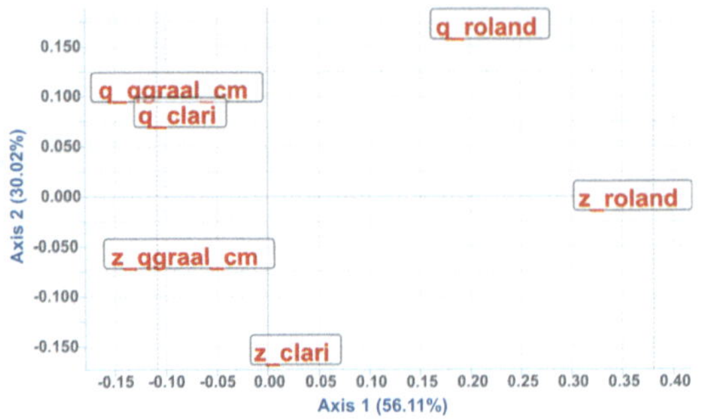

Fig. 5 CA on the three texts *Chanson de Roland* (roland), *Conquête de Constantinople* (clari) and *Queste del Saint Graal* (qgraal) (6 DUs × 33 POS) (TXM)

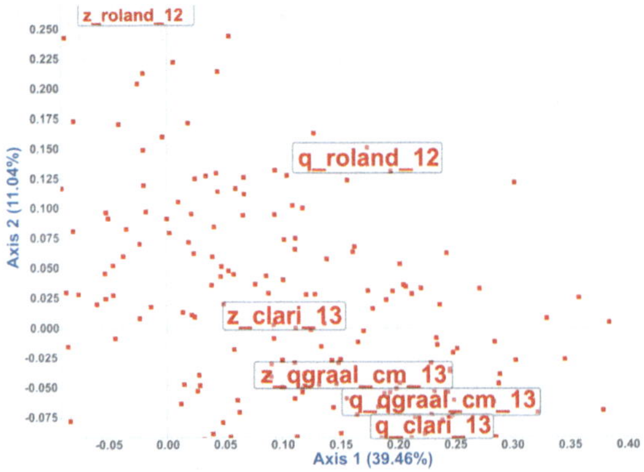

Fig. 6 Position of the DUs for the three texts *Chanson de Roland* (roland), *Conquête de Constantinople* (clari) and *Queste del Saint Graal* (qgraal) in the CA computed on the full corpus at the text level (247 DUs × 33 POS) (see § 4.2) (TXM)

the other hand. Biber [15] also insisted on this grammatical expression of the first dimension through the opposition of a clausal style and a phrasal style. Brunet [22, pp. 147–148] shared the same observations recurrently, pointing out the striking polarization of discourse between the "verb clan" and the "noun clan."

5 Complementary Tools to Analyze 1D Gradient in Textual Data

We now test new means to gain insight into the causation of this gradient in our data.

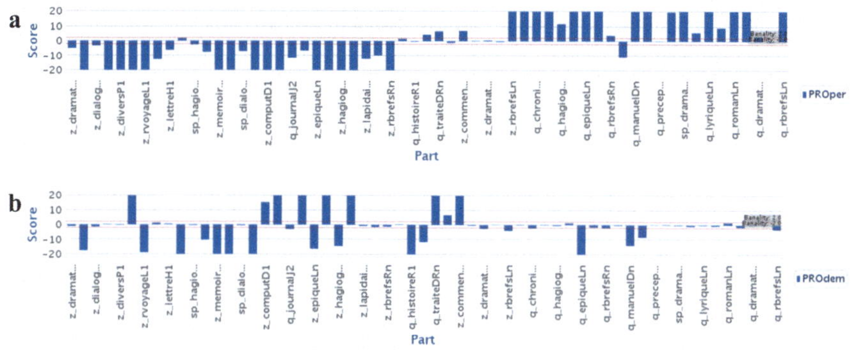

Fig. 7 Gradient-ordered specificity barplot (TXM) for: **a** personal pronoun, as example of a POS which is correlated to the first axis, and **b** demonstrative pronoun, as example of a POS which is not correlated to the first axis. For readability reasons, the height of specificity bars is capped at 20

5.1 *Gradient-Ordered Barplot*

The first method we propose is to visualize the evolution of POS frequencies according to the orality gradient using a specificity barplot chart where the DU order on the x-axis is given by the DU order on the first CA axis: this display visually reveals how much a POS is correlated with more-oral or less-oral features and details its affinity with each DU. For instance, personal pronouns are typical for the more-oral pole: this is displayed as a rising profile (Fig. 7a), and one can easily find out which DUs have an outlying use of this POS. By contrast, a POS that is not correlated to the orality gradient, such as a demonstrative pronoun (Fig. 7b), presents a chart with no overall pattern.

This kind of visualization enriches qualitatively the factorial analysis output. Whereas, the coordinate on the dimension is a unique quantitative score that tells us *whether* a POS (or a feature) is more-oral or less-oral oriented, and with which intensity, the gradient-oriented barplot accounts for *how* this polarity is implemented throughout the corpus.

5.2 *Back-to-Text Close Reading by Getting Representative Words for Each Side of the First Axis*

The second methodological innovation concerns obtaining lexical information about orality characteristics in our texts. We selected two sets of DUs based on their cosine squared scores for the first CA axis in order to represent the more-oral ($\cos^2 1 > 0.4$) and less-oral ($\cos^2 1 > 0.29$) poles. The $\cos^2$ thresholds were adjusted to get two balanced sets with enough different DUs to get an adequate representativeness. Less-oral pole comprises 10 DUs: z_journalJ2, z_plaidP1,

Table 3 Typical demonstrative pronouns for each pole (selected by highest specificity score)

Less-oral	**icellui** (141.4), ce (115.3), **icelle** (104.8), **iceulx** (84.8), **ceulx** (79.9), **icelles** (66.4), cen (28.3), **ycelle** (26.2), **ycellui** (23.6), CE (22.7), **chil** (22.5), **ceulz** (18.8), **yceulx** (16.3), chou (15.3), **ycelles** (15.0), **ciaus** (14.0), cecy (9.2), **ceaux** (8.5), Cen (6.8), **celuy** (6.5), **celles** (6.4), cela (5.9), **celx** (5.7), **chiax** (5.1)...
More-oral	Ce (84.8), C' (52.1), çou (36.0), c' (23.2), che (20.0), Che (18.5), Ch' (12.8), ice (7.7), **cist** (6.7), Ice (6.7), chou (6.1), **ceste** (5.9), **cestui** (5.8), **cesti** (5.4), ç' (5.3), cela (5.2)...

Each word is followed by its specificity score in brackets. Bolded forms are *cil* or *cist* variants

z_diversP1, z_registreP2, z_commentaireD1, z_memoiresHn, z_lettreH1, z_rvoyageL1, z_dramatiqueLn, and z_chroniqueHn. More-oral pole comprises 12 DUs: q_romanLn, sp_dramatiqueR1, q_rbrefsLn, q_lyriqueLn, q_romanDn, q_bestiaireD2, sp_dramatiqueLn, z_lyriqueLn, q_hagiographieRn, q_chroniqueHn, sp_lyriqueLn, and q_memoiresHn. Then, a specificity computation [2] revealed lexical features for each pole, showing typical words as they can be read in texts.

Our example sheds light on the uses of demonstrative pronouns, which are not related to the orality gradient as a category (Fig. 7b) but have strong associations with it at a lexical level (Table 3): speech chunks make much use of the demonstrative *cist*, which is more related to the speaker and more used as a deictic pronoun; the demonstrative *cil* is more frequent in less-oral pole in anaphoric and recognitional uses [33, pp. 335–338].

Viewing in context words which correspond to features associated with one pole or the other, as in Biber's methodology, is enlightening, even crucial. Indeed, it helps to understand in any text passage which concrete phenomena and expressions coincide with the more-orality or less-orality trend of this passage. But our method introduces an intermediary step, based on representative subcorpora, which makes it more powerful for two reasons. Firstly, thanks to the statistical computation, linguistic realizations are ordered: one immediately gets a synthetic view of the most characteristic word forms. And secondly, characterizations found at the lexical level are complementary to characterizations at the descriptive level (POS, or other features): characteristic words are not all implementations of characteristic features (see the example of demonstrative pronouns above), and conversely every implementation of a characteristic feature may not necessarily correspond to a polarized word form. For instance, *Sire* is a common noun, which is a POS identified as typically less-oral, that mainly occurs in DUs identified as more-oral and as a term of address introducing speech turns: so it would be somehow confusing to find *Sire* underlined in less-oral DUs (as it is a noun) and not underlined in more-oral DUs (despite being one of the most typical words in these DUs).

6 Conclusion

In this contribution, we have shown several ways to take into account the limits of real data, especially textual data: managing the POS tags' reliability (Sect. 2), validation process to identify where data is lacking (3), cross-checking genre-level results with finer text-level experiments (4), and refining the analysis based on morphosyntactic tags with lexical information (5). However, our main objective has been to establish a methodology in order to reveal and study any gradient-like deep structuration of data. A simple seriation (as illustrated in [29] could provide the same results for the first step, as it generates the same ordered view of the data. Yet, CA gives much more information, qualifying the relation of each variable to the gradient with indicators such as contributions and cosines squared. Interpretation can go further: CA coordinates are controlled with bootstrap and confidence ellipses, gradient-ordered barplot visualizations are efficient to analyze the relationship of any individual variable to the overall gradient, and the gradient poles can be illustrated by words, which add a concrete textual account for the deep structure. Thus, in our corpus of French medieval texts, we discover that orality is the main contrastive dimension and it characterizes quoted speech as well as text genres. We discuss what contributes to the prevalence of such feature: its affordance to a one-dimensional representation and its relevance at the corpus scale. The methodology could be applied to other data and is entirely implemented using tools freely available to the scientific community.

We believe that the methodology we elaborated provides an inspiring framework for the analysis of textual data. It allowed us to account for fine-grain and gradual relations between linguistic units while revealing the fundamental morphosyntactic opposition between the noun on the one hand, and the verb and personal pronoun on the other hand. Brunet worded this ubiquitous internal contrast in corpora with a witty metaphor [22, end of Chap. 11]: "Put together peaceful and indistinct players on a plain and neutral ground and give them a round and smooth ball, then come back a few moments later. You will see teams fighting against each other."

Acknowledgements This research has benefited from the PaLaFra ANR-DFG project (ANR-14-FRAL-0006). We would like to thank L. Lebart and C. Poudat for their inspiring comments on previous versions of this paper.

References

1. McEnery T, Hardie A (2012) Corpus linguistics: method, theory and practice. Cambridge University Press, Cambridge
2. Lebart L, Salem A, Berry L (1998) Exploring textual data. Kluwer Academic, Dordrecht
3. Koch P, Österreicher W (2001) Gesprochene Sprache und geschriebene Sprache. Langage parlé et langage écrit. In: Holtus G, Metzeltin M, Schmitt C (eds) Lexikon der romanistischen Linguistik 1-2. Niemeyer, Tübingen, pp 584–627

4. Marchello-Nizia C (2012) L'oral représenté: un accès construit à une face cachée des langues 'mortes'. In: Guillot C et al (eds) Le changement en français. Études de linguistique diachronique. Peter Lang, Bern/Berlin/Bruxelles, pp 247–264
5. Glikman J, Mazziotta N (2013) Représentation de l'oral et structures syntaxiques dans la prose de la Queste del saint Graal (1225-1230). In: Lagorgette D, Larrivée P (eds) Représentations du sens linguistique 5. Université de Savoie, Chambéry, pp 43–64
6. Guillot C, Lavrentiev A, Pincemin B, Heiden S (2013) Le discours direct au Moyen Âge: vers une définition et une méthodologie d'analyse. In: Lagorgette D, Larrivée P (eds) Représentations du sens linguistique 5. Université de Savoie, Chambéry, pp 17–41
7. Guillot C, Heiden S, Lavrentiev A, Pincemin B (2015) L'oral représenté dans un corpus de français médiéval (9e-15e): approche contrastive et outillée de la variation diasystémique. In: Kragh KJ, Lindschouw J (eds) Les variations diasystémiques et leurs interdépendances dans les langues romanes -Actes du Colloque DIA II. Éditions de linguistique et de philologie, Strasbourg, pp 15–28
8. Koch P (1993) Pour une typologie conceptionnelle et médiale des plus anciens documents/monuments des langues romanes. In: Selig M, Frank B, Hartmann J (eds) Le passage à l'écrit des langues romanes. Narr, Tübingen, pp 39–81
9. Guillot-Barbance C, Pincemin B, Lavrentiev A (2017) Représentation de l'oral en français médiéval et genres textuels. Langages 208:53–68
10. Biber D (1988) Variation across speech and writing. Cambridge University Press
11. Biber D (2009) Multidimensional approaches. In: Lüdeling A, Kytö M (eds) Corpus linguistics: an international handbook. Walter de Gruyter, Berlin, pp 822–855
12. Biber D (1995) Dimensions of register variation: a cross-linguistic comparison. Cambridge University Press
13. Biber D (2001) Dimensions of variation among eighteenth-century speech-based and written registers. In: Conrad S, Biber D (eds) Variation in English: multi-dimensional studies. Longman, London, pp 200–214
14. Biber D, Finegan E (2001) Diachronic relations among speech-based and written registers in English. In: Conrad S, Biber D (eds) Variation in English: multi-dimensional studies. Longman, London, pp 66–83
15. Biber D (2014) The ubiquitous oral versus literate dimension: a survey of multidimensional studies. In: Connor-Linton J, Amoroso LW (eds) Measured language: quantitative studies of acquisition, assessment, and variation. Georgetown University Press, Washington DC, pp 1–20
16. Crossley SA, Louwerse MM (2007) Multi-dimensional register classification using bigrams. Int J Corpus Linguistics 12:453–478
17. Benzécri J-P et al (1973) L'Analyse des Données, tome 2. L'Analyse des Correspondances. Dunod, Paris
18. Husson F, Lê S, Pagès J (2017) Exploratory multivariate analysis by example using R, 2nd edn. Chapman and Hall/CRC, Boca Raton
19. Lebart L, Pincemin B, Poudat C (2019) Analyse des données textuelles. Presses de l'université du Québec, Québec
20. Née É (dir) (2017) Méthodes et outils informatiques pour l'analyse des discours. Presses Universitaires de Rennes
21. Poudat C, Landragin F (2017) Explorer un corpus textuel. De Boeck, Louvain-la-Neuve
22. Brunet É (2016) Tous comptes faits, Écrits choisis tome III, Questions linguistiques. Pincemin B (ed). Champion, Paris
23. Leblanc J-M, Pérès M (2014) Modèles tridimensionnels pour la représentation de l'état des connaissances et propositions de visualisation pour l'analyse des corpus textuels. In: Née E, Daube J-M, Valette M, Fleury S (eds) Proceedings of the 12th international conference on textual data statistical analysis JADT 2014. Université Paris 3 & INaLCO, Paris, pp 373–384
24. Viprey J-M (2006) Ergonomiser la visualisation AFC dans un environnement d'exploration textuelle: une projection « géodésique » . In: Viprey J-M et al (eds) Proceedings of 8th international conference on textual data statistical analysis JADT'06, vol II. Presses universitaires de Franche-Comté, Besançon, pp 989–1000

25. Salem A (1991) Les séries textuelles chronologiques. Histoire et Mesure 6(1):149–175
26. Heiden S (2010) The TXM platform: building open-source textual analysis software compatible with the TEI encoding scheme. In: Otoguro R et al (eds) 24th Pacific Asia conference on language, information and computation—PACLIC24. Institute for Digital Enhancement of Cognitive Development. Waseda University, Sendai, pp. 389–398
27. Lebart L, Piron M (2016) Exploring numerical and textual data in practice with Dtm-Vic. L2C. http://www.dtmvic.com/06_ManualE.html. Accessed 22 Oct 2019
28. Schmid H (1994) Probabilistic part-of-speech tagging using decision trees. In: International conference on new methods in language processing, pp 44–49
29. Dupuis F, Lebart L (2008) Visualisation, validation et sériation. Application à un corpus de textes médiévaux. In Heiden S, Pincemin B (eds) Actes JADT 2008. Presses universitaires de Lyon, pp 433–444
30. Lebart L (2004) Validation techniques in text mining (with application to the processing of open-ended questions). In: Sirmakessis S (ed) Text mining and its applications. Springer, Heidelberg, pp 169–178.
31. Benzécri J-P et al (1981) Pratique de l'Analyse des données, tome 3. Linguistique & lexicologie. Dunod, Bordas, Paris
32. Mazziotta N, Glickman J (2019) Oral représenté et narration en ancien français. Spécificités syntaxiques dans trois textes de genres distincts. Linx 78. https://doi.org/10.4000/linx.3151
33. Guillot-Barbance C (2017) Le démonstratif en français: étude de sémantique grammaticale diachronique (9ème-15ème siècles). Peeters, Louvain

Discursive Functions of French Modal Forms: What Can Correspondence Analysis Tell Us About Genre and Diachronic Variation?

Corinne Rossari, Ljiljana Dolamic, Annalena Hütsch, Claudia Ricci, and Dennis Wandel

Abstract Our aim is to describe discursive functions of a set of French modal forms by establishing their combinatory profiles based on their co-occurrence with different connectors. We then compare these profiles using correspondence analysis in order to find evidence of genre and diachronic variation. The use of these forms is explored in contexts of informative discourse within two distinctly different genres—contemporary written press and encyclopedic discourse—as well as within two diachronic spans.

1 Introduction

Our aim is to analyze the genre and diachronic variation of discursive functions of French modal forms (MFs) in different discourse genres: written press and encyclopedia. Our methodology uses quantitative analysis based on statistical tools to detect similarities and differences in the use of terms in different corpora, as previously done for prepositions in Blumenthal [3], Charolles et al. [5] and Vigier [15]. By discursive function, we mean the rhetorical aim of the utterance in which a form occurs: counter-argument, argument, or conclusion (cf. [14]). We name the above-defined sequences *discourse movements*. We would like to see if the values that forms of a same paradigm can endorse are sensitive to diachronic or genre variation. On the other hand, we aim to verify whether the core meaning of a form entails a different behavior in relation to discursive sequences, and if so, if this behavior is in turn sensitive to diachronic and genre variation. For this purpose, we use two statistical tools: an association measure, log-likelihood [8], and correspondence analysis (CA) (see Désagulier [6],

This paper is an extended version of the paper *Discursive Functions of Epistemic Adverbs: What can Correspondence Analysis tell us about Genre and Diachronic Variation?* By the same authors, published in *Actes JADT 2018*, Roma: UniversItalia, pp. 668–675.

C. Rossari (✉) · L. Dolamic · A. Hütsch · C. Ricci · D. Wandel
University of Neuchâtel, Neuchâtel, Switzerland
e-mail: corinne.rossari@unine.ch

D. F. Iezzi et al. (eds.), *Text Analytics*, Studies in Classification, Data Analysis, and Knowledge Organization,
https://doi.org/10.1007/978-3-030-52680-1_12

Greenacre [9]). Firstly, we take into account a paradigm of MFs conveying different nuances linked to the core meaning of possibility, which consists of the following epistemic adverbs[1]: *certainement*, *certes* ('certainly'), *peut-être* ('maybe'), *probablement* ('probably'), *sans doute* ('without a doubt') and *sûrement* ('surely'). Secondly, we compare forms conveying two different core meanings: possibility (P) and necessity (N). Our paradigm consists of the following items: *devoir*, *falloir* ('must'/'have to,' 'be necessary') (N meaning) and *pouvoir* ('can'/'may'), *peut-être* ('maybe')[2] (P meaning). The functions of all MFs are explored in contexts of informative discourse within two distinctly different genres: contemporary written press and encyclopedic discourse. The former is represented by three daily newspapers: *Le Monde* (2008, 20,410,766 tokens), *Le Figaro* (2008, 10,795 373 tokens) and *Sud-Ouest* (2002, 29,763 988 tokens). For the latter, we consider two diachronic spans: the eighteenth century, represented by Diderot & D'Alembert's *Encyclopédie* (DDA, 29 940 181 tokens) and the twenty-first century, represented by the 2005 edition of *Encyclopédie Universalis* (UNI, 49 859 864 tokens) and by a random sample of the 2015 version of *Wikipédia* (WIKI, 50 396 345 tokens).[3]

We first carry out an analysis based on the combinatory profile of each MF constituting our first paradigm of forms (Sect. 2) within our corpus of contemporary written press, and, after having pinpointed what such an analysis can and cannot show, we use a more holistic approach based on correspondence analysis (Sect. 3). The same approach is chosen for our second paradigm. Our purpose is to verify if this type of approach allows to extend the analysis of modals, bringing out differences in behavior between the modals conveying necessity (N-MFs) and those conveying possibility (P-MFs).

2 Analysis of Combinatory Profiles of Epistemic Adverbs (E-ADVs)

In order to identify the discursive functions of the E-ADVs considered here, we searched connectors (C) specifically co-occurring with each of these E-ADVs within a 20-token span. We have chosen a 20-token span rather than a sentence span because a connector's combinatory profile can go beyond sentence boundaries. We define

[1]Our selection of forms is based on Roulet's [13] paradigm of epistemic assertive adverbs. The English translations of all modal forms and connectors considered in this paper do not necessarily represent all the possible renderings of the French forms. They are literal translations, which we provide for the sake of clarity.

[2]*Peut-être* belongs to both paradigms because, for reasons of representativeness, we decided to include two forms for each core meaning: two N forms (*devoir*, *falloir*) and two P forms (*pouvoir*, *peut-être*).

[3]All the corpora used were supplied by the platform *BTLC* (*Base Textuelle Lexico-statistique de Cologne*), conceived by Diwersy [7], and were constituted within the French-German projects *Presto* (http://presto.ens-lyon.fr) and Emolex (http://emolex.u-grenoble3.fr).

connectors as linguistic forms linking segments of discourse. Such a functional category is not part of the tagset of the platform used. We therefore made our query by searching for three different categories which are tagged: adverbs, subordinating conjunctions, and coordinating conjunctions. We then manually filtered the resulting forms by keeping those which proved to function as connectors.

For all our subcorpora, each E-ADV is thus specifically assigned a series of connectors within constructions of the type 'E-ADV…C1/C2/Cn' and 'C1/C2/Cn…E-ADV,' which, taken as a whole, represent their *discursive combinatory profile*.[4] We call each sequence within a combinatory profile a *discourse movement* as we consider it to have specific, rhetorically motivated discursive aims. These aims (mentioned in Sect. 1) are signaled by the connectors co-occurring specifically with an E-ADV: *néanmoins* ('however') and *mais* ('but') present the content of the utterance preceding them as a counter-argument to that of the utterance they introduce, *donc* ('therefore') and *finalement* ('finally'/'after all') signal that the utterance they introduce is a conclusion, and *car* and *parce que* ('because') signal that the utterance they introduce is an argument in favor of the utterance preceding them.

The tables below show the discursive combinatory profiles in three subcorpora of contemporary press (*Le Monde 2008*; *Le Figaro 2008*; *Sud-Ouest 2002*). The significance of each co-occurrence of a connector with an E-ADV is calculated using the significance measure log-likelihood (LL) (Table 1).

The data lead to the following observations: (i) Although the E-ADVs belong to the same semantic class, each of them has its own specific combinatory profile. (ii) Certain E-ADVs share comparable combinatory profiles: *sans doute* and *peut-être* share an almost identical set of specific connectors, whereas more frequently E-ADVs essentially only share one or more specific connectors (for instance, the connector *mais* for *certainement*, *sûrement*, *peut-être*, and *sans doute*). (iii) Certain E-ADVs stand out for their unique combinatory features: *certes* is almost exclusively associated with *mais*, but only with *mais*_R, and with a notably higher log-likelihood score than the other E-ADVs. *Probablement* is also associated with only a few connectors, but with a low log-likelihood score, close to the threshold of 10.83. (iv) There is homogeneity in the significant associations for each E-ADV in the three subcorpora of contemporary press. However, preceding studies—Rossari et al. [11] and Rossari and Salsmann [12]—show that an E-ADV's combinatory profile varies depending on different genres and diachronic periods. This finding contrasts with what is observed here for the press genre: in the encyclopedias DDA and UNI, the association *peut-être…mais* is less significant than the association *mais…peut-être*; in addition, in DDA, no significant association *certes…mais* is observed, while the association *sans doute…mais* in the same corpus proves to be highly significant.

The analysis of combinatory profiles (based on log-likelihood scores; cf. [4] allows for one-to-one comparison of different sequences of the type E-ADV…C and C…E-ADV. Thus, the associations of each E-ADV with each connector can easily be compared across subcorpora representing different newspapers, but also across different genres and diachronic periods. It is also possible to compare the associations

[4] We adapt the term *combinatory profile* used by Blumenthal et al. [4] and Blumenthal [1, 2].

Tables 1 Log-likelihood scores (threshold: 10.83; all scores equal or above are marked in bold)

Le Monde (2008)	Certainement (385)		Certes (1943)		Peut-être (2900)		Probablement (723)		Sans doute (2482)		Sûrement (307)	
	L	R	L	R	L	R	L	R	L	R	L	R
Car (6 706)	0.08 (3)	0.08 (3)	−0.27 (11)	0 (13)	**49.30** (57)	−0.37 (12)	**18.99** (17)	0.92 (7)	**16.09** (35)	0.02 (17)	1.51 (4)	−0.64 (1)
Donc (8 276)	2.09 (6)	2.09 (6)	−2.47 (10)	2.9 (23)	−1.45 (18)	**24.09** (51)	−0.68 (4)	1.43 (9)	0.03 (21)	4.18 (30)	0.10 (3)	0.10 (3)
Finalement (1 559)	−1.18 (0)	−0.01 (1)	−1.77 (1)	1.14 (5)	3.60 (9)	**15.43** (15)	−0.01 (1)	0.58 (2)	0.01 (4)	6.96 (10)	−0.94 (0)	6.08 (3)
Mais (51 544)	**29.83** (47)	**27.94** (46)	**3248** (979)	−22.55 (52)	**371.45** (423)	**107.28** (284)	10.30 (57)	10.30 (57)	**205.88** (310)	**32.85** (193)	**30.90** (41)	**65.68** (55)
Néanmoins (968)	−0.73 (0)	−0.73 (0)	5.84 (6)	**14.22** **(9)**	0.02 (3)	−0.23 (2)	0.12 (1)	−1.38 (0)	−0.06 (2)	−0.06 (2)	−0.58 (0)	−0.58 (0)
Parce que (2 514)	0.88 (2)	2.81 (3)	1.78 (8)	−4.47 (1)	**62.15** (37)	9.75 (17)	2.03 (4)	−3.58 (0)	**86.58** (41)	0.12 (7)	3.78 (3)	0.07 (1)
Le Figaro (2008)	Certainement (268)		Certes (1084)		Peut-être (1851)		Probablement (441)		Sans doute (1240)		Sûrement (211)	
	L	R	L	R	L	R	L	R	L	R	L	R
Car (3 922)	−0.57 (1)	−0.57 (1)	3.89 (14)	−2.35 (4)	**28.08** (37)	−0.48 (11)	**14.27** (12)	1.95 (6)	8.45 (19)	4.43 (16)	7.53 (6)	−3.08 (0)
Donc (4 763)	−1.02 (1)	−1.02 (1)	−0.03 (9)	1.81 (14)	−20.39 (2)	3.16 (24)	−0.22 (3)	6.74 (10)	−0.37 (9)	0.37 (13)	−3.74 (0)	−0.48 (1)
Finalement (1 150)	−1.14 (0)	−1.14 (0)	−0.04 (2)	−0.95 (1)	−3.15 (1)	6.52 (10)	0 (1)	0 (1)	1.67 (5)	−1.34 (1)	−0.90 (0)	−0.90 (0)

(continued)

Tables 1 (continued)

Le Figaro (2008)	Certainement (268)		Certes (1084)		Peut-être (1851)		Probablement (441)		Sans doute (1240)		Sûrement (211)	
	L	R	L	R	L	R	L	R	L	R	L	R
Mais (28 552)	**36.23** (41)	**14.25** (30)	**1757.55** (545)	−1.39 (49)	**245.20** (281)	**93.30** (204)	0.56 (27)	2.38 (31)	**86.88** (151)	**34.59** (117)	**17.23** (27)	**87.11** (52)
Néanmoins (580)	1.07 (1)	−0.58 (0)	10.02 (6)	0.49 (2)	0.44 (3)	−3.84 (0)	−0.95 (0)	−0.95 (0)	−2.67 (09	−2.67 (0)	−0.45 (0)	−0.45 (0)
Parce que (1 435)	0.10 (1)	−1.43 (0)	−1.65 (1)	−0.30 (2)	**31.90** (22)	−2.25 (2)	6.88 (5)	−0.03 (1)	4.79 (8)	0.14 (4)	0.28 (1)	2.22 (2)
Sud-Ouest (2002)	Certainement (1277)		Certes (2795)		Peut-être (4950)		Probablement (812)		Sans doute (3930)		Sûrement (684)	
	L	R	L	R	L	R	L	R	L	R	L	R
Car (12 434)	10.78 (23)	6.53 (20)	2.89 (32)	5.92 (36)	**44.71** (91)	−0.29 (38)	1.34 (10)	−1.35 (4)	**45.13** (78)	−0.26 (30)	6.87 (13)	−0.09 (5)
Donc (19 185)	−3.00 (10)	5.72 (27)	−6.45 (22)	0.10 (38)	−1.98 (53)	1.55 (74)	−1.32 (7)	9.77 (22)	−1.61 (42)	9.50 (74)	−0.41 (7)	0.15 (10)
Finalement (2 858)	−0.09 (2)	0.11 (3)	3.19 (10)	0.07 (6)	7.35 (19)	3.68 (16)	−0.23 (1)	−0.23 (1)	2.23 (12)	0.72 (10)	−0.08 (1)	0.31 (2)
Mais (77 108)	**28.87** (113)	**64.77** (139)	**6962.42** (1778)	−5.72 (118)	**520.54** (682)	**211.15** (513)	10.35 (64)	9.48 (63)	**209.38** (434)	**123.59** (376)	7.09 (52)	**162.80** (130)
Néanmoins (1 698)	−0.16 (1)	−0.16 (1)	9.25 (10)	−0.01 (3)	5.39 (12)	−0.54 (4)	0.93 (2)	−1.85 (0)	6.72 (11)	1.20 (7)	0.06 (1)	0.06 (1)
Parce que (5 981)	−4.77 (11)	−13.45 (6)	**12.51** (15)	8.48 (13)	**814.50** (135)	**104.89** (33)	**16.58** (13)	−0.57 (2)	**233.04** (108)	0.00 (16)	−1.04 (8)	−0.48 (9)

‘L’ (‘left’) and ‘R’ (‘right’) indicate on which side of the pivot a collocate appears

of different E-ADVs with one or a few connectors. However, this method has certain insufficiencies when it comes to simultaneously comparing all of these variables in a holistic view. This type of analysis of combinatory profiles never takes into account all variables at the same time (e.g., frequencies, log-likelihood scores, paradigm of E-ADVs, and paradigm of connectors). Moreover, using a threshold (in our case 10.83) in order to decide whether an association is significant is useful for traditional collocation analysis. But our goal is to also represent the use of each E-ADV in its typical discourse movements in contrast to its non-typical discourse movements. It thus seems counterproductive that all sequences (E-ADV...C/C...E-ADV) which are not statistically significant for certain E-ADVs as such are not taken into account when establishing their combinatory profiles, since these non-significant cases play an important role in characterizing the overall use of the E-ADVs and connectors. In order to allow for a holistic approach, we propose to use correspondence analysis (CA) [9].

3 Correspondence Analysis (CA) of E-ADVs

The correspondence analysis presented in this section was performed using the R software and the package 'CA' [10].

(1) In DDA, representing the eighteenth century, *certes* has a use which stands out. *Certes* left and right of *mais* differs clearly from all other E-ADVs as to its association with the connectors. *Certes* is not typically used with any other connector analyzed and, most importantly, its association is not stronger with *mais* on its right than it is with *mais* on its left. Conversely, in all other five corpora (encyclopedic and press corpora), which represent the twenty-first century, there is an important difference between the use of *certes* right and left of *mais*: while *certes*_L is strongly linked to *mais*, *certes*_R is not.
(2) In all six corpora, *mais* appears to be in contrast to all other connectors when it comes to its associations with E-ADVs. Its central position appears to be linked to its high frequency, indicating its important contribution to the horizontal axis, which is confirmed by the correspondence analysis indicators.
(3) An association between *sans doute*_L and *parce que* can be observed in DDA and WIKI, whereas in UNI, the adverb and the connector appear to be in the inverted relation. This behavior indicates that variation has to be expected even among subcorpora within the encyclopedic genre, concerning at least two parameters.
(4) The results of the correspondence analysis show that in all subcorpora of one particular genre, in most cases, the same E-ADVs are strongly associated with the same connector or group of connectors (*donc* and *finalement*; *car* and *parce que*; *mais*). This phenomenon is particularly pronounced in the subcorpora representing written press. The connector *mais* differs the most from the other connectors with regard to the strength of its associations. Although *mais* is associated with most E-ADVs, its association appears to be strong with only a few of them

in all corpora (*certes*_L being the only constant), while most other connectors have a higher number of strong associations. This indicates that certain discourse movements (such as E-ADV…*car/parce que*) seem to be rather regular, whereas *certes…mais* proves to be a special association, although only in the twenty-first century corpora (Fig. 1).

4 Correspondence Analysis (CA) of Necessity and Possibility Modal Forms

Before presenting the plots for these cases, some preliminary remarks are necessary. Interpreting the association of connectors with verbs requires manual sorting for the connectors which are mobile, i.e., whose position within the sentence may vary. A left syntactic position can in many cases correspond to a right semantic position: for instance, in a clause such as *il faut donc agir* (*so we must act*), the right scope of the connector *donc* includes in fact the modal verb. In our analysis, such a syntactic position is treated like a right semantic position of the modal form in relation to the connector, and thus is counted among the occurrences of the modal form in the right position. For *donc*, all the occurrences have been sorted. The other mobile connectors, *finalement* and *néanmoins*, which assume the same discursive function as *donc* and *mais* and would require the same sorting, have not been taken into account for this research.

We present symmetric plots for visual clarity, but our interpretations are based on the asymmetric plots, which maintain the distances between the forms (Fig. 2).

(1) The results of the correspondence analysis show that in all corpora some modal forms are grouped in relation to their core meaning. *Falloir* and *devoir*, which have a core meaning of necessity, are systematically grouped with regard to their right and left position. This shows that they have a very similar behavior in relation to the connectors considered here. The fact that such a clustering is transversal to genres and diachronic spans indicates that the core meaning of the forms is preponderant for the selection of the discursive sequence by the speaker.
(2) Connectors with the same function are grouped: *car* and *parce que* show a similar behavior in relation to the four modal forms taken into account. This confirms that the modal forms are sensitive to the function of the connectors, and consequently to the discursive sequence they indicate, independently of diachronic span or genre represented by the corpus.
(3) We cannot see regularity in the behavior of modal forms conveying a core meaning of possibility. *Pouvoir* and *peut-être* in the left position are grouped in some corpora, such as DDA and UNI, but not in others. This is due to the fact that there is a strong attraction between *peut-être* and *parce que*, because the modal can modify the causal relation in itself, in examples such as: *Paul est parti tôt, peut-être parce qu'il ne connaissait personne* (Paul left early, maybe because he didn't know anyone). In such a case, *peut-être* modifies the causal relation

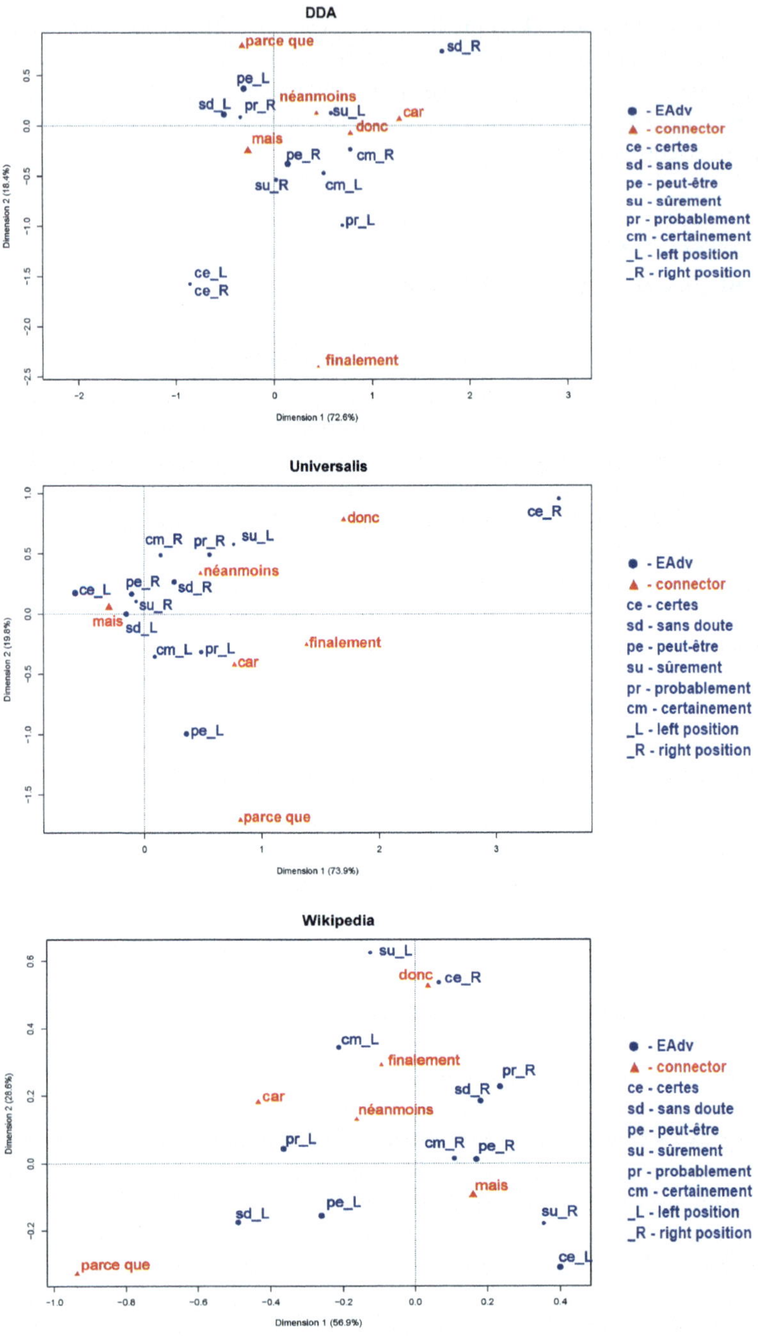

Fig. 1 Correspondence analysis scatter plots for the six corpora

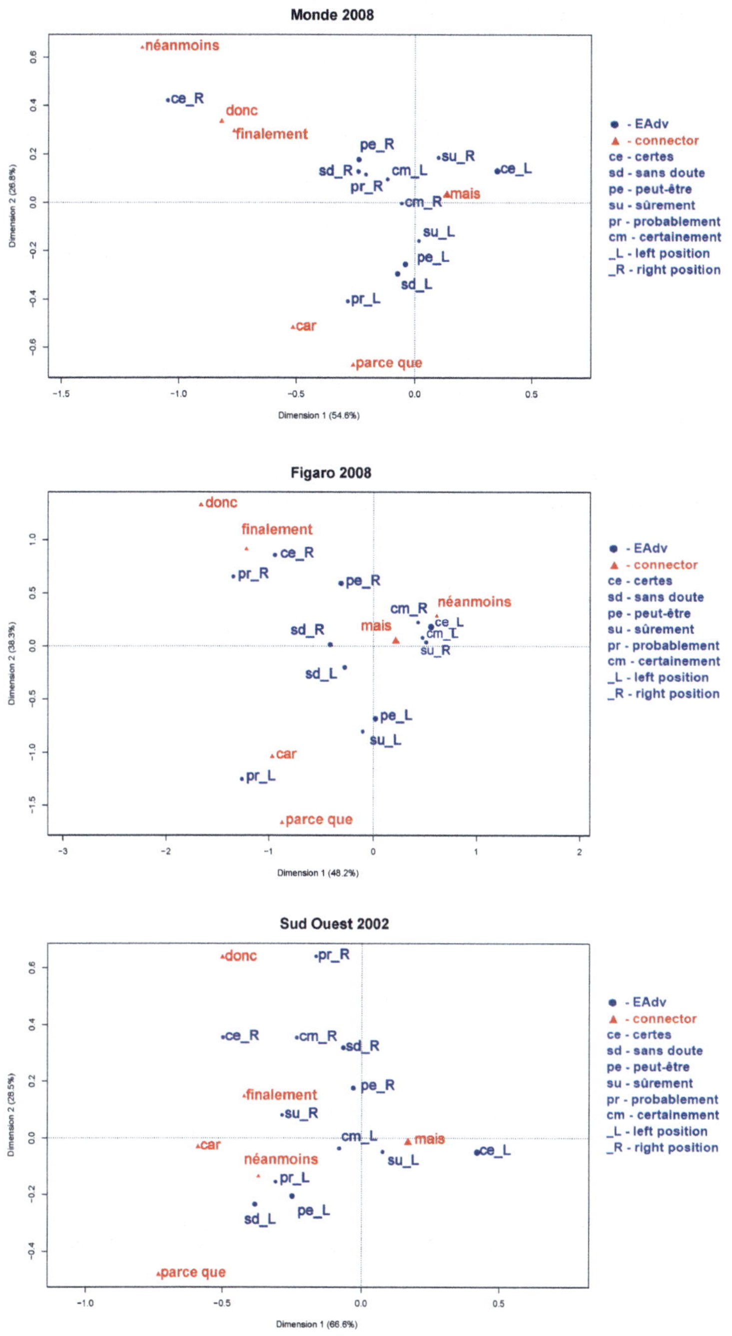

Fig. 1 (continued)

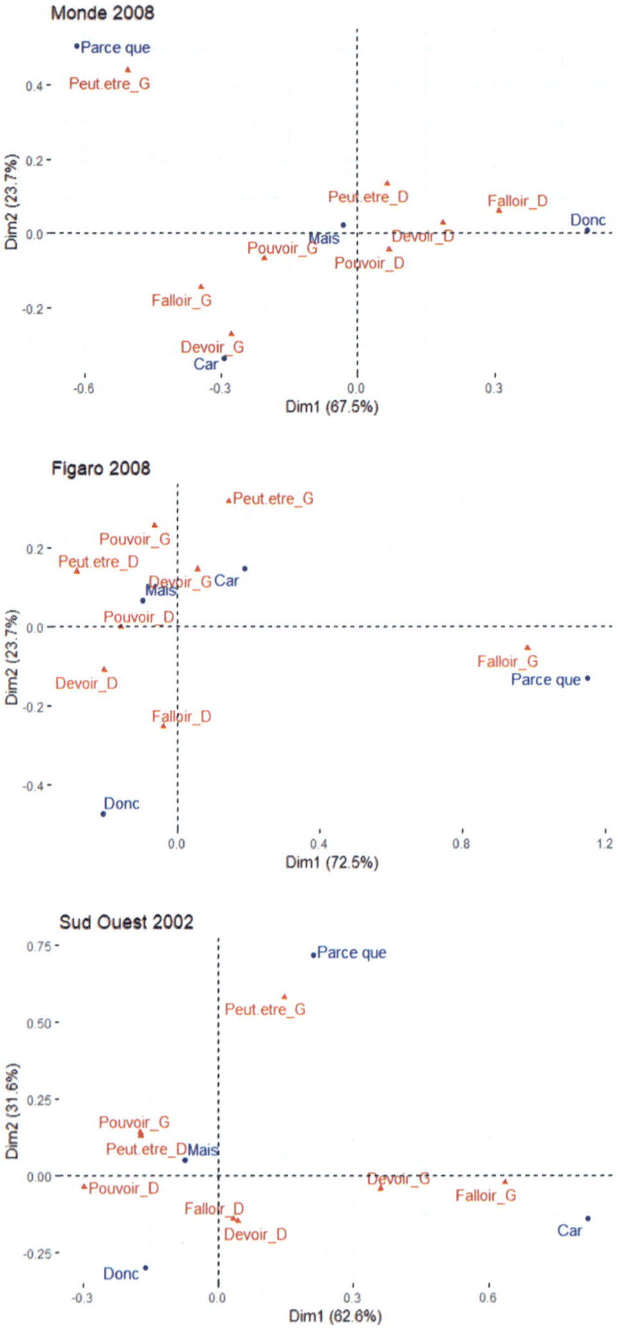

Fig. 2 Correspondence analysis scatter plots for modal forms of necessity and possibility within the six corpora (_G: left position; _D: right position)

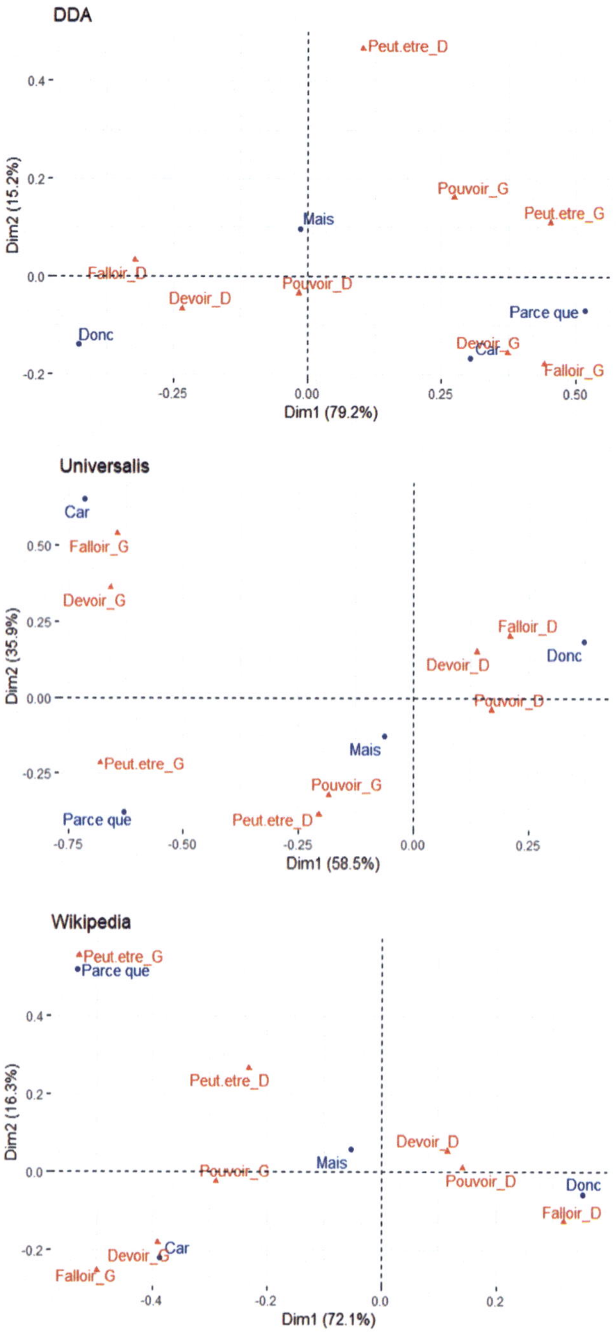

Fig. 2 (continued)

marked by *parce que* and not the content *il ne connaissait personne*. These uses are very frequent and explain the proximity between *peut-être* and *parce que*, proximity that distances the connector from the other modals conveying a core meaning of possibility: *pouvoir* cannot have such a use.

(4) One connector shows remarkable homogeneity in all corpora in relation to its association with modals: it is *mais*, which is systematically associated with *pouvoir* in the left position and with *peut-être* in the right position. The result concerning *peut-être* cannot be taken into account because of the modal's attraction to *parce que*, as indicated above. But the result concerning *pouvoir* is meaningful: it shows that, within a concessive sequence, a modal conveying possibility as its core meaning is preferred. This is consistent with the concessive interpretation that the modal verb can trigger (*il peut être riche, mais il n'est pas intelligent*, he may be rich, but he is not smart). Such an interpretation is not accessible with modal verbs having a core meaning of necessity: *il doit être riche, mais il n'est pas intelligent/il faut qu'il soit riche, mais il n'est pas intelligent*.

(5) No corpus indicates a particular use of a N-MF or of a P-MF; conversely, we have seen that the plots concerning epistemic adverbs make emerge some change concerning *certes* and *sans doute*, (see § 1 and 3 in Sect. 3). The interpretation that we can advance is that the parameter relating to the core meaning, i.e., the difference in behavior between P modal forms and N modal forms, is less subject to change—we notice that the same clustering of N-MFs can be observed in all corpora—than the parameter relating to the different values that the forms of a same paradigm can convey—we notice that some E-ADVs show different degrees of attraction in relation to connectors.

5 Perspectives

Our attempt to use correspondence analysis to study different discursive movements has provided promising results regarding the genre and diachronic variation of discursive functions of French modal forms. It illustrates that the genre and time factors play a role in the use of forms conveying different values stemming from the same core meaning, as observed in Sect. 3. But the core meaning in itself shows strong stability and determines the nature of the discursive sequence in which the form appears, as observed in Sect. 4.

It seems therefore that the values that the modal forms of a paradigm can convey are context-dependent, but the core meaning attached to the form is context-free and can thus be considered as a cognitive primitive.

To test this hypothesis, we can go in two directions: (i) we can extend the analysis to other languages to assess stability and variation in relation to values or to the core meaning of the form; (ii) we can extend the analysis to other corpora representing different diachronic spans as well as to corpora resorting to different communication channels such as those that favor a spontaneous use of language (spoken language, chat, blogs).

Acknowledgments We would like to thank Mitra Akasereh for her precious help in the constitution and the analysis of the modal verbs plots, as well as Jessica Chessex for her help in collecting and sorting data.

References

1. Blumenthal P (2008) Combinatoire des prépositions: approche quantitative. Langue française 157:37–51
2. Blumenthal P (2012) Particularités combinatoires du français en Afrique: essai méthodologique. Le français en Afrique 27:55–74
3. Blumenthal P (2017) D'une encyclopédie à l'autre. L'ascension des locutions prépositionnelles. Langages 206(2):123–138
4. Blumenthal P, Diwersy S, Mielebacher J (2005) Kombinatorische Wortprofile und Profilkontraste. Berechnungsverfahren und Anwendungen. Zeitschrift für romanische Philologie 121:49–83
5. Charolles M, Diwersy S, Vigier D (2017) Evolution des emplois des marqueurs de topiques de discours dans Le Figaro de la fin du XIXe et du début du XXIe siècles. Langages 206:85–104
6. Désagulier G (2017) Corpus linguistics and statistics with R. Introduction to quantitative methods in linguistics. Springer, New York
7. Diwersy S (2014) Corpus diachronique de la presse française: base textuelle créée dans le cadre du projet ANR-DFG PRESTO. Institut des Langues Romanes, Université de Cologne
8. Evert S (2008) Corpora and collocations. In: Lüdeling A, Kytö M (eds) Corpus Linguistics. An international handbook. Mouton de Gruyter, Berlin, New York, pp 1212–1248
9. Greenacre MJ (2017) Correspondence analysis in practice, 3rd edn. Chapman, Boca Raton
10. Nenadić O, Greenacre MJ (2007) Correspondence Analysis in R, with two- and three-dimensional graphics: the ca package. J Stat Softw 20(3):1–13
11. Rossari C, Hütsch A, Ricci C, Salsmann M, Wandel D (2016) Le pouvoir attracteur de *mais* sur le paradigme des adverbes épistémiques: du quantitatif au qualitatif. In: Mayaffre D et al (eds) Proceedings of 13th international conference on statistical analysis of textual data, II, pp 819–823
12. Rossari C, Salsmann M (2017) Étude quantitative des propriétés dialogiques des adverbes épistémiques. Actes des 9èmes Journées Internationales de la Linguistique de corpus 87–93
13. Roulet E (1979) Des modalités implicites intégrées en français contemporain. Cahiers Ferdinand de Saussure 33:41–76
14. Roulet E, Auchlin A, Moeschler J, Schelling M, Rubattel C (1991) L'articulation du discours en français contemporain, 3rd edn. Lang, Bern
15. Vigier D (2017) La préposition dans au xvie siècle. Apports d'une linguistique instrumentée. Lang, 206(2):105–122

Multilingual Text Analysis

How to Think About Finding a Sign for a Multilingual and Multimodal French-Written/French Sign Language Platform?

Cédric Moreau

Abstract This article examines the access to the signs in French sign language (LSF) within a corpus taken from the collaborative platform Ocelles, from a multilingual French bijective/LSF perspective. There is currently no monolingual dictionary in SL, so deaf users must necessarily master the written language of by relying on the maximal resemblance sequence of signs/experience, or use the lexical unit without resemblance with the referent. This model, which is also integrative, therefore takes into account the diachronic link existing within language under the influence of pressures between transfer structures and lexical units. The morphemic approach to the study of lexical units is in this case legitimate since their compositionality does not rely on strict phonology but, in the first place, on complex morphology. First of all, we shall present our paradigm and the origins of the Ocelles multilingual and multimodal platform (written, oral, and signed languages), out of which our French-written/LSF corpus is built. We will then describe a process likely to enable users to search for an LSF signifier and to relate this result to that of the corresponding written French signifier.

1 Introduction

When a sign language speaker wishes to access a resource in his or her language, particularly to search for a definition in a sign language (SL) dictionary, he or she is confronted with two obstacles. The first is that very few tools presented as digital sign language dictionaries are just lexicons. Among 105 sites listed on the Web, one capitalization uses the term "dictionary," yet only 17 of them have written definitions with only seven giving definitions in SL. The number of SL dictionaries is therefore extremely small. Moreover, the number of definitions does not exceed 5000, so we are very far from the 135,000 proposed by the Larousse online dictionary [12]. The second obstacle concerns the difficulty for the deaf user to access the very content of

C. Moreau (✉)
Grhapes EA 7287, INSHEA, UPL, Suresnes, France
e-mail: cedric.moreau@inshea.fr

D. F. Iezzi et al. (eds.), *Text Analytics*, Studies in Classification, Data Analysis, and Knowledge Organization,
https://doi.org/10.1007/978-3-030-52680-1_13

such a dictionary. Indeed, we found that in the vast majority of cases, the proposed entries are closely linked with the knowledge of the national written language. A necessary prerequisite is therefore mastery of the written language, presenting a major obstacle for deaf people who have SL as their first language and the written language, often poorly mastered, as their second one. From the seven sites mentioned above, only two offer an entry via the linguistic parameters of the SL [12].

This question also takes on a particular resonance when we question the mode of transmission of SL. This is not a hereditary mode of transmission, since about 95% of deaf people have hearing parents who, for the most part, do not practice SL. Language learning therefore takes place in a variety of contexts, at any age, often without the fixed reference of an adult family member.

The remaining percentage (about 5%) is therefore made up of deaf people from deaf parents. Parents mostly are in the previous category from hearing families. Only 0.02% of the significant deaf population is made up of a generic pain with three successive generations of deaf signers. The SL learning standard cannot therefore be compared to those of the learners [4].

In addition, French sign language (LSF), which has been banned as a language of instruction for more than a century, has only been recognized as a language of the Republic since 2005. In this context, the multilingual and multimodal collaborative project Ocelles[1] was born, which aims to define all concepts, in all fields of knowledge and in all languages (written, oral, or signed) [14].

2 Confrontation of Two Paradigms

2.1 A Phonological Hypothesis of SL

Basing on a study of preschool communication between deaf small children and their hearing environment, Susan Goldin-Meadow highlighted the creation of gestures called "home signs" [8, 7]. To try to communicate with their surroundings, these children realize them in the perceptual-practical universe. These productions make it possible to hypothesize pre-linguistic conceptual stabilizations, unlike the productions of hearing children of the same age, for which the link between language and this perceptual-practical knowledge does not exist. Once in school, these children then come into contact with an institutionalized sign language. According to Golwin-Meadow as the meaningful forms of institutionalized sign languages have a phonological status, the components of the child's home signs would then lose their morpheme status to become phoneme equivalents. This hypothesis can be considered as a starting point for the confrontation of two paradigms. Iconicity is then compared to illustrative co-verbal gestures, relegated to the rank of pantomime outside any linguistic phenomenon.

[1] https://ocelles.inshea.fr. Project under the egis and with the help of the Délégation générale à la langue française et aux langues de France (DGLFLF) and the Ministry of National Education.

In this paradigm, most sign language "dictionaries" currently fit. Their entries are mainly defined on the basis of a hypothetical conceptual equivalence between the words of the dominant vocal languages and those of the lexematic units (LU) of the signed languages [6]. The origin of this methodology is rooted in work asking deaf speakers to translate a lexeme out of context from the vocal language into a signed language. This corpus is then used in the development of an inventory of minimum pairs, in which configurations (hand shapes), their locations, and movements are widely represented [9].

2.2 A Morphemic Hypothesis of SL

Our work is a part of a completely different paradigm in which the consequence of deafness is no longer a simple channel change effect. The possibility to say and show which used to be the only fact of the visuo-gestuel channel has given sign languages the architecture different from that of vocal languages.

According to Cuxac [3], two discursive enunciation strategies coexist in LSF. The signer via the visuo-gestuel channel chooses to say without showing or to say by showing. It can thus "show" the experience based on the maximum similarity of sign sequence/experience, or use LU without similarity to the referent. The semiological model [4] has its origins in the description of a typology and in the identification of the three main transfer structures:

- entity volumes (Transfers of Form and Size (TFS)),
- the movements of actors in relation to stable rentals, in the image of a four-dimensional environment (the three of space and time) recreated in front of the speaker (Situational Transfers (ST)),
- the entity desired by the speaker, who then becomes that entity (Personal Transfers (PT)).

(Cuxac [3] and Sallandre [15]). Imaginary or real experiences are thus anamorphosed by the speaker.

The semiological model therefore takes into consideration the diachronic link that exists within the language under the influence of pressures between transfer structures and LU. This link is sometimes found in the etymology of some of the LUs. The morphemic approach to the study of LUs is in this case legitimate, since their compositionality does not depend on a phonology in the strict sense but in the first place, on a complex morphology.

When a sign is made (transfer or LU), the whole body of the speaker takes on a semantic value via an organization of the morphemic elements that compose it facial expression, posture, facial orientation, configuration, movement, location [16], orientation [5, 10, 11, 18].

3 Significant Elements in the Search for a Sign for a Written French Multilingual and Multimodal Platform/LSF

3.1 Context of a Sign Search in a Bilingual Corpus Written Language/SL

The Ocelles collaborative project makes it possible to link, over the course of the contributions, definitions of concepts to several signifiers being in textual, oral or signed forms. The entries are not constrained by the original language, and the architecture is deployed as users contribute. The text entry can therefore take the form of a word or expression if the origin of the deposit comes from a sign language transfer structure. The current reflection therefore focuses on the type of possible indexation of signs that is essential to the sign search process in the context of a bilingual corpus of written language/SL.

3.2 Automation of Indexing

The indexing of a sign is done via the corresponding text entry. There is currently no automatic indexing of dynamic collaborative corpus of SL signs that could serve as a basis for a LU search engine or a transfer directly from the linguistic parameters of the SL. The very nature of the video signal, which is very complex to analyze, does not allow automatic indexing. In addition to the loss of three-dimensional information related to projections from 3D space to 2D video, this work would require fine tools for analysis and recognition of the different body components, occurring in parallel, at very different spatial and temporal scales, developed for vocal, linear, and single source languages but not for SL [2].

3.3 Current Situation and Limit

Today, the entry of SL signs from the linguistic parameters is mainly made from the configuration. From the 105 sites listed offering signs in SL, only 18 offer the possibility of directly accessing a sign from the linguistic parameters of sign language, without using a written language. From these 18, 17 offer an input from the configuration (the number of these manual inputs varies from 9 to 211 depending on the site), 6 offer the input from the movement, 10 from the location, and 1 for symmetry, lip image, and facial expression [12]. This phonological indexing of SL, with such a gap in the possible number of configurations from 9 to 211 for example, questions the management of the potential error of the speaker who is looking for a sign that he would have perceived in speech (which is the case in most cases, given the oral nature of SL). In addition, on a choice between 211 configurations, the speaker has one chance in 211 to choose the right one or 210 risks in 211 to make mistakes.

3.4 Description and Search Criteria

Indexing cannot therefore be based only on a strictly phonological approach and must take into account the management of possible errors. Our hypothesis is based on the importance for the speaker of certain linguistic units in a morphemic approach used when formulating a sign [12].

Our approach is based on the principle of collaborative indexing to reflect speakers' perceptions. The principle is based on the following process:

- taking into account the type(s) of transfer(s) used (ST/PT/TFS) in the realization of a sign unless the lexemic unit can possibly find its origin in one of these transfers,
- iteration in the choice of key images being a basis for a description of the significant linguistic units is based [17],
- a more detailed description of the units selected is then proposed.

While today linguistic structures cannot be accepted as familiar to all contributors, their consideration cannot be ignored. Two approaches are being considered. A first one, inherent to the main objective of the platform, is based on the proposal of a definition of these concepts in order to gradually familiarize the speakers with their uses. A succession of possible anamorphoses that are more and more precise is then proposed. This approach is consistent with the use of any tool for which a minimum of prerequisites are required, such as the alphabet for a dictionary. A second approach is based on taking these gaps into account by integrating the process into a continuum, which allows for a possible contribution based on the selection and description of images that are representative of the sign from the user's point of view. It is therefore the set of macro-microscopic descriptions of each contributor serving as a basis for the significant linguistic units weighting. This data will then be reused as a search criterion for a sign.

4 Conclusion

Indexing signs of sign languages in a research based on the linguistic parameters of these languages are currently an important issue for deaf education in the perspective of an inclusive society in which written and signed languages coexist. The fine analysis of 5 min of video recording of an SL production takes about half a day for the researcher. Given the scale of the task and the urgency of the situation, the work of describing linguists must be done in parallel with that of speakers. In addition, the generation of this type of corpus, based on the principle of spontaneous deposits, highlights the need for multidisciplinary work from the point of view of facilitating both uses and data processing. The choice of key images for indexing a sign, for example, is confronted with a large amount of information that is visually and morphologically close. Also, the calculation of a consideration of this question must be questioned in linguistics, computer science, ergonomics, or mathematics in order to support users in the best way and optimize the resulting treatments. Finally, this approach could be combined with automatic indexation to improve existing models [1].

References

1. Belissen V, Gouiffes M, Braffort A (2019) Optimization of a gesture representation network for sign language analysis. Palaiseau, France. https://hal.archives-ouvertes.fr/hal-02146369. Accessed 20 Dec 2019
2. Braffort A, Dalle P (2012) Accessibilité du cyberespace: langues des signes. In: Vannini L, Le Crosnier H (eds) NET.LANG Réussir le cyber espace multiling. C&F éditions éd, pp 269–289. https://net-lang.net. Accessed 20 Dec 2019
3. Cuxac C (2000) La Langue des Signes Française: les voies de l'iconicité. OPHRYS, Paris
4. Cuxac C, Pizzuto EA (2010) Émergence, normes et variation en langues des signes: vers une redéfinition conceptuelle. Sourds et langue des signes. Norme et variations, Langage et Société (131):37–53
5. Friedman LA (1977) On the other hand: new perspectives on American sign language. Academic Press, New York
6. Fusellier-Souza I (2006) Processus de création et de stabilisation lexicale en Langues des Signes (LS) à partir d'une approche sémiogénétique, vol 7. GLOTTOPOL, Janvier, pp 72–95
7. Goldin-Meadow S (2003) The resilience of language: what gesture creation in deaf children can tell us about how all children learn language. In: Werker J, Wellman H (eds) Essays in developmental psychology series. Psychology Press, New York
8. Goldin-Meadow S, Mylander C (1991) Home sign systems in deaf children: the development of morphology without a conventional language model. Theoretical issues in sign language research. Psychology, vol 2, pp 41–63
9. Klima ES, Bellugi U (1979) The signs of language. Harvard University Press, Cambridge, MA
10. Liddell SK (1980) American sign language syntax. Mouton De Gruyter, The Hague
11. Moody B (1980) Histoire et Grammaire, tome I. Ellipses, Paris
12. Moreau C (2012) Stratégies de reconstruction du sens en langue des signes française à partir de données incomplètes en et hors contexte. Perspectives pour la constitution d'un lexique/dictionnaire à entrée directe en langue des signes. Thèse de doctorat sous la direction de C. Cuxac, Université Paris VIII, Saint-Denis
14. Moreau C (2017) How to think of an inclusive educational tool based on "special" needs. Kazan, Университет управления «ТИСБИ», p 372
15. Sallandre M-A (2003) Les unités du discours en Langue des Signes Française. Tentative de catégorisation dans le cadre d'une grammaire de l'iconicité. Thèse de doctorat sous la direction de C. Cuxac, Université Paris VIII, Saint-Denis
16. Stokoe WC, Casterline DC, Cronenberg CG (1965) A dictionary of American Sign Language on linguistic principles. Gallaudet College Press, Washington D.C.
17. Thom R (1988) Esquisse d'une sémiophysique: physique aristotélicienne et théorie des catastrophes. InterEditions, Paris
18. Yau S-C (1992) Création Gestuelle et début du Langage Création de langues gestuelles chez les sourds isolés. Langages Croisés, Hong Kong

Corpus in "Natural" Language Versus "Translation" Language: LBC Corpora, A Tool for Bilingual Lexicographic Writing

Annick Farina and Riccardo Billero

Abstract The aim of this paper is to describe the work done to exploit the LBC database for the purpose of translation analysis as a resource to edit the bilingual lexical sections of our dictionaries of cultural heritage (in nine languages). This database, made up of nine corresponding corpora, contains texts whose subject is cultural heritage, ranging from technical texts on art history to books on art appreciation, such as tour guides, and travel books highlighting Italian art and culture. We will illustrate the different questions with the SketchEngine LBC French corpus, made up at the moment of 3,000,000 words. Our particular interest here is in research that not only orients lexical choices for translators but that also precedes the selection of bilingual quotations (from our Italian/French parallel corpus) and that we rely on for editing an optional element of the file called "translation notes." We will rely on this as much for works on "universals of translation" already described by Baker (Corpus linguistics and translation studies. Implications and applications. In Baker M et al (eds) Text and technology. Benjamins, Amsterdam/Philadelphia, pp 233–250 (1993)) as for studies aimed at improving translation quality assessment (TQA). We will show how a targeted consultation of different corpora and subcorpora that the database allows us to distinguish ("natural language" vs "translation," "technical texts" vs "popularization texts" or "literary texts") can help us identify approximations or translation errors, so as to build quality comparative lexicographical information.

Riccardo Billero wrote parts 1 and 2 of this article: Parts 3, 4, and 5 were written by Annick Farina. The IT support for the project (the database and the statistical results) was produced by Riccardo Billero.

A. Farina (✉) · R. Billero (✉)
Università Degli Studi di Firenze, Florence, Italy
e-mail: annick.farina@unifi.it

R. Billero
e-mail: riccardo.billlero@gmail.com

D. F. Iezzi et al. (eds.), *Text Analytics*, Studies in Classification, Data Analysis, and Knowledge Organization,
https://doi.org/10.1007/978-3-030-52680-1_14

1 Introduction

The *Lessico dei Beni Culturali* project and the research group by the same name have existed since 2013 within the Educational Department of Languages, Interculture and Psychology (FORLILPSI) of the University of Florence, with the following objectives:

- to create a reference platform for the use and study of the heritage lexicon;
- to create text corpora for all the languages of the project;
- to provide tools for consulting these corpora to enable and promote research in different fields such as linguistics and the Humanities in general;
- to offer access to corpora to all types of users, by offering a dialogue with all those who want to contribute in a constructive way to enriching the platform;
- to write multilingual dictionaries of art and heritage words using corpora as a resource.

This work reflects the wide variety of cultural and scientific interests of the members of our research unit (specialists in linguistics, lexicography, computational linguistics, translation, literatures from the different countries affected by the project, history of art, etc.) all of which converge on the study of the heritage lexicon in a broad spectrum perspective that favors opening up to the world and its diversity.

The ultimate goal of the *Lessico dei Beni Culturali* project is to create monolingual dictionaries of nine different languages according for the specific use of describing (and translating descriptions) the Tuscan heritage mainly in texts for popular utilization (tourist guides, museum sites, etc.). To do this, we have created textual databases, which we are completing as a work in progress,[1] which will be used for various tasks, from creating lists of entries to drafting monolingual lexicographic/terminological files and translation files linking the entries of the different languages to each other (for the description of these databases see Billero et al. [3]).

It is the use of these textual databases for the drafting of bilingual translation files that we will illustrate here,[2] based on the analysis of different queries on NoSketchEngine (mainly statistics and context) from our French LBC corpus, currently composed of approximately three million words. In particular, we will compare data from several comparable subcorpora of texts "in natural language" and texts "in translation." We will also propose a first comparison of results from a subset of the Italian corpus with a subset containing the French translations of the same texts, which constitute fragmentary material for the moment because we are still working on the insertion of the texts with the aim of creating parallel bases of translation from Italian into all the languages of the project.

[1]Consultable at https://corpora.lessicobeniculturali.net/.

[2]Apropos of using our databases to create monolingual dictionaries, see the article by Nicolás and Lanini in this volume. We are currently building the list of entries of the different languages following the model they have defined for Italian. The bilingual link between these different entries will only be possible once we have established our parallel databases.

We will show how a targeted consultation of the various corpora and subcorpora that the database allows us to distinguish ("natural language" Italian vs "natural language" French, French "in translation" vs "natural language" French, "specialized texts" French vs "popularization" French vs "literary" French) can help us identify approximations or translation errors, and hence provide highly accurate comparative lexicographic information.

2 The LBC Corpora

One of the main resources of the LBC Research Unit is the LBC corpora. Begun in 2016, they have been designed from the start as a tool to be used not only by the group for writing dictionaries but that should also be accessible for other types of uses and users, as a work tool for translators or as a documentary source for specialists in art history or literature.

For this reason, in designing the corpus and corpus interrogation model for extensive use by specialists, we have defined three basic needs:

1. to furnish useful linguistic material to create and serve as a document base for writing multilingual entries in LBC dictionaries;
2. to furnish linguistic material that can serve as a basis for linguistic, literary, or more broadly cultural studies;
3. to furnish useful textual material for spreading knowledge of the Tuscan heritage to the general public.

In creating the LBC corpora, we have therefore chosen texts that have as their main subject Italian artistic heritage and, more in particular, that convey a perspective "from without" on Florence and Tuscany, and descriptions that are together objective and personal. We have since started to broaden the type of texts, integrating works on Italy.

Some of the texts that appear in our corpora are:

- Biographies, for example, Benvenuto Cellini's autobiography (*La vita* and its translations in the different languages of the project) or Dumas' *Une année à Florence*;
- Art texts or having art as their subject, for example, Vasari (*Le Vite* and its translations) or Alarcón's *Vidas de los pintores y estatuarios*;
- Novels, for example, Forster's *A Room with a View* or Fernandez' *Dans la main de l'Ange*;
- Essays, for example, Feuillet's *L'art italien* or Burckhardt's *Geschichte der Renaissance*;
- Tourist guides such as the *Routard Guide* or *GeoGuide*;
- Specialized dictionaries, such as Viollet Le Duc's *Dictionnaire de l'architecture.*

These texts originate from different eras, from the Renaissance to the present day. For each text, both its dates of composition and its year of publication, of translation when applicable, and publication of the translation are recorded. For us, the most important dates are those of its composition, as regards "natural" language texts, and that of the translation, as regards translation texts. What interests us is the representative character of these texts in relation to their use of language and not for the type of discourse pertaining to heritage. However, we also leave the possibility for users of the corpora to search with criteria other than ours.

The different texts were collected and underwent a work process illustrated in detail in Billero [2]: these texts, accompanied by bibliographic type metadata, have been integrated into the different corpora published online through NoSketchEngine.

Because of the use we want to make of corpora for our project, we needed the proper software for processing and analyzing data from our corpus. Software belonging to this category can be of two types: what can only be installed on a computer and what can be accessed from an online service (and so available on the internet).

A research on the most recent techniques allowed us to identify three of the main software types that interested us—and which therefore also allowed us to make our corpus accessible to persons outside the group: CWB (IMS Open Corpus Work-Bench), INL BlackLab, and SketchEngine. While the first two are available free of charge, SketchEngine must normally be purchased, though it is free for members of European research institutions. There is also a free version, NoSketchEngine, which lacks some of the commercial version functions, which are mainly useful for lexicographic studies such as thesauruses and combinatorial word sketch maps.

All of the above-mentioned software has a number of basic functions that are fundamental for querying, allowing both simple searches and searches using regular expressions or the "Corpus Query Language" (CQL acronym), a query language defined according to graphic or lemmatized forms or grammatical categories, created at the base for CWB, to then be integrated into the other software mentioned. They also allow the use of lemmatization tools such as TreeTagger, Freeling, RfTagger … These three software programs also have a programming interface app for software-to-software communication (API) that allows them to perform query operations or introduce data through additional software created specifically for the purposes of such research.

CWB and SketchEngine by now enjoy excellent documentation and decent recognition in the university community. This is not the case with BlackLab, which has shortcomings in this regard.

SketchEngine offers some combinatorial word functions that are very useful for creating lists of specific terms of the various arts and that can be used to relate words that refer to the same art objects, to delimit areas of knowledge or references, for example, to monuments. They can also be used to differentiate objects from people or to disambiguate proper names that refer to these two categories. However, these possibilities are not available in the free NoSketchEngine version.

In the end, we chose SketchEngine as the corpus management software in our research activities, while we use NoSketchEngine to online consultation of our corpus.

3 Comparison Between "Natural Language" and "Translation" Corpus: A Perspective Halfway Between Descriptive and Prescriptive Translatology

We rely on analyses which do not consider the language of translation as a "third code" [6], in the belief that translated texts find their place perfectly within a single textual base of the same language, alongside texts "in natural language." However, on the model of proposals for using a translation corpus for a didactic purpose, for teaching both languages and translation, it seems necessary to offer the possibility of consulting the database in separate subcorpora that group texts of both types and define criteria for evaluating translated texts to be integrated into the database. To do this, we have created separate subcorpora of texts translated into all the languages of the project. These corpora are useful to us as translation memory tools for working on the bilingual part of our lexicographical files in a perspective that is more prescriptive than descriptive.

As will be shown in comparing results from our French LBC database "in natural language" and "in translation" with a corpus of almost 100,000 words currently not integrated into our database (and so not available online) consisting of translations of "popular" works into French (in particular tourist guides of Tuscany and museum brochures), some of the texts that interest us present features similar to "translationese," which could only distort queries of the base that aim at attesting forms or structures typical of French as it is written and spoken by most native speakers without interference from another language. We intend to define acceptability thresholds for texts "in translation" so as to introduce into our published corpus only texts that we will have recognized as similar to "natural language" texts, while limiting their proportion in our corpus so that the latter category remains preponderant.

3.1 Descriptive and Prescriptive Information in LBC Dictionaries: Universals and Deviations

Following Baker [1], we start from the principle that there are translation universals which will serve as a framework for illustrating the different types of queries carried out within our subcorpora and for comparing the results obtained. It is on these universals that we depend to provide the descriptive part of the detailed comparative lexicographic information present in the bilingual part of our dictionaries.[3] This information corresponds first to the observation of parallel corpora, which certify the translation of the lemmas (words or collocations) described by the dictionary,

[3]For a more precise description of this comparative linguistic information inside our dictionaries, as we defined it in a first proposal for a drafting protocol which we intend to review during the year 2020, we refer to [4].

by appearing in bilingual quotations within the bilingual part of the article. We will analyze in particular:

- the simplification (mainly, as regards our corpus, the choice of hyperonyms to translate certain more specific terms) which we will depend on in the drafting phase of our dictionaries to provide any added semantic information that accompanies the suggested translation: distinctive features specific to the lemma, which are not rendered by the translation, will be indicated, with or without parentheses, after the translation (e.g., *tavola* translated as *peinture (sur bois)-painting (on wood)*—and *tavoletta* translated as *(petite) peinture (sur bois)*—*(small) painting (on wood)*)
- leveling (non-respect of register, for example, the choice of technical rather than everyday terms, and vice versa). All entries have a usage mark indication. In the case of a translation that involves a shift of register, this change will be noted in the "translation note" part or will appear in the part reserved for distinctive semantic indicators, where several translations of the same lemma are possible, with or without loss of register. This is the case, for example, of the (unmarked) Italian word *tondo* compared to (unmarked) *médaillon* and the Italianism *tondo* (technical term used mainly in French in art historians).

Baker also analyzes the explicitation which is particularly frequent in the texts which interest us because it is almost systematically employed when using an Italianism, in particular for *realia*, which are given a special treatment in our dictionaries (cf. [4, 5]. It would be possible to systematically search for this type of data in our corpus by extracting all the occurrences of "type of" or "sort of" or the elements indicated in parentheses, but we have deliberately left out this category, which is too strongly linked to the subject described by our texts and to stylistic choices shared between the authors of "natural language" texts and translators in the context of our database, and therefore would not allow us to illustrate by comparing the two types of linguistic constraint resources related to translation operations.[4] We have also left aside "normalization" or "conservatism" which is ill-suited to our material, little conducive to variation or to lexical and stylistic exploration.

Unlike Baker ([1]: 243), who defines translation universals as "features which typically occur in translated text rather than original utterances and which are not the result of interference from specific linguistic systems," we have adopted a fairly prescriptive, or rather didactic perspective, taking into consideration the phenomena of interference (influence of the source language on the target language) frequent in translation operations between two close languages like Italian and French, and in texts whose quality is far from homogeneous. Interference is, in our view, the source not only of many cases of simplification and leveling deviations found in our comparisons but of other manifestations which can be likened to losses resulting from the translation operation, or even to errors or translation inaccuracies. The translation quality assessment (TQA) model and, in particular, the different types of quality

[4]The abundant use of Italianisms is a dominant feature of the tourist guides analyzed, comparable to a desire by their authors to give these texts a «hint of Italianness» (Farina [5]: 61).

measures which can guide the translator toward an improvement in fluidity and precision can serve as a reference for this (cf. "Multidimensional Quality Metrics" [11]. These analyses mainly orient us toward the choosing a position that may seem to run counter to utilizing a descriptive corpus like Baker's. In fact, it presents itself as an accessory allowing the user of our dictionaries to make choices, on the basis of a descriptive exploitation of the resources consulted, as we have already described it, and the indication of statistical data resulting from frequency analyses like the ones we will present below. The editor of the lexicographical files will be able to decide, if necessary and when our analyses of these data urge him or her to identify errors or deviations that could be reduced, not to suggest a form that appears in the base as a translation (indicating in the part of the article providing statistical information on the translations found) or to write the "translation note" part, optional in our bilingual articles, to advise users on their choices by explaining why certain forms may be preferred to others.

4 Natural Language Versus Translated Language: Observation of the Corpus

The difference in frequency of words or collocations present in comparable corpora containing French texts in "natural language" and texts which come from a French translation can allow us to identify forms chosen under the influence of the source language.

4.1 Zero Frequency in Natural Language Texts

We compared the list of words present in the LBC subcorpus of popular texts written in French containing 270,000 words with a non-integrated corpus in the database for the moment of the same text types but in translation (93,000 words) by performing a list of words found exclusively in the "in translation" subcorpus.

- errors

Most of the forms encountered can be considered errors, such as the absence of circumflex accents (*cloitre*), the influence of Italian spelling on French spelling (*baroche*), excessive "Frenchization" in spelling (*Caliari*) or by using a French translation whose usage would call for the Italian form (*Sainte-Réparate* is used in French to designate the person or the cathedral of Nice but not the Santa Reparata church of Florence, for which the French form is not attested anywhere in the LBC database) or vice versa (*Giove* is never used in Italian in our corpus, but is always translated as *Jupiter*), the use of words that have nothing to do with the description of Florentine heritage, probably because they correspond to a meaning of the source word that applies to other contexts (e.g., *coursive* in a description of Florentine monuments,[5] or *panonceau* to refer to the panels of the Gates of Paradise). This kind of error does not give rise to the production of targeted information inside dictionaries except, in the case of a high frequency of the error (as with *panonceau*, which is found in several sources and with a total of 28 occurrences but not for *coursive*, which is attested in just one text) to justify the insertion of an Italian entry and its French translation even in cases where the frequency of the form in the corpus would not justify it. The presence, even in a small proportion of errors of this type in a text, is enough for us to dismiss the work from the published corpus.

- leveling

We can distinguish forms that correspond to a "pragmatic" or stylistic difference between French and Italian which do not interest lexicographically, such as the use of *mentionnons* in several translation texts, which is not found in any of the texts from the complete database, or certain simple past forms (*décora, succéda*) which are not used in popular texts in "natural" French. These are forms that correspond to different standards relating to the text types in the corpus: a more in-depth analysis would probably show little or no attested use of "nous" in French tourist guides, and the infrequent use of forms in the simple past in comparison to the compound past or present, etc.

What interests us much more in this comparison is identifying forms that, while being perfectly "correct" in French, can be considered out of context with respect to the usages attested in the same type of context in natural language. The difference in the use of a word not attested can have the effect of an "anachronism" (difference in frequency of use in synchrony). This is the case, for example, of the *grand-ducal* adjective and the past participle *paraphé*, whose Italian equivalents are more frequent in the present-day language than are their literal French translations. The difference in register can also apply in the case of a difference in "technicity." The adjective *autographe* present in several popularization sources in translation is absent from texts of the same type in our natural language corpus, but we find some occurrences of it in more specialized texts of the general corpus. The difference in register will give rise to a differentiated marking between the lemma in the source language and its attested translation.

[5]In the sentences: "*Vers le début du XVIème siècle on pensa décorer le tambour externe de la coupole par une coursive en marbre qui en fasse tout le tour.*"; "*Écoutant l'avis autorisé de Michel-Ange qui définit cette coursive une cage à grillons, les autorités firent cesser les travaux.*"; and "*Sous les petites arcades de la coursive du corps du palais se trouvent de nombreux emblèmes liés à la vie politique de la commune de Florence*" in Riccardo NESTI, *Florence - Histoire, art, folklore, tous les chefs-d'œuvre* (Florence, ATS Italia Editrice, nd [2008]), the word *coursive* probably suggested itself as a translation for *ballatoio*, which refers to a gallery or a balcony in reference to buildings. This word is equivalent to *ballatoio* in French only in the context of describing a ship.

4.2 *Frequency Difference in Source Texts Compared to Target Texts*

To illustrate phenomena of simplification, we have queried two subcorpora of our LBC database made up of 51 lives from Giorgio Vasari's *Le vite de 'più eccellenti pittori, scultori e architettori* (1568) and their French translations into (Leclanché-Weiss translation, 1900).

Unable to base ourselves on statistics from parallel translation databases (for the description of these databases, see [12], we focused on French words that had a high frequency, by comparing this frequency with that of the closest common word in Italian (same meaning, same distinctive features). This allowed us to note frequency differences that will lead us to a more in-depth study to define analog networks in both languages which enable us to suggest translation links that help to avoid a loss of precision (Figs. 1 and 2).

Tableau, for example, has a frequency of 2232 per million words in our French subcorpus, while *quadro* has a frequency of 793 per million words in the Italian subcorpus containing the same texts in the original language. A large number of *quadro* hyponyms are in effect translated by *tableau* into French. If this loss is probably compensated by the addition of distinctive features that accompany the word, we think that the translator can only gain in precision if we propose other forms to him or her to render the meanings of these various hyponyms.

5 Conclusion

The comparison of results concerning the frequency of forms within the LBC corpus allowed us to illustrate the use of different subcorpora to orient both descriptive and normative information that we wish to provide in the bilingual part of our LBC dictionaries.

> We will consider, even if it remains to seen [...], that an over- or under-representation of a given linguistic phenomenon can correspond to a violation of the constraint of usage [...] and that a good translation must tend towards a blending between the original language and the translated language. [9]

The application of methods aimed at verifying the quality of translations and the creation of tools based on critical analyses of existing translations, by comparing

Fig. 1 Search for the word "tableau" in Vasari's *Vies* subcorpus translated into French

them, in particular, with productions which do not go through the mediation of another language, should yield an optimization of the natural character of the translated texts and of their precision, an essential aim for the disseminating quality information.

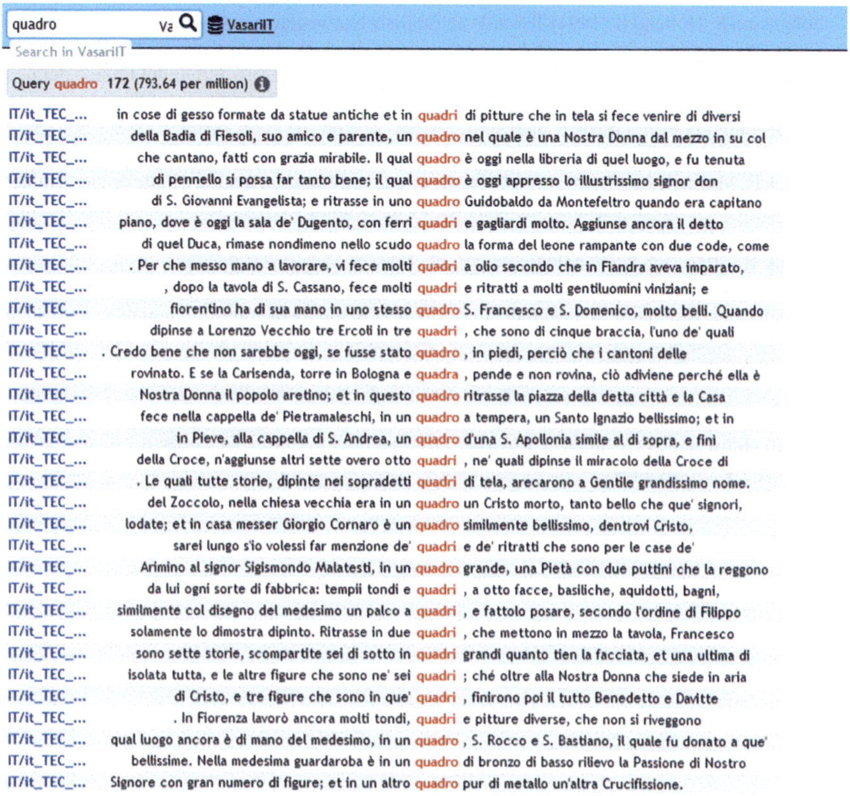

Fig. 2 Search for the word “quadri” in the Italian subcorpus of Vasari’s *Vite*

References

1. Baker M (1993) Corpus linguistics and translation studies. Implications and applications. In: Baker M et al (eds) Text and technology. Benjamins, Amsterdam/Philadelphia, pp 233–250
2. Billero R (2020) Cultural heritage lexicon: a case study. In: Alamán AP, Zotti V (eds) The language of art and culture heritage: a pluringual and digital perspective. Cambridge Scholars Publishing, pp 86–103. ISBN: 978-1-5275-4567-0
3. Billero R, Nicolas Martine MC (2017) Nuove risorse per la ricerca del lessico del patrimonio culturale: corpora multilingue LBC. CHIMERA Romance Corpora and Linguistic Studies 4(2):203–216. ISSN : 2386-2629
4. Farina A (2016) Le portail lexicographique du Lessico plurilingue dei Beni Culturali, outil pour le professionnel, instrument de divulgation du savoir patrimonial et atelier didactique, PUBLIF@RUM, vol 24. https://publifarum.farum.it/ezine_articles.php?id=335
5. Farina A (2014) Descrivere e tradurre il patrimonio gastronomico italiano: le proposte del Lessico plurilingue dei Beni Culturali. In Chessa F, De Giovanni C (eds) La terminologia dell’agroalimentare. Franco Angeli, Milano, pp 55–66
6. Frawley W (1984) Prolegomenon to a theory of translation. In: Frawley W (ed) Translation: literary, linguistic and philosophical perspectives. University of Delaware Press, Newark, pp 159–175

7. Johansson S, Hofland K (1994) Towards an English-Norwegian parallel corpus. In Fries U et al (eds) Creating and using English language corpora. Rodopi, Amsterdam, pp 25–37
8. Lanini L, Nicolas Martinez MC (2018) Verso un dizionario *corpus-based* del lessico dei beni culturali: procedure di estrazione del lemmario. In: JADT'18 Proceedings of the 14th international conference on statistical analysis of textual data, pp 108–116, ISBN 978-88-3293-137-2, UniversItalia
9. Loock R, Mariaule M, Oster C (2013) Traductologie de corpus et qualité : étude de cas. Tralogy - Session 5 - Assessing Quality in MT/Mesure de la qualité en TA. https://lodel.irevues.inist.fr/tralogy/index.php?id=188
10. Loock R (2016) La Traductologie de corpus. Presses Universitaires Septentrion, Villeneuve-d'Ascq
11. Uszkoreit H, Burchardt A, Lommel A (2013) A new model of translation quality assessment tralogy—Session 5—Assessing Quality in MT/Mesure de la qualité en TA. https://lodel.irevues.inist.fr/tralogy/index.php?id=319
12. Zotti V (2017) L'integrazione di corpora paralleli di traduzione alla descrizione lessicografica della lingua dell'arte : l'esempio delle traduzioni francesi delle Vite di Vasari. In Zotti V, Pano A (eds) Informatica Umanistica. Risorse e strumenti per lo studio del lessico dei beni culturali. Firenze University Press.

The Conditional Perfect, A Quantitative Analysis in English-French Comparable-Parallel Corpora

Daniel Henkel

Abstract The frequency of the conditional perfect in English and French was observed in a corpus of almost 12-million words corpus consisting of four 2.9-million-word comparable and parallel subcorpora, tagged by POS and lemma, and analyzed using regular expressions. Intralinguistically, authors and translators were compared using the Wilcoxon-Mann–Whitney test and Cliff's delta as a measurement of effect size, while potential interlinguistic influences were assessed by means of Spearman's correlation test. French-translated-from-English was found to be distinct from original French in its use of all conditional perfect forms, but significant differences between translators and authors were observed in both languages when COULD, MIGHT and POUVOIR were used as auxiliaries.

1 Introduction

Since corpus-based translation studies (CBTS) first began to gain momentum around the turn of the twenty-first century, differences have consistently been shown between corpora of translated English, French and other languages in comparison with untranslated reference corpora in the same languages. The hybrid nature of translated texts is now thus widely acknowledged among specialists[1] in the field, so much so that any further proof might seem superfluous. These studies have focused on phenomena such as the use of '*that*' to introduce subordinate clauses [2], features specific to the target-language such as sufficiency verbs in Finnish [3], manner-of-motion verbs [4], existential predications [5] or metadiscursive markers [6]. Such comparisons have provided valuable insights about the languages involved and the

[1] Albeit with some divergence of opinion as to whether such differences are best interpreted as evidence of source-language interference or as consequences of the translation process regardless of the source language, i.e., characteristics inherent in the 'third code' or 'translationese' [1].

D. Henkel (✉)
Université Paris 8 Vincennes St-Denis, Saint-Denis, France
e-mail: daniel.henkel@univ-paris8.fr

D. F. Iezzi et al. (eds.), *Text Analytics*, Studies in Classification, Data Analysis, and Knowledge Organization,
https://doi.org/10.1007/978-3-030-52680-1_15

translation process itself but have focused most often on particular linguistic features in terms of their overall frequency,[2] without taking into account that in any corpus of texts multiple occurrences will be found within the same text, thus violating the assumptions of certain statistical tests,[3] that outliers may distort totals or averages, especially in small corpora, and that a great deal of variation may exist within texts in the same corpus. Little consideration has been given so far to the fact that each language-system consists of many individual styles or idiolects[4] which gravitate around a common center, but individually exhibit widely differing characteristics. In other words, while the variation from one author or translator to another is inherent in the very nature of corpus linguistics, this dimension remains absent from the equation in many corpus-based translation analyses.

Following a research design described by Johansson [13] as a "bidirectional" model, this study will demonstrate how the quantitative analysis of comparable corpora (i.e., corpora in different languages composed of untranslated texts sharing common characteristics) and parallel corpora (i.e., corpora consisting of source- and target-texts) together can yield greater insight than either approach on its own. Although the English/French language pair has been the object of intense speculation, since the time of Vinay and Darbelnet [14], bidirectional studies of English and French (as with many other language pairs) have been limited by a scarcity of available resources. Cappelle and Loock [5], for example, in order to compensate for the lack of a corpus of French-translated-from-English, resort to the use of self-collected corpora of fewer than 300 k words taken in each case from three different authors. Similarly, Granger's [6] more recent study of metadiscursive markers in English-translated-from-French versus original English based on corpora of European parliament debates of around 1.2–1.4 m words once again lacks a corpus of French-translated-from-English. The corpora compiled for this study, comprising almost 3 million words each (12 million total) from 108 different authors or translators (4 corpora of 27 authors/translators each), while still quite small compared to monolingual reference corpora such as Frantext, BNC or CoCA, nonetheless represent an advance in terms of size, diversity and balance compared those used in recent studies, especially with respect to French-translated-from-English.

Furthermore, in order to overcome the shortcomings of studies based on average frequencies, several statistical methods, which treat each corpus, not as an amalgam of

[2]Olohan [7] apparently subscribes to Stubbs' [8] view that "*corpus linguistics [...] investigates relations between frequency and typicality, and instance and norm. It aims at a theory of the typical,*" (while nonetheless encouraging investigation of individual translators' styles in her conclusion), and the predominance of this approach is confirmed again over a decade later by Loock [9] who observes that "*many studies within the CBTS framework still solely rely on overall quantitative analyses to establish differences between original and translated languages*" (emphasis added).

[3]Moreover, probability statistics, such as *p*-values or log-likelihood, are often misinterpreted (an all-too-common error, which seems to know no disciplinary boundaries) as if they were measurements of effect size rather than an indication of the amount of evidence available.

[4]It should nonetheless be acknowledged, *pace* Loock [9], that this problem has been recognized before and that attempts have been made to take into account the amount of variation between authors or texts in the same corpus, by calculating variance [10], subjective appraisal of frequency data from selected translators [11], or student's *t*-test [12].

occurrences, but as a sample of authors or translators, will be used to determine, first of all, whether sufficient evidence of a difference exists (Wilcoxon-Mann-Whitney), and how great the difference between authors and translators in comparison with the variation between authors or translators within each sample (Cliff's delta), after which the degree of influence between original and translated texts will be assessed by means of a correlation test (Spearman's rho).

The conditional perfect[5] (En. 'would have done', Fr. 'aurait fait') was chosen as an indicator in this study, given that the syntactic constituents of the corresponding constructions in English and French are organized differently and subject to different kinds of variation. The English construction, as illustrated in (1) and (2), invariably consists of two auxiliaries: The first of which is one of the preterit modals WOULD, COULD, MIGHT, SHOULD (or OUGHT), while the second is invariably HAVE, and the lexical verb is invariably in the past participle (e.g., 'done,' 'gone,' 'spoken,' etc.).

1. But if it had not been for the black dog, he **would**have gone back. (H.G. Wells, *The War in the Air*) ⇒ N'eût été le chien noir, il serait retourné sur ses pas.
2. The key had no duplicate; it was impossible that the gray woman **could**have gone that way. (E. Wallace, *Green Archer*) ⇒ La clef n'avait pas de double; il était impossible que la dame aux cheveux gris fût partie par là.

In French, the corresponding constructions take a single auxiliary verb, AVOIR for the majority of lexical verbs, as in (3):

3. Des meubles l'encombraient, trop nombreux pour la pièce. Si j'avais soulevé leurs housses, j'aurais vu que le damas de l'étoffe et la dorure des bois dénonçaient l'opulence ancienne. (P. Bourget, *La duchesse bleue*) ⇒ It was full of furniture, too full in fact. If I had raised the covers from the furniture I should have seen that the quality of the upholstery and the gilded wood betokened former opulence.

A subset of intransitive (or "unaccusative") verbs, such as ALLER, VENIR, PARTIR, SORTIR, TOMBER, DEVENIR, take ÊTRE systematically as their auxiliary (4a), as well as all pronominal verbs (e.g., SE REGARDER, SE PARLER), followed by the lexical verb again in the past participle (4b):

4. Ah ! si je l'avais vue, je serais allée(a) à elle, je me serais livrée(b) à elle … (O. Mirbeau, *Le journal d'une femme de chambre*) Ah! if I had seen her, I would have gone(a) to her, I would have delivered myself(b) to her.

When the semi-auxiliaries POUVOIR and DEVOIR are included, however, AVOIR is used as the first auxiliary regardless of the lexical verb,[6] followed by POUVOIR or DEVOIR in the past participle, while the lexical verb is in the infinitive, as in (5):

---

[5]The term 'conditional perfect' is used as a generic lable for the corresponding syntactic constructions in English and French, even though, from a semantic point of view, not all of the occurrences are necessarily 'conditional' or 'perfect' *sensu stricto*.

[6]Unlike Italian which maintains the distinction between AVERE/ESSERE even with POTERE/DOVERE.

5. C'était la faute de Frédéric; il avait laissé passer le bon moment, il aurait dû venir plus tôt, se remuer. (G. Flaubert, *L'éducation sentimentale*) ⇒ It was all Frederick's fault; he had let the lucky moment pass by; he should have come sooner and stirred himself.

The English construction thus varies only as far as the choice of the first auxiliary is concerned, with no syntactic constraints on the compatibility between auxiliaries and lexical verbs, whereas, in French, the first auxiliary is subject to syntactic constraints, and the addition of a semi-auxiliary affects the form of the lexical verb. These differences between the two constructions are difficult for language learners to assimilate, and it was thought that they might also be a source of translation difficulties, or divergences between translated and original texts.

2 Methods

Two important terminological distinctions must be made at the outset. The first is between *ex nihilo*, a.k.a. 'original,' English (En0) and French (Fr0), i.e., discourse in each language produced independently of any known prior influence, as opposed to English-translated-from-French (EtrF) and French-translated-from-English (FtrE), which will be used to refer to translations into each language, based on a pre-existing work in the other language, and therefore potentially subject to interlinguistic influences. The second distinction is between two sorts of bilingual corpora, 'comparable' and 'parallel'. In keeping with the clarification offered by McEnery and Xiao [15], the term 'comparable corpus' will hereafter refer to a bilingual corpus consisting of two subcorpora of *ex nihilo* English and French texts, which are therefore not translations of one another, but which share a certain number of common characteristics, whereas the term 'parallel corpus' will designate a corpus made up of one subcorpus of *ex nihilo* works in a source-language and another subcorpus consisting of the translations of those same works into the target language.

The corpora used in this study were compiled from public domain works available in electronic format (.epub,.mobi,.html or.txt), the translations of which were also available in electronic format via publicly available sources (primarily Project Gutenberg, but also Project Gutenberg Australia, Wikisource, Bibliothèque numérique romande, Bibliothèque électronique du Québec, and Ebooksgratuits). Common criteria[7] based on size and date were then used to select 27 works by 27 different authors in En0 and the same number in Fr0, so as to obtain, first of all, two reference subcorpora comparable in terms of date, size, discourse type and diversity (Table 1). The translations of these works were then compiled into two subcorpora of EtrF

[7]Whenever several works by the same author were available, preference was given either to the most recent or the one with the highest word count. In general date was given precedence over size, except in cases where a major difference in word count was found between works published within a relatively close interval.

Table 1 Summary of characteristics for comparable En0 and Fr0 subcorpora

	Subcorpus 1 En0 ($n = 27$)	Subcorpus 2 Fr0 ($n = 27$)
Word counts[a]	**Max.** 201,305 (Wallace, *Ben Hur*) **Min.** 50,096 (Fitzgerald, *The Great Gatsby*) **Median** 94,265 (Reid, *The Finger of Fate*) **Total** 2,891,531	**Max.** 239,132 (Rolland, *Jean Christophe*) **Min.** 38,233 (Audoux, *Marie Claire*) **Median** 93,435 (Vandérem, *Les deux rives*) **Total** 2,957,738
Dates	**Max.** 1928 (Woolf, *Orlando*) **Min.** 1868 (Collins, *The Moonstone*) **Median** 1901 (Kipling, *Kim*)	**Max.** 1921 (Leblanc, *Les dents du tigre*) **Min.** 1869 (Flaubert, *L'éducation sentimentale*) **Median** 1901 (Bazin, *Les Oberlé*)

[a]Word counts were estimated using the text editor Geany, after replacing punctuation with whitespaces, given that punctuation was found to artificially inflate word counts in French as compared to English

and FtrE, so as to produce an 11.6 m-word 'super-corpus'[8] consisting of four 2.9 m-word subcorpora, designed to be both comparable and parallel (i.e., "bidirectional" in Johansson's [13] terminology), and thereby provide a basis for three types of comparisons:

- between En0 and Fr0, in order to establish benchmark data for each language,
- between EtrF and En0, so as to ascertain whether the linguistic indicator under investigation, i.e., the conditional perfect, has a similar distribution in EtrF compared to En0, and likewise for FtrE in comparison with Fr0,
- between source- and target-texts, to determine whether correlations exist between the parallel subcorpora (i.e., EtrF ~ Fr0 and FtrE ~ En0) which could be taken as evidence of interlinguistic interference.

All of the texts were cleaned of metatext, tagged for POS and Lemma in *TreeTagger* [16] (with the backslash "/" as the tag marker), and interrogated in *AntConc* [9] using the following regular expressions to target the conditional perfect.

English (all verbs):

- \S + /(([wc]|sh)ould|(mi|ou)ght)(\S + /RB[RS]?/\S +)*(to/\S +)? \S + /(have|of)(\S + /RB[RS]?/\S +)* \S + /V[VBH][ND]/\S + [9]

French (verbs taking AVOIR as an auxiliary, verbs taking ÊTRE, reflexive constructions):

- \S + /VER:cond/avoir(\S + /ADV/\S +)* \S + /VER:pper

[8]The corpus used in this analysis is expanded version of the 8 m-word corpus of 80 authors and translators (4 × 20) on which the original conference paper was based.

[9]The form "would of" was included as an idiomatic spelling of "would've": *If he'd of lived he'd of been a great man.* (F.S. Fitzgerald, *The Great Gatsby*) ⇒ *S'il avait vécu, il serait devenu un grand homme.*

- \S + /VER:cond/être(\S + /ADV/\S +)* \S + /VER:pper/(r[eé])?(aller|(ad|de|inter|par|pro|sur)?venir|rester|demeurer|(ap|dis)?paraître|naître|mourir|décéder|arriver|partir|tomber|monter|descendre|passer|rentrer |retourner|sortir)
- ((je/\S + (\S + /ADV/\S +)* m[e']/\S +)|(tu/\S + (\S + /ADV/\S +)* t[e']/\S +)|(nous/\S + (\S + /ADV/\S +)* nous/\S +)|(vous/\S + (\S + /ADV/\S +)* vous/\S +)|(s[e']/\S +)) (en|y/\S +)* \S + /VER:cond/être(\S + /ADV/\S +)* \S + /VER:pper/

After this initial inventory of all conditional perfect constructions, these formulas were adapted to target WOULD, COULD, MIGHT, POUVOIR and DEVOIR specifically.

For English:

- \S + /would(\S + /RB[RS]?/\S +)* \S + /(have|of)(\S + /RB[RS]?/\S +)* \S + /V[VBH][ND]/\S +
- \S + /could(\S + /RB[RS]?/\S +)* \S + /(have|of)(\S + /RB[RS]?/\S +)* \S + /V[VBH][ND]/\S +
- \S + /might(\S + /RB[RS]?/\S +)* \S + /(have|of)(\S + /RB[RS]?/\S +)* \S + /V[VBH][ND]/\S +

For French:

- \S + /VER-cond/avoir(\S + /ADV/\S +)* \S + /VER-pper/pouvoir
- \S + /VER-cond/avoir(\S + /ADV/\S +)* \S + /VER-pper/devoir

The results obtained from these queries were converted into frequencies per 10,000 words (freq./10 k) for each author or translator and analyzed in R [17] using the Wilcoxon-Mann-Whitney test, Cliff's delta, and Spearman's correlation test as described in the following section.

3 Results and Analysis

The data collected from each subcorpus are presented in Tables 2, 3, 4 and 5 and summarized in Fig. 1.

As is readily apparent from Fig. 1, the conditional perfect is used more frequently in En0 (median = 11/10 k) than in Fr0 (median = 5.6/10 k). EtrF and FtrE occupy an intermediate zone, with practically identical medians (8.6/10 k in EtrF, 8.9/10 k in FtrE). Interlinguistically, taking Cliff's delta as the estimate for effect size, the difference between En0 and Fr0 is in the range conventionally described as "large" ($d = 0.68$). Intralinguistically, the difference in French between authors (Fr0) and translators (FtrE) is also in this range ($d = 0.57$), while in English, it is much smaller ($d = 0.29$). In fact, according to the Wilcox-Mann-Whitney test, the evidence of a difference between authors and translators as distinct subtypes in English is not

Table 2 Conditional perfect frequencies in En0

Author, Title	Word count	Would have	Could have	Might have	Should have
Bellamy, Looking Back	78,414	$n = 69$ (8.8/10 k)	$n = 32$ (4.08/10 k)	$n = 17$ (2.17/10 k)	$n = 26$ (3.32/10 k)
Benson, Lord of the World	101,303	$n = 19$ (1.88/10 k)	$n = 8$ (0.79/10 k)	$n = 12$ (1.18/10 k)	$n = 8$ (0.79/10 k)
Buchan, Greenmantle	100,621	$n = 82$ (8.15/10 k)	$n = 29$ (2.88/10 k)	$n = 16$ (1.59/10 k)	$n = 12$ (1.19/10 k)
Burnett, The Secret Garden	83,065	$n = 39$ (4.7/10 k)	$n = 15$ (1.81/10 k)	$n = 11$ (1.32/10 k)	$n = 10$ (1.2/10 k)
Collins, The Moonstone	199,978	$n = 112$ (5.6/10 k)	$n = 54$ (2.7/10 k)	$n = 91$ (4.55/10 k)	$n = 52$ (2.6/10 k)
Conan Doyle, The Lost World	77,152	$n = 41$ (5.31/10 k)	$n = 41$ (5.31/10 k)	$n = 14$ (1.81/10 k)	$n = 28$ (3.63/10 k)
Cox, Out of the Silence	114,354	$n = 80$ (7/10 k)	$n = 28$ (2.45/10 k)	$n = 21$ (1.84/10 k)	$n = 14$ (1.22/10 k)
Eliot, Middlemarch	164,459	$n = 167$ (10.15/10 k)	$n = 37$ (2.25/10 k)	$n = 63$ (3.83/10 k)	$n = 45$ (2.74/10 k)
Fitzgerald, The Great Gatsby	50,096	$n = 20$ (3.99/10 k)	$n = 5$ (1/10 k)	$n = 12$ (2.4/10 k)	$n = 3$ (0.6/10 k)
Hardy, Tess of d'Ubervilles	153,079	$n = 119$ (7.77/10 k)	$n = 27$ (1.76/10 k)	$n = 65$ (4.25/10 k)	$n = 35$ (2.29/10 k)
Hope, Rupert of Hentzau	85,029	$n = 58$ (6.82/10 k)	$n = 21$ (2.47/10 k)	$n = 23$ (2.7/10 k)	$n = 10$ (1.18/10 k)
James, Washington Square	65,220	$n = 64$ (9.81/10 k)	$n = 4$ (0.61/10 k)	$n = 23$ (3.53/10 k)	$n = 17$ (2.61/10 k)
Joyce, Dubliners	69,204	$n = 14$ (2.02/10 k)	$n = 5$ (0.72/10 k)	$n = 6$ (0.87/10 k)	$n = 0$ (0/10 k)
Kipling, Kim	108,400	$n = 58$ (5.35/10 k)	$n = 15$ (1.38/10 k)	$n = 16$ (1.48/10 k)	$n = 19$ (1.75/10 k)
Lewis, Free Air	83,803	$n = 37$ (4.42/10 k)	$n = 10$ (1.19/10 k)	$n = 7$ (0.84/10 k)	$n = 2$ (0.24/10 k)
London, Martin Eden	142,463	$n = 80$ (5.62/10 k)	$n = 29$ (2.04/10 k)	$n = 10$ (0.7/10 k)	$n = 11$ (0.77/10 k)
Mansfield, The Garden Party	59,773	$n = 18$ (3.01/10 k)	$n = 23$ (3.85/10 k)	$n = 17$ (2.84/10 k)	$n = 4$ (0.67/10 k)
Morrow, The Ape, the Idiot and Other People	57,062	$n = 34$ (5.96/10 k)	$n = 12$ (2.1/10 k)	$n = 15$ (2.63/10 k)	$n = 7$ (1.23/10 k)
Reid, The Finger of Fate	94,265	$n = 75$ (7.96/10 k)	$n = 29$ (3.08/10 k)	$n = 72$ (7.64/10 k)	$n = 23$ (2.44/10 k)
Stevenson, Kidnapped + David Balfour	186,177	$n = 85$ (4.57/10 k)	$n = 45$ (2.42/10 k)	$n = 64$ (3.44/10 k)	$n = 42$ (2.26/10 k)
Stoker, Dracula	163,104	$n = 53$ (3.25/10 k)	$n = 27$ (1.66/10 k)	$n = 17$ (1.04/10 k)	$n = 30$ (1.84/10 k)

(continued)

Table 2 (continued)

Author, Title	Word count	Would have	Could have	Might have	Should have
Twain, The Prince and the Pauper	70,647	$n = 18$ (2.55/10 k)	$n = 9$ (1.27/10 k)	$n = 5$ (0.71/10 k)	$n = 6$ (0.85/10 k)
Wallace, Green Archer	122,339	$n = 52$ (4.25/10 k)	$n = 36$ (2.94/10 k)	$n = 24$ (1.96/10 k)	$n = 17$ (1.39/10 k)
Wallace, Ben Hur	201,305	$n = 110$ (5.46/10 k)	$n = 43$ (2.14/10 k)	$n = 46$ (2.29/10 k)	$n = 7$ (0.35/10 k)
Wells, The War in the Air	99,499	$n = 34$ (3.42/10 k)	$n = 6$ (0.6/10 k)	$n = 15$ (1.51/10 k)	$n = 1$ (0.1/10 k)
Wilde, The Picture of Dorian Gray	80,410	$n = 40$ (4.97/10 k)	$n = 8$ (0.99/10 k)	$n = 7$ (0.87/10 k)	$n = 21$ (2.61/10 k)
Woolf, Orlando	80,310	$n = 39$ (4.86/10 k)	$n = 10$ (1.25/10 k)	$n = 16$ (1.99/10 k)	$n = 11$ (1.37/10 k)

statistically significant ($p = 0.069$),[10] while for French the probability that a similar disparity could occur by chance alone ($p = 0.0003$) leaves little doubt that FtrE and Fr0 should be seen as distinct subspecies as far as their use of the conditional perfect is concerned.

More striking differences between authors and translators appear when specific auxiliaries are examined. In English, while the difference between authors' and translators' use of WOULD HAVE ($d = 0.09, p = 0.63$), SHOULD HAVE ($d = 0.11, p = 0.67$) and OUGHT TO HAVE ($d = 0.05$, $p = 0.76$) is both statistically insignificant and negligible in terms of effect size, the divergence is quite noticeable (Fig. 2) for COULD HAVE ($d = 0.36$, $p = 0.02$) and MIGHT HAVE ($d = 0.52$, $p < 0.001$), with median frequencies in translation that are roughly only half of those observed among authors.

In French, when AVOIR is the auxiliary, the effect size ($d = 0.59$) is quite close to the overall result for all verbs regardless of auxiliary ($d = 0.57$), as would be expected given that constructions with AVOIR make up the bulk of these occurrences (87% in Fr0, 91% in FtrE). There is, however, quite surprisingly, no significant difference between authors and translators when ÊTRE is the auxiliary, both for intransitive verbs ($d = 0.05$, $p = 0.74$) and pronominal constructions ($d = 0.19$, $p = 0.23$) (Fig. 3).

When the semi-auxiliary POUVOIR is included, the difference between authors and translators becomes even more manifest ($d = 0.68$, $p < 0.001$) (Fig. 4):

[10]This p-value is closer to the conventional threshold for statistical significance ($p < 0.05$) than in the original study based on samples of 20 authors and translators, which suggests that for an even larger sample it would eventually reach "significance" as p-values tend to decrease with an increasing number of observations. Even with the present set of data, a "significant" result could be obtained by performing the same test with a one-sided alternative hypothesis. The fact remains, however, that whatever difference there may be is small in terms of effect size, hence, the importance of never interpreting results on the basis of p-values alone.

Table 3 Conditional perfect frequencies in EtrF

Translator	Word count	Would have	Could have	Might have	Should have
Tr.Audoux	39,947	$n = 6$ (1.5/10 k)	$n = 5$ (1.25/10 k)	$n = 2$ (0.5/10 k)	$n = 11$ (2.75/10 k)
Tr.Barbusse	116,181	$n = 26$ (2.24/10 k)	$n = 4$ (0.34/10 k)	$n = 3$ (0.26/10 k)	$n = 12$ (1.03/10 k)
Tr.Bazin	76,051	$n = 42$ (5.52/10 k)	$n = 10$ (1.31/10 k)	$n = 6$ (0.79/10 k)	$n = 12$ (1.58/10 k)
Tr.Benoît	64,337	$n = 12$ (1.87/10 k)	$n = 5$ (0.78/10 k)	$n = 12$ (1.87/10 k)	$n = 11$ (1.71/10 k)
Tr.Bourget	75,804	$n = 48$ (6.33/10 k)	$n = 4$ (0.53/10 k)	$n = 8$ (1.06/10 k)	$n = 25$ (3.3/10 k)
Tr.Daudet	70,539	$n = 30$ (4.25/10 k)	$n = 6$ (0.85/10 k)	$n = 10$ (1.42/10 k)	$n = 4$ (0.57/10 k)
Tr.Flaubert	175,686	$n = 91$ (5.18/10 k)	$n = 4$ (0.23/10 k)	$n = 14$ (0.8/10 k)	$n = 10$ (0.57/10 k)
Tr.France	76,832	$n = 44$ (5.73/10 k)	$n = 7$ (0.91/10 k)	$n = 10$ (1.3/10 k)	$n = 5$ (0.65/10 k)
Tr.Gaboriau	148,493	$n = 123$ (8.28/10 k)	$n = 41$ (2.76/10 k)	$n = 21$ (1.41/10 k)	$n = 25$ (1.68/10 k)
Tr.Gourmont	69,372	$n = 42$ (6.05/10 k)	$n = 9$ (1.3/10 k)	$n = 6$ (0.86/10 k)	$n = 19$ (2.74/10 k)
Tr.Gyp	65,299	$n = 8$ (1.23/10 k)	$n = 6$ (0.92/10 k)	$n = 9$ (1.38/10 k)	$n = 13$ (1.99/10 k)
Tr.Hémon	46,702	$n = 19$ (4.07/10 k)	$n = 6$ (1.28/10 k)	$n = 4$ (0.86/10 k)	$n = 11$ (2.36/10 k)
Tr.Hugo	125,430	$n = 49$ (3.91/10 k)	$n = 19$ (1.51/10 k)	$n = 28$ (2.23/10 k)	$n = 5$ (0.4/10 k)
Tr.Huysmans	130,183	$n = 19$ (1.46/10 k)	$n = 5$ (0.38/10 k)	$n = 10$ (0.77/10 k)	$n = 9$ (0.69/10 k)
Tr.Leblanc	128,495	$n = 65$ (5.06/10 k)	$n = 18$ (1.4/10 k)	$n = 12$ (0.93/10 k)	$n = 14$ (1.09/10 k)
Tr.Leroux	98,447	$n = 85$ (8.63/10 k)	$n = 21$ (2.13/10 k)	$n = 9$ (0.91/10 k)	$n = 11$ (1.12/10 k)
Tr.Loti	66,142	$n = 9$ (1.36/10 k)	$n = 5$ (0.76/10 k)	$n = 1$ (0.15/10 k)	$n = 0.01$ (0/10 k)
Tr.Louÿs	60,014	$n = 25$ (4.17/10 k)	$n = 11$ (1.83/10 k)	$n = 8$ (1.33/10 k)	$n = 12$ (2/10 k)
Tr.Malot	87,359	$n = 75$ (8.59/10 k)	$n = 16$ (1.83/10 k)	$n = 12$ (1.37/10 k)	$n = 19$ (2.17/10 k)
Tr.Massenet	57,738	$n = 29$ (5.02/10 k)	$n = 6$ (1.04/10 k)	$n = 6$ (1.04/10 k)	$n = 1$ (0.17/10 k)
Tr.Mirbeau	101,961	$n = 40$ (3.92/10 k)	$n = 13$ (1.27/10 k)	$n = 2$ (0.2/10 k)	$n = 16$ (1.57/10 k)

(continued)

Table 3 (continued)

Translator	Word count	Would have	Could have	Might have	Should have
Tr.Proust	198,723	$n = 239$ (12.0/10 k)	$n = 41$ (2.06/10 k)	$n = 63$ (3.17/10 k)	$n = 59$ (2.97/10 k)
Tr.Rolland	232,901	$n = 216$ (9.27/10 k)	$n = 45$ (1.93/10 k)	$n = 12$ (0.52/10 k)	$n = 21$ (0.9/10 k)
Tr.Souvestre	102,962	$n = 64$ (6.22/10 k)	$n = 16$ (1.55/10 k)	$n = 26$ (2.53/10 k)	$n = 11$ (1.07/10 k)
Tr.Vanderem	96,391	$n = 43$ (4.46/10 k)	$n = 15$ (1.56/10 k)	$n = 18$ (1.87/10 k)	$n = 4$ (0.41/10 k)
Tr.Verne	165,434	$n = 164$ (9.91/10 k)	$n = 53$ (3.2/10 k)	$n = 10$ (0.6/10 k)	$n = 8$ (0.48/10 k)
Tr.Zola	205,505	$n = 78$ (3.8/10 k)	$n = 22$ (1.07/10 k)	$n = 31$ (1.51/10 k)	$n = 35$ (1.7/10 k)

Compared to authors in Fr0, translators use conditional perfect constructions with POUVOIR around twice as often.

To what extent such differences may be attributed to the influence of the analogous forms in the source-texts can be assessed visually in the form of a scatter plot, as illustrated in Fig. 5a, b, where each point represents a pair of source- and target-texts and shows the frequency of occurrences in the target-text on the y-axis in relation to the frequency of occurrences in the original source-text on the x-axis:

In both cases, Spearman's[11] correlation test reveals a statistically significant ($p < 0.05$) positive correlation of moderate strength ($\rho = 0.47$ for EtrF/Fr0, $\rho = 0.37$ for FtrE/En0). None of the other correlation tests performed on COULD HAVE, MIGHT HAVE and POUVOIR, however, were statistically significant.

4 Discussion

These correlations suggest that translators do tend to imitate the source-texts, to a certain extent, by translating occurrences of the conditional perfect in one language by the analogous form in the other language, but that this factor is insufficient as an explanation to account for the disparity in the use of the conditional perfect between translators and authors in the same language.

In addition to this interlinguistic influence or "shining through" [18], the divergence in the use of the conditional perfect is also a consequence of the inadequation or partial correspondence between the analogous constructions in the two language-systems, and their partial correspondence with other forms or syntactic constructions.

[11] Spearman's was used in preference to Pearson's r (which yields somewhat stronger coefficients and lower p-values) due to the presence of outliers, and because not all of the series of results were found to be normally distributed.

Table 4 Conditional perfect frequencies in Fr0

Author, Title	Word count	Conditional Avoir	Cond. Être Intrans	Cond. Être Pronom	Cond. Avoir + Pouvoir	Cond. Avoir + Devoir
Audoux, Marie Claire	38,233	$n = 36$ (9.42/10 k)	$n = 3$ (0.78/10 k)	$n = 0$ (0/10 k	$n = 5$ (1.31/10 k)	$n = 1$ (0.26/10 k)
Barbusse, Le Feu	119,442	$n = 46$ (3.85/10 k)	$n = 1$ (0.08/10 k)	$n = 1$ (0.08/10 k	$n = 5$ (0.42/10 k)	$n = 2$ (0.17/10 k)
Bazin, Les Oberlé	78,366	$n = 39$ (4.98/10 k)	$n = 1$ (0.13/10 k)	$n = 1$ (0.13/10 k	$n = 8$ (1.02/10 k)	$n = 6$ (0.77/10 k)
Benoît, L'Atlantide	67,638	$n = 29$ (4.29/10 k)	$n = 1$ (0.15/10 k)	$n = 3$ (0.44/10 k	$n = 7$ (1.03/10 k)	$n = 2$ (0.3/10 k)
Bourget, La Duchesse Bleue	88,872	$n = 71$ (7.99/10 k)	$n = 2$ (0.23/10 k)	$n = 2$ (0.23/10 k	$n = 10$ (1.13/10 k)	$n = 5$ (0.56/10 k)
Daudet, L'Immortel	65,954	$n = 34$ (5.16/10 k)	$n = 3$ (0.45/10 k)	$n = 0$ (0/10 k	$n = 8$ (1.21/10 k)	$n = 1$ (0.15/10 k)
Flaubert, L'Éducation sentimentale	149,768	$n = 98$ (6.54/10 k)	$n = 10$ (0.67/10 k)	$n = 1$ (0.07/10 k	$n = 8$ (0.53/10 k)	$n = 14$ (0.93/10 k)
France, Les dieux ont soif	71,967	$n = 16$ (2.22/10 k)	$n = 2$ (0.28/10 k)	$n = 2$ (0.28/10 k	$n = 2$ (0.28/10 k)	$n = 1$ (0.14/10 k)
Gaboriau, L'argent des autres	173,224	$n = 44$ (2.54/10 k)	$n = 10$ (0.58/10 k)	$n = 7$ (0.4/10 k	$n = 0$ (0/10 k)	$n = 2$ (0.12/10 k)
Gourmont, Sixtine	72,967	$n = 52$ (7.13/10 k)	$n = 3$ (0.41/10 k)	$n = 5$ (0.69/10 k	$n = 7$ (0.96/10 k)	$n = 5$ (0.69/10 k)
Gyp, Bijou	58,636	$n = 19$ (3.24/10 k)	$n = 2$ (0.34/10 k)	$n = 1$ (0.17/10 k	$n = 1$ (0.17/10 k)	$n = 5$ (0.85/10 k)
Hémon, Chapdelaine	49,352	$n = 35$ (7.09/10 k)	$n = 2$ (0.41/10 k)	$n = 0$ (0/10 k	$n = 6$ (1.22/10 k)	$n = 0$ (0/10 k)
Hugo, Quatrevingt-treize	122,395	$n = 13$ (1.06/10 k)	$n = 4$ (0.33/10 k)	$n = 1$ (0.08/10 k	$n = 0$ (0/10 k)	$n = 1$ (0.08/10 k)
Huysmans, La Cathédrale	132,784	$n = 19$ (1.43/10 k)	$n = 3$ (0.23/10 k)	$n = 0$ (0/10 k	$n = 5$ (0.38/10 k)	$n = 1$ (0.08/10 k)
Leblanc, Les Dents du Tigre	130,095	$n = 46$ (3.54/10 k)	$n = 1$ (0.08/10 k)	$n = 0$ (0/10 k	$n = 11$ (0.85/10 k)	$n = 3$ (0.23/10 k)
Leroux, Rouilletable chez le Tsar	102,905	$n = 65$ (6.32/10 k)	$n = 8$ (0.78/10 k)	$n = 7$ (0.68/10 k	$n = 12$ (1.17/10 k)	$n = 2$ (0.19/10 k)
Loti, Les Derniers Jours de Pékin	72,326	$n = 13$ (1.8/10 k)	$n = 0$ (0/10 k)	$n = 2$ (0.28/10 k	$n = 4$ (0.55/10 k)	$n = 0$ (0/10 k)
Louÿs, Aphrodite	55,481	$n = 32$ (5.77/10 k)	$n = 1$ (0.18/10 k)	$n = 4$ (0.72/10 k	$n = 8$ (1.44/10 k)	$n = 0$ (0/10 k)

(continued)

Table 4 (continued)

Author, Title	Word count	Conditional Avoir	Cond. Être Intrans	Cond. Être Pronom	Cond. Avoir + Pouvoir	Cond. Avoir + Devoir
Malot, Conscience	105,655	$n = 105$ (9.94/10 k)	$n = 8$ (0.76/10 k)	$n = 9$ (0.85/10 k	$n = 9$ (0.85/10 k)	$n = 6$ (0.57/10 k)
Massenet, Mes Souvenirs	76,634	$n = 40$ (5.22/10 k)	$n = 3$ (0.39/10 k)	$n = 2$ (0.26/10 k	$n = 8$ (1.04/10 k)	$n = 2$ (0.26/10 k)
Mirbeau, Le journal d'une femme de chambres	116,755	$n = 48$ (4.11/10 k)	$n = 7$ (0.6/10 k)	$n = 4$ (0.34/10 k	$n = 9$ (0.77/10 k)	$n = 6$ (0.51/10 k)
Proust, Du côté de chez Swann	181,464	$n = 267$ (14.71/10 k)	$n = 20$ (1.1/10 k)	$n = 10$ (0.55/10 k	$n = 57$ (3.14/10 k)	$n = 9$ (0.5/10 k)
Rolland, Jean-Christophe 1–4	239,132	$n = 80$ (3.35/10 k)	$n = 11$ (0.46/10 k)	$n = 6$ (0.25/10 k	$n = 29$ (1.21/10 k)	$n = 7$ (0.29/10 k)
Souvestre, Fantômas	91,924	$n = 45$ (4.9/10 k)	$n = 4$ (0.44/10 k)	$n = 5$ (0.54/10 k	$n = 5$ (0.54/10 k)	$n = 3$ (0.33/10 k)
Vandérem, Les deux rives de la Seine	93,435	$n = 39$ (4.17/10 k)	$n = 2$ (0.21/10 k)	$n = 4$ (0.43/10 k	$n = 4$ (0.43/10 k)	$n = 1$ (0.11/10 k)
Verne, L'île mystérieuse	209,872	$n = 143$ (6.81/10 k)	$n = 18$ (0.86/10 k)	$n = 8$ (0.38/10 k	$n = 36$ (1.72/10 k)	$n = 8$ (0.38/10 k)
Zola, Les trois villes Paris	192,462	$n = 125$ (6.49/10 k)	$n = 8$ (0.42/10 k)	$n = 8$ (0.42/10 k	$n = 9$ (0.47/10 k)	$n = 14$ (0.73/10 k)

The examples given in the introduction in which the source- and target-segments contain corresponding conditional perfect constructions, with the exception of (2), contain an *if*-protasis and a conditional apodosis in which the verb in the conditional perfect is interpreted as an affirmative retrospective hypothesis. When no protasis is expressed or implied, when the English preterit modals retain a temporal interpretation as past tenses, or when negation comes into play, however, the correspondence begins to break down. The following examples, which are far from exhaustive, illustrate several such cases of non-correspondence between conditional perfect constructions with COULD, MIGHT and POUVOIR, i.e., those for which the level of divergence between authors and translators in each language is greatest:

6. And I heard that far-off crying again, just as we heard it the other night. There isn't a wind today, so you see it couldn't have been the wind. (F.H. Burnett, *The Secret Garden*) ⇒ Et j'ai de nouveau entendu quelqu'un pleurer comme l'autre soir. Il n'y a pas de vent aujourd'hui, ainsi vous voyez que ça ne pouvait pas être le vent.
7. … but our certain knowledge stopped with Rischenheim's start for the capital and Rupert's presence there at three o'clock. The pair might have met or might have missed. We had to act as though they had missed … (A. Hope, *Rupert of*

Table 5 Conditional perfect frequencies in FtrE

Author, Title	Word count	Conditional Avoir	Cond. Être Intrans	Cond. Être Pronom	Cond. Avoir + Pouvoir	Cond. Avoir + Devoir
Tr.Bellamy	73,359	$n = 59$ (8.04/10 k)	$n = 3$ (0.41/10 k)	$n = 3$ (0.41/10 k	$n = 14$ (1.91/10 k)	$n = 1$ (0.14/10 k)
Tr.Benson	96,638	$n = 57$ (5.9/10 k)	$n = 3$ (0.31/10 k)	$n = 1$ (0.1/10 k	$n = 14$ (1.45/10 k)	$n = 2$ (0.21/10 k)
Tr.Buchan	104,989	$n = 59$ (5.62/10 k)	$n = 8$ (0.76/10 k)	$n = 2$ (0.19/10 k	$n = 11$ (1.05/10 k)	$n = 7$ (0.67/10 k)
Tr.Burnett	80,582	$n = 65$ (8.07/10 k)	$n = 2$ (0.25/10 k)	$n = 7$ (0.87/10 k	$n = 15$ (1.86/10 k)	$n = 5$ (0.62/10 k)
Tr.Collins	198,786	$n = 111$ (5.58/10 k)	$n = 12$ (0.6/10 k)	$n = 15$ (0.75/10 k	$n = 18$ (0.91/10 k)	$n = 7$ (0.35/10 k)
Tr.ConanDoyle	73,815	$n = 64$ (8.67/10 k)	$n = 0$ (0/10 k)	$n = 0$ (0/10 k	$n = 13$ (1.76/10 k)	$n = 5$ (0.68/10 k)
Tr.Cox	130,792	$n = 175$ (13.38/10 k)	$n = 8$ (0.61/10 k)	$n = 12$ (0.92/10 k	$n = 37$ (2.83/10 k)	$n = 18$ (1.38/10 k)
Tr.Eliot	168,019	$n = 109$ (6.49/10 k)	$n = 7$ (0.42/10 k)	$n = 4$ (0.24/10 k	$n = 23$ (1.37/10 k)	$n = 4$ (0.24/10 k)
Tr.Fitzgerald	56,280	$n = 51$ (9.06/10 k)	$n = 2$ (0.36/10 k)	$n = 2$ (0.36/10 k	$n = 8$ (1.42/10 k)	$n = 5$ (0.89/10 k)
Tr.Hardy	151,247	$n = 203$ (13.42/10 k)	$n = 7$ (0.46/10 k)	$n = 9$ (0.6/10 k	$n = 48$ (3.17/10 k)	$n = 23$ (1.52/10 k)
Tr.Hope	82,833	$n = 89$ (10.74/10 k)	$n = 7$ (0.85/10 k)	$n = 3$ (0.36/10 k	$n = 20$ (2.41/10 k)	$n = 3$ (0.36/10 k)
Tr.James	75,497	$n = 124$ (16.42/10 k)	$n = 2$ (0.26/10 k)	$n = 9$ (1.19/10 k	$n = 12$ (1.59/10 k)	$n = 13$ (1.72/10 k)
Tr.Joyce	72,684	$n = 47$ (6.47/10 k)	$n = 2$ (0.28/10 k)	$n = 0$ (0/10 k	$n = 12$ (1.65/10 k)	$n = 1$ (0.14/10 k)
Tr.Kipling	124,770	$n = 59$ (4.73/10 k)	$n = 4$ (0.32/10 k)	$n = 5$ (0.4/10 k	$n = 13$ (1.04/10 k)	$n = 10$ (0.8/10 k)
Tr.Lewis	96,142	$n = 69$ (7.18/10 k)	$n = 7$ (0.73/10 k)	$n = 4$ (0.42/10 k	$n = 13$ (1.35/10 k)	$n = 5$ (0.52/10 k)
Tr.London	131,127	$n = 126$ (9.61/10 k)	$n = 5$ (0.38/10 k)	$n = 4$ (0.31/10 k	$n = 18$ (1.37/10 k)	$n = 8$ (0.61/10 k)
Tr.Mansfield	68,591	$n = 78$ (11.37/10 k)	$n = 2$ (0.29/10 k)	$n = 2$ (0.29/10 k	$n = 20$ (2.92/10 k)	$n = 6$ (0.87/10 k)
Tr.Morrow	60,439	$n = 15$ (2.48/10 k)	$n = 5$ (0.83/10 k)	$n = 3$ (0.5/10 k	$n = 1$ (0.17/10 k)	$n = 0$ (0/10 k)
Tr.Reid	92,925	$n = 106$ (11.41/10 k)	$n = 4$ (0.43/10 k)	$n = 10$ (1.08/10 k	$n = 33$ (3.55/10 k)	$n = 2$ (0.22/10 k)
Tr.Stevenson	190,455	$n = 157$ (8.24/10 k)	$n = 10$ (0.53/10 k)	$n = 11$ (0.58/10 k	$n = 27$ (1.42/10 k)	$n = 20$ (1.05/10 k)

(continued)

Table 5 (continued)

Author, Title	Word count	Conditional Avoir	Cond. Être Intrans	Cond. Être Pronom	Cond. Avoir + Pouvoir	Cond. Avoir + Devoir
Tr.Stoker	176,515	$n = 157$ (8.89/10 k)	$n = 7$ (0.4/10 k)	$n = 3$ (0.17/10 k	$n = 34$ (1.93/10 k)	$n = 8$ (0.45/10 k)
Tr.Twain	85,737	$n = 59$ (6.88/10 k)	$n = 1$ (0.12/10 k)	$n = 0$ (0/10 k	$n = 6$ (0.7/10 k)	$n = 3$ (0.35/10 k)
Tr.WallaceE	87,130	$n = 84$ (9.64/10 k)	$n = 9$ (1.03/10 k)	$n = 3$ (0.34/10 k	$n = 20$ (2.3/10 k)	$n = 7$ (0.8/10 k)
Tr.WallaceL	145,902	$n = 162$ (11.1/10 k)	$n = 7$ (0.48/10 k)	$n = 8$ (0.55/10 k	$n = 49$ (3.36/10 k)	$n = 6$ (0.41/10 k)
Tr.Wells	108,417	$n = 66$ (6.09/10 k)	$n = 2$ (0.18/10 k)	$n = 6$ (0.55/10 k	$n = 16$ (1.48/10 k)	$n = 14$ (1.29/10 k)
Tr.Wilde	82,291	$n = 59$ (7.17/10 k)	$n = 2$ (0.24/10 k)	$n = 1$ (0.12/10 k	$n = 8$ (0.97/10 k)	$n = 9$ (1.09/10 k)
Tr.Woolf	87,349	$n = 55$ (6.3/10 k)	$n = 2$ (0.23/10 k)	$n = 0$ (0/10 k	$n = 19$ (2.18/10 k)	$n = 5$ (0.57/10 k)

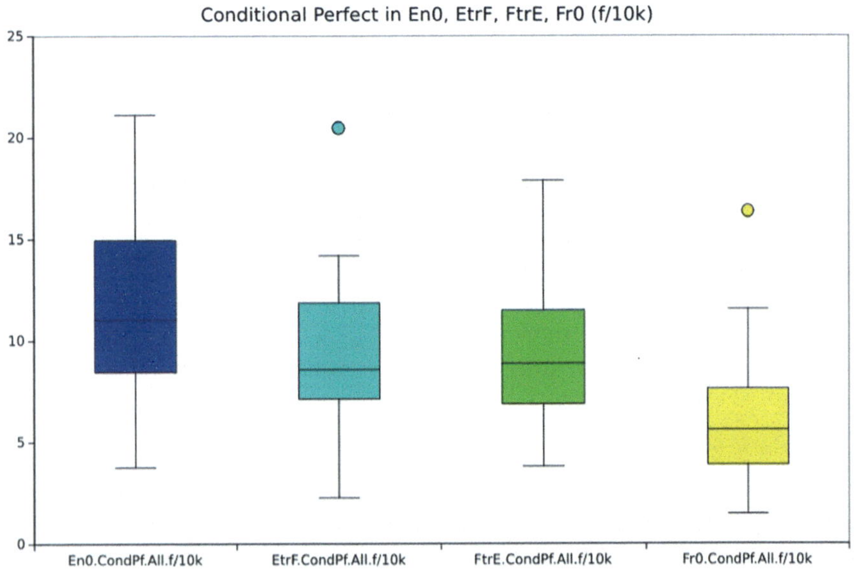

Fig. 1 Distributions of conditional perfect frequencies (f/10 k) for all verbs and auxiliaries in En0, EtrF, FtrE and Fr0

Hentzau) ⇒ … mais nous n'avions la certitude que de deux choses: le départ de Rischenheim pour la capitale et la présence de Rupert dans cette ville à cinq heures. Les deux cousins pouvaient s'être rencontrés ou manqués. Nous devions agir comme s'ils s'étaient manqués …

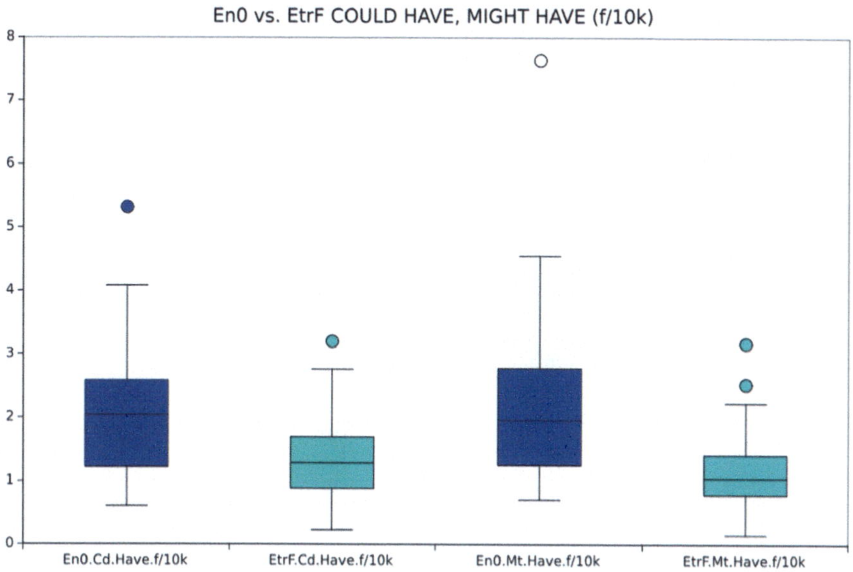

Fig. 2 Distributions of conditional perfect frequencies (f/10 k) for COULD HAVE and MIGHT HAVE in En0 versus EtrF

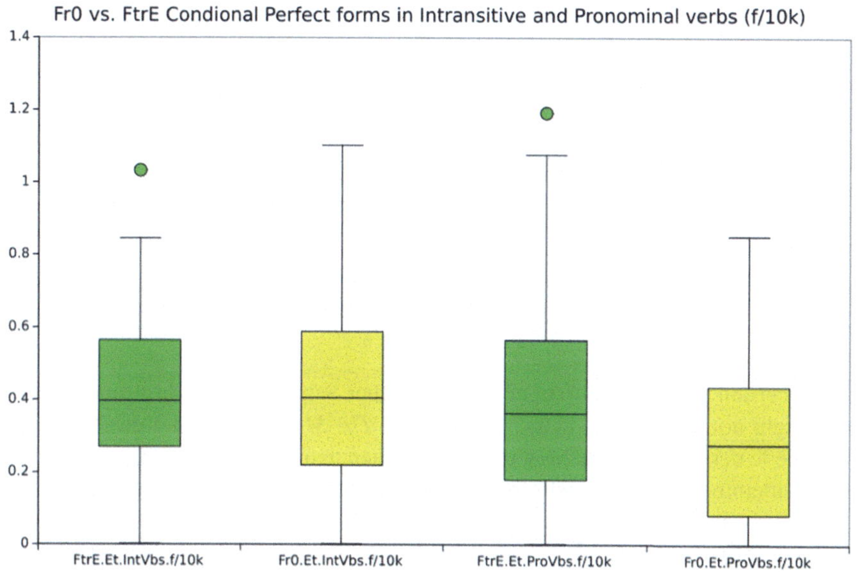

Fig. 3 Distributions of conditional perfect frequencies (f/10 k) for intransitive (IntVbs) and pronominal verbs (ProVbs) in Fr0 and FtrE

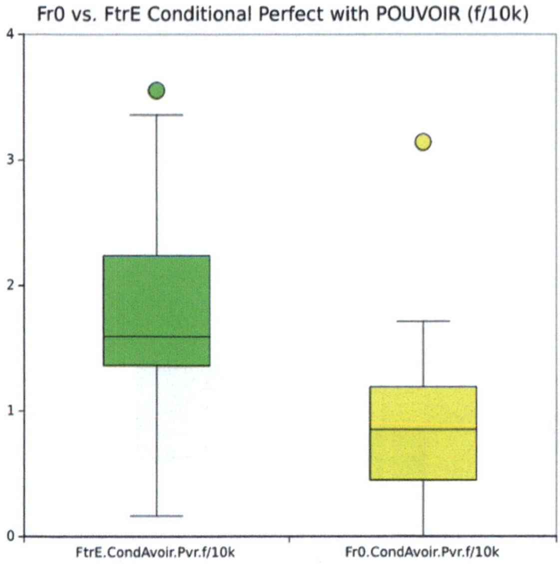

Fig. 4 Distributions of conditional perfect frequencies (f/10 k) including POUVOIR in Fr0 and FtrE

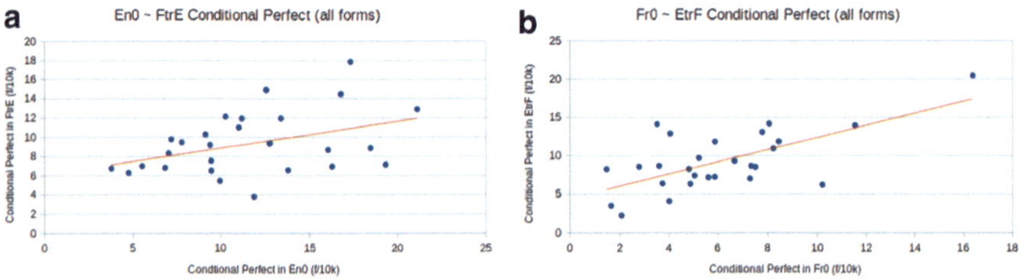

Fig. 5 **a** Frequency of conditional perfect forms in FtrE versus En0. ($\rho = 0.37$, $p = 0.05$). **b** Frequency of conditional perfect forms in EtrF versus Fr0. ($\rho = 0.47$, $p = 0.01$)

8. A certain doubt came up in my mind whether Sergeant Cuff's last random shot might not have hit the mark. (W. Collins, *The Moonstone*) ⇒ J'eus la pensée que le dernier effort de M. Cuff pourrait bien avoir atteint son but …
9. Évidemment, si Nab avait eu quelque fait important à faire connaître à son maître, il ne pouvait employer(a) un plus sûr et plus rapide messager, qui pouvait passer là où ni les colons ni Top lui-même n'auraient peut-être pu(b) le faire. (J. Verne, *L'île mystérieuse*) ⇒ Neb could not have chosen(a) a better messenger, who could get through obstacles which none of the others could have surmounted(b).
10. Monsieur Rouletabille, on peut avoir tenté d'empoisonner mon père et n'être point venu par la fenêtre ! (G. Leroux, *Rouilletable chez le Tsar*) ⇒ Monsieur

Rouletabille, someone might have tried to poison my father, and not have come by way of the window.

11. How strange and well-nigh incredible does it seem that so prodigious a moral and material transformation as has taken place since then could have been accomplished in so brief an interval! (E. Bellamy, *Looking Back*) ⇒ Qu'il nous paraît étrange, presque incroyable qu'une aussi prodigieuse transformation morale et matérielle ait pu avoir lieu en si peu de temps !
12. Et don Luis demeurait confondu devant la révélation d'un forfait dont il ne semblait pas que la conception eût pu germer dans un cerveau d'homme. (M. Leblanc, *Les dents du tigre*) ⇒ And Don Luis remained stupefied at the revelation of a crime which looked as if no human brain could have conceived it.

Translators may opt to use POUVOIR in the imperfect (6)–(7) when COULD or MIGHT[12] retain a temporal dimension, and/or in order to place the modal verb directly within the scope of the negation (6). When MIGHT expresses retrospective uncertainty about previous events, there seems to be a tendency to translate it as POUVOIR followed by a perfect infinitive, a construction in which the constituents follow the same linear order as in English, and which is much less common in original French but nonetheless attested (10). The same processes can be seen at work in the opposite direction in (9a)–(10). Finally, examples (11)–(12) show that, in both directions, the French subjunctive should be taken into account as another potential translational equivalent.

5 Conclusion

This study demonstrates how quantitative bidirectional analysis of translated parallel corpora in comparison with untranslated comparable corpora, can be used both to identify disparities between target-texts and target-language norms as represented in an *ex nihilo* corpus of original texts, and to assess the influence of source-texts on target-texts. Taking the conditional perfect as an indicator, French-translated-from-English is demonstrably different from original French, and in both languages, translated texts can be identified as a distinct subspecies when COULD, MIGHT and POUVOIR are included in such constructions, with frequencies that are approximately twice as high in FtrE compared with Fr0, while the frequencies in EtrF are roughly half of those observed in En0. In sum, EtrF is lacking in occurrences of COULD HAVE and MIGHT HAVE, while FtrE overuses the conditional perfect, with POUVOIR in particular.

Because the conditional perfect is more frequent in original English than in original French, it was thought, at first, that the intermediate frequency observed in translated texts in both directions might be a consequence of "shining through," i.e., imitation or interlinguistic influence, but the degree of correlation between

[12] As a result of indirect discourse or free indirect speech.

source- and target-text frequencies appears to be insufficient to account for the difference that exists between translated and original texts. Many diverse cases of non-correspondence can be found in translation which seem to reflect the differing internal logic of the constructions, and in particular the place and interpretation of the respective modal auxiliaries. These findings both lend support to the idea that the difference in the way these analogous constructions are organized internally might be a source of divergence in translation, and call for further investigation into the ways the corresponding modal auxiliaries in each language interact with tense and aspect, as well as with the French subjunctive.[13]

References

1. Koppel M, Ordan N (2011) Translationese and its dialects proceedings of the 49th annual meeting of the association for computational linguistics (pp 1318–1326), June 19–24
2. Olohan M, Baker M (2000) Reporting that in translated English. Evidence for subconscious processes of explicitation? Across Lang Cult 1(2):141–158
3. Tirkkonen-Condit S (2004) Unique items-over-or under-represented in translated language? Benjamins Transl Libr 48:177–186
4. Cappelle B (2012) English is less rich in manner-of-motion verbs when translated from French. Across Lang Cultures 13(2):173–195
5. Cappelle B, Loock R (2013) Is there interference of usage constraints?: A frequency study of existential there is and its French equivalent il ya in translated vs. non-translated texts. Target. Int J Transl Stud 25(2):252–275
6. Granger S (2018) Tracking the third code: a cross-linguistic corpus-driven approach to metadiscursive markers. The corpus linguistics discourse. John Benjamins, Amsterdam & Philadelphia, pp 185–204
7. Olohan M (2002) Comparable corpora in translation research. In: LREC language resources in translation work and research workshop proceedings (pp 5–9)
8. Stubbs M (2001) Texts, corpora, and problems of interpretation: a response to widdowson. Appl Linguist 22(2):149–172
9. Loock R (2013) Close encounters of the third code. In Lefer MA, Vogeleer S (eds) Interference and normalization in genre-controlled multilingual corpora. Belgian J Linguist 27:61–86
10. Laviosa S (1998). Core patterns of lexical use in a comparable corpus of English narrative prose. Meta: Journal des traducteurs/Meta: Translators' Journal, 43(4):557–570
11. Baker M (2004) A corpus-based view of similarity and difference in translation. Int J Corpus Linguist 9(2):167–193
12. Peterlin AP (2010) Hedging devices in Slovene-English translation: a corpus-based study. Nordic J Engl Stud 9(2):171–193
13. Johansson S (2007) Seeing through multilingual corpora. In corpus linguistics 25 years on. Brill Rodopi, pp. 51–71
14. Vinay JP, Darbelnet J (1958) Stylistique comparée de l'anglais et du français. Didier, Paris
15. McEnery T, Xiao R (2007) Parallel and comparable corpora. Corpus-based perspectives in linguistics, pp 131–145
16. Schmid H, TreeTagger, Universität Stuttgart, https://www.cis.uni-muenchen.de/~schmid/tools/TreeTagger/

[13]Indeed, one obvious question concerns the extent to which the overuse of the conditional perfect in FtrE, with or without POUVOIR, may come at the expense of the subjunctive.

17. R Core Team (2017) R: a language and environment for statistical computing. R Foundation for Statistical Computing, Vienna, Austria. URL https://www.R-project.org/
18. Teich E (2003) Cross-linguistic variation in system and text: a methodology for the investigation of translations and comparable texts. Berlin: Mouton de Gruyter

Repeated and Anaphoric Segments Applied to Trilingual Knowledge Extraction

Lionel Shen

Abstract In a context of globalized societies, multilingualism is becoming an economic and social phenomenon. Translation constitutes a crucial element for communication. A good translation guarantees the quality of the transmission of all information. However, can translation on its own be used to face the challenge of multilingual information monitoring? With the advent of the digital age and the integration of many new technologies, corporate governance is undergoing a complete metamorphosis. One of the priorities remains the efficient exploitation of accumulated big data. This paper endeavors to shed light on the need for multilingual monitoring and on current or future developments of text mining, for three major languages (French, English, and Chinese), in crucial areas for the world's future and to describe the specificity and efficiency of anaphora.

1 Introduction

The world, which has used hundreds of languages for thousands of years, has formalized words and grammars to transcribe, teach, and transmit via media, knowledge, facts, and thoughts. From hieroglyphics to ideograms, via alphabets, these representations thus spread the image of the world through periods, evolutions, mores, and currents of thought. Today, this represents hundreds of billions of words in different corpuses, with varying occurrences. It is not possible for human beings to deal with the mass of archived or circulating publications unaided.

Only the use of computers, can now, in the context of globalization, allow the massive scanning of the sequences of corpora necessary to study the occurrences and uses of words, at least in the essential languages that disseminate knowledge, information, and communication between humans. The usefulness of this research is extensive [3], ranging from social needs, human, scientific, and economic wars, to media and strategic policy issues. It is the ability to rapidly detect, record, analyze,

L. Shen (✉)
Sorbonne Nouvelle University, Paris, France
e-mail: lionel.shen@sorbonne-nouvelle.fr

D. F. Iezzi et al. (eds.), *Text Analytics*, Studies in Classification, Data Analysis, and Knowledge Organization,
https://doi.org/10.1007/978-3-030-52680-1_16

and understand information that will enable the various forces of power to anticipate decisions and act effectively.

In a context of globalized societies, we can talk about multilingualism or even plurilingualism. Today, the frenzy around the big data phenomenon and multilingualism are transforming all the services and behaviors of our time. Translation thus becomes a crucial element for communication. A good translation guarantees that information is transmitted completely and at a high quality. However, given the challenge of the project of carrying out multilingual monitoring, can translation on its own suffice?

When faced with information from multiple languages, their meanings must be reconciled by identifying, detecting, and highlighting the coexistence of identical or similar concepts. In this case, we are trying to bring together French, English, and Chinese concepts. This article is based on explorations of trilingual thematic corpuses applied to information mining and attempts to highlight the specificity of anaphors [1] and recurrences in three languages.

2 Definitions and Conceptual Scope

To clarify our point, it is important to recall the definitions concerning the two aspects of the anaphora.

> In rhetoric, A process aimed at achieving an effect of symmetry, insistence, etc., by repeating the same word or group of words at the beginning of several successive sentences or proposals.
>
> In grammar, A process of recalling a word or group of words previously stated by a grammatical term.[1]

It should be noted that French linguists, such as Kleiber, Charolles, and Schnedecker, have shown the full scope of the notion of anaphora, which extends well beyond a single word, forming what Catherine Schnedecker has called a chain of reference. While the present study focuses on the aspect of the notion of repetition of an identical form, the recurrence of the notion (or concept) and their semantic substitutions in corpora composed of di-versus journalistic texts, a type of discourse responding to particular specificities and introducing a referent in discourse in the following of our analyses. In the following paragraphs, this paper will establish a fresh approach: considering the chain of references beyond the pronominal anaphors through all the texts of the corpora studied. The results of this will generate notable benefits in terms of its practical application.

[1] Source : CNRTL, https://www.cnrtl.fr/definition/anaphore.

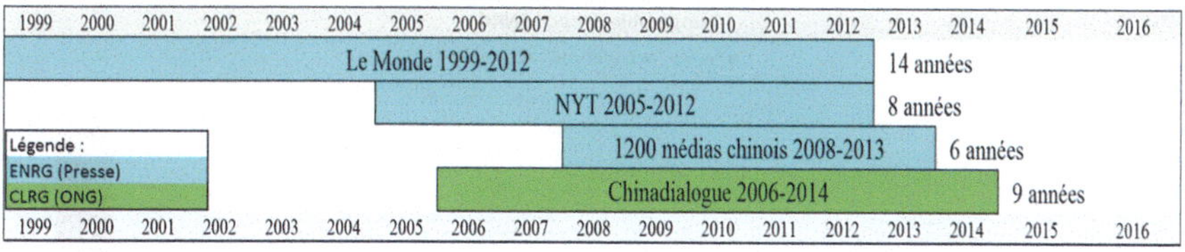

Fig. 1 Time aspect of the data

3 Corpus

For the sake of complementarity, parallel information, as part of the continuity of comparable studies, provides very interesting scientific data for the multilingual monitoring we are carrying out, and this has led to us building a parallel corpus. It should be remembered that parallel corpora are made up of source texts and translations, and make it possible to approach translinguistic phenomena with much greater accuracy, while also working on materials made up of translations into several languages, in order to better understand the problems of comparison.

In an attempt to answer this question, two trilingual thematic corpuses were compiled from newspaper articles from three spheres of communication, namely Le Monde for France, the New York Times, abbreviated to NYT, for the United States, and about 1200 different media for China, the best known being The People's Daily. The two information sites, www.qq.com and www.sina.com.cn, are the main sources of the Chinese subcorpus. The texts of the parallel bilingual English–Chinese (NGO) corpus are taken from the Web site www.chinadialogue.net. The construction of these two corpora was based on three topical themes, namely the environment, energy, and climate change, in three languages: French, English, and Chinese. These two corpuses, a comparable corpus called ENRG and a parallel corpus called CLRG, are composed of textual data extracted from press speeches, as well as those of NGOs. Two textometric tools were used, Lexico[2] and Le Trameur[3] for poly-cooccurrences.

The temporal aspect of the data differs according to the sources and covers varying periods of time: 14 years for Le Monde, 8 years for the NYT, and 6 years for the Chinese media, while for the parallel corpus, the articles date from 2006 to 2014, or 9 years (Fig. 1).

As for the volume of the ENRG corpus:

- ENRG_FR, for France, Le Monde, 4817 articles, or about 2.7 million occurrences (hits),
- ENRG_US, for the USA, the NYT, 3993 articles, or about 2.7 million occurrences,
- ENRG_CN, for China, about 1200 media, 14,509 articles, 10.4 million occurrences (Figs. 2 and 3).

[2]https://www.lexi-co.com/.

[3]https://www.tal.univ-paris3.fr/trameur/.

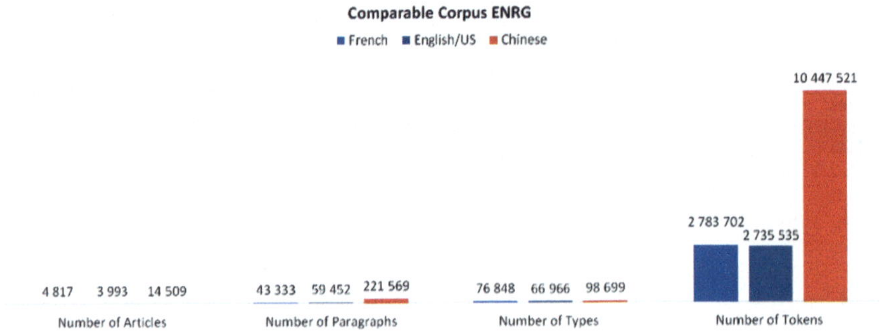

Fig. 2 ENRG multilingual corpus

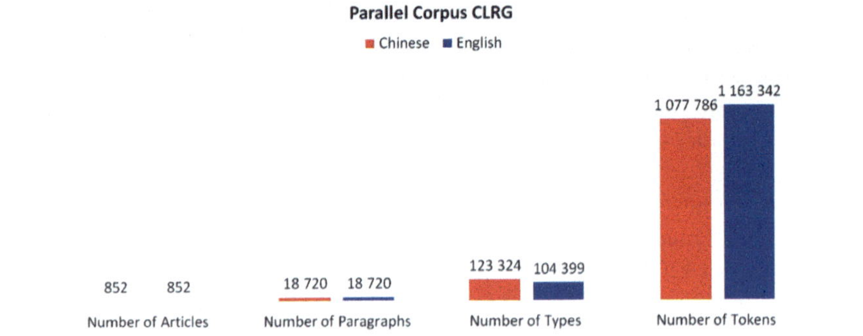

Fig. 3 CLRG multilingual corpus

We find that the number of articles recovered from the Chinese media (ENRG_CN) is more than three times higher than that of Le Monde (ENRG_FR) and that of the NYT. It should be noted that two text sources in Chinese have been mobilized.

The volume of the parallel corpus, named CLRG on Chinadialogue, amounts to 852 articles per component, CLRG_CN, CLRG_EN, 1704 articles in total, or about 1.1 million occurrences in each component, Chinese versus English.

Corpus analysis

The analyses are carried out using textometric tools, in particular factor analyses of correspondences, the specificity of the hypergeometric model, repeated segments [10], cooccurring networks and poly-current networks [4] or section maps. Subsequently, the global, local, convergences, divergences, and specificities of each corpus were highlighted during the explorations of this paper.

4 Repetitions in ENRG_FR and ENRG_US

Through the studies of our two corpora, relatively simple and obvious linguistic facts seem to emerge, such as the notion of repetitions and anaphors. But in reality, it is not that simple, particularly if we look at them in a translinguistic way. From one language to another, the phenomenon changes. We have therefore decided to quantify some of these repetitions. Here is an example concerning French and American.

In the comparable corpus, ENRG_FR and ENRG_US, we have selected four pairs of words (writing form), "énergie VS energy," "climat VS climate," "nucléaire VS nuclear," and "réchauffement VS warming" (Fig. 4).

We can see that, for the pairs of words "énergie VS energy," "climat VS climate," and "réchauffement VS warming," the number of repetitions is higher in English than in French. As for the following pair of words "nucléaire VS nuclear," this phenomenon of language in French is dominant, by almost three and a half times. The question is whether the form "climate" in English can act as an adjective. Indeed, the association of terms such as "climate change or climate system" remains relatively recurrent in English. The same phenomenon occurred for the form "nuclear" in French with a much higher number. This is because the "nuclear" form can be both a noun and an adjective.

To support our point, we have chosen another term that is relatively difficult to transform into an adjective, namely the form "réchauffement(s) VS warming." We get more instances of Warming in English.

Nous pouvons donc en déduire, d'une part, le phénomène de la répétition est plus prononcé en anglais (qu'en français), d'autre part, le mécanisme des anaphores est plus employé en français. Quant à la ventilation de la notion "réchauffement climatique" du corpus ENRG_FR, on s'aperçoit que, cette forme est parfois substituée par "réchauffement planétaire ou réchauffement de la planète," mais le segment le plus employé reste "changement climatique" en français (Fig. 5). We can therefore

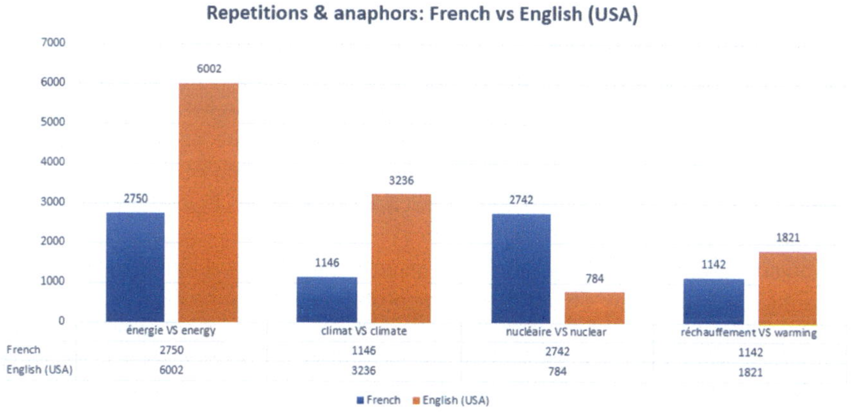

Fig. 4 Repetitions and anaphors: French versus English (USA)

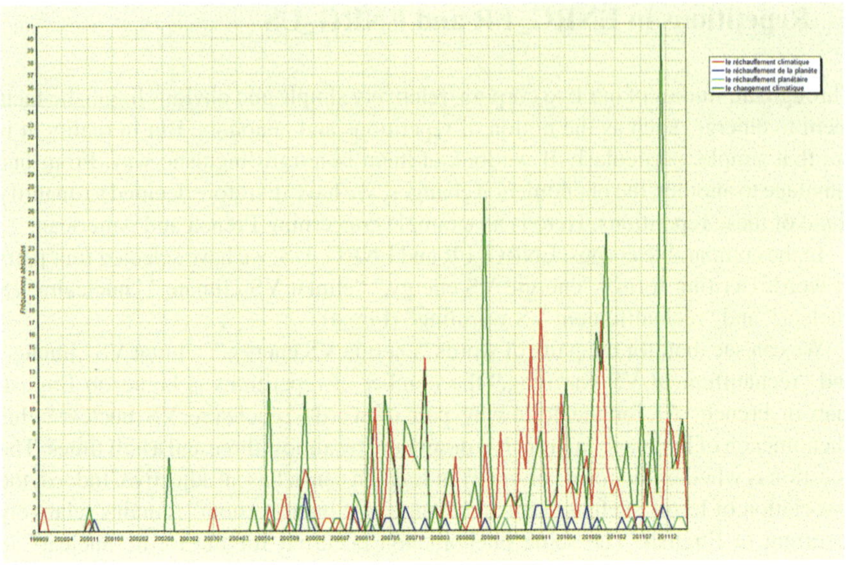

Fig. 5 Anaphors in ENRG_FR

deduce, on the one hand, that the phenomenon of repetition is more pronounced in English (than in French), and on the other hand, that the anaphora mechanism is more commonly used in French. As for the breakdown of the concept "réchauffement climatique/climate warming" in the ENRG_FR corpus, we see that this form is sometimes substituted by "réchauffement planétaire ou réchauffement de la planète/global warming or global warming," but the most commonly used segment is still "climate change" in French (Fig. 5).

The results of the segments "global warming and climate change/global warming and climate change" from the NYT corpus are displayed in Fig. 6. It can be seen that the anaphoric deictic of these sequences is less important in English than in French.

The statistical breakdown (Fig. 7) shows a strong presence of the two segments repeated between March and October 2011, particularly the month of March, very affected by the Fukushima event, the months of July and August of the same year, marked by inventories, increased control audits, etc., of the nuclear fleet in the most nuclearized country in the world. A monthly projection of the two repeated segments on the section map provides direct access to articles that report events and information.

The article containing only the third-generation nuclear reactor segment is very revealing, as it relates the situation of all the problems concerning the EPR project in France and in the world. Indeed, it is thanks to the repetition of the anaphoric form [5] of the EPR that we were able to identify it (Fig. 8).

We can reaffirm this observation by quoting another article that clearly shows that the anaphoric deictic represents a strong trend in the proper use of French. We can deduct from this that while the phenomenon of repetition is more pronounced

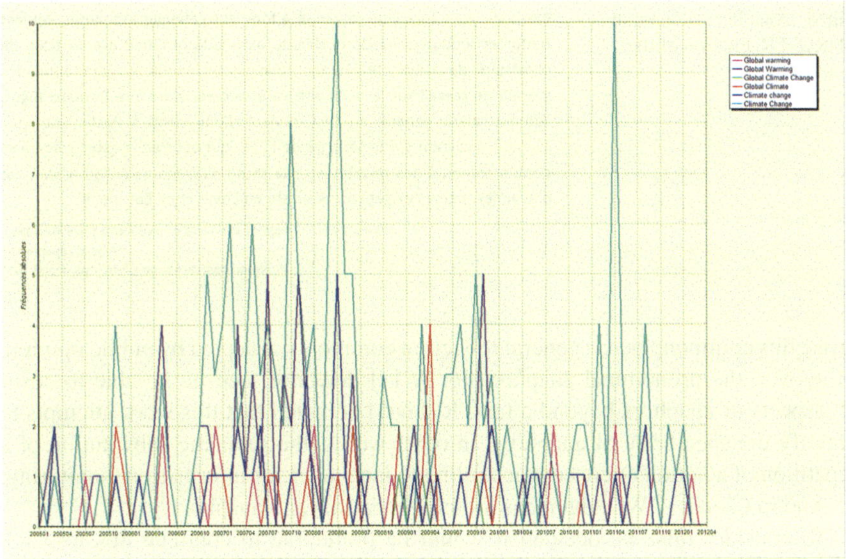

Fig. 6 Anaphores in ENRG_US

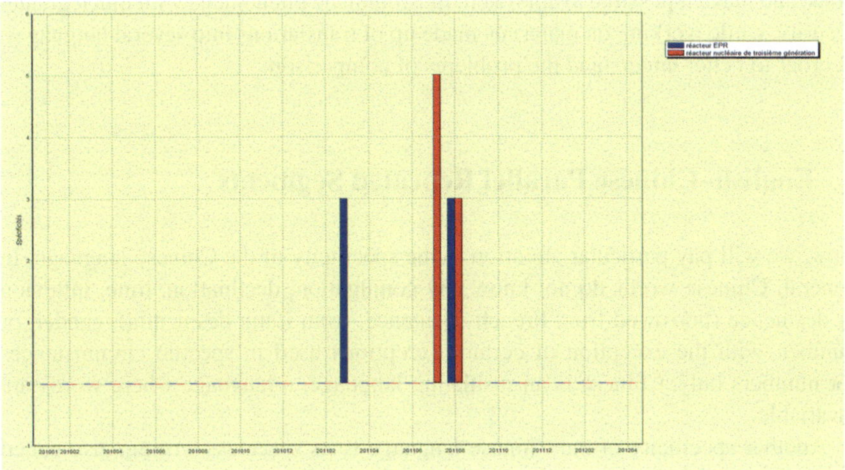

Fig. 7 Anaphoric deictic in ENRG_FR

in English than in French, the mechanism of anaphors is applied more in French. In this specific case, it is a question of observing phenomena specific to the different languages: in French, we learned at school that it is customary to change the form of repetition of a referent in order to avoid repetitions, for example, a proper name or a defined description will be repeated by a pronoun in the rest of a text, whereas in English, repetition is customary. One of our objectives was to establish, evaluate and

Fig. 8 Anaphoric deictic in ENRG_FR, citation of the article

Ainsi, les deux pays, qui ont misé à fond sur l'énergie atomique, veulent renforcer leur partenariat nucléaire, alors que la Chine est en train de construire deux EPR, le réacteur français de troisième génération. Un groupe de travail sur le nucléaire va être créé, alors que Bernard Bigot, administrateur général du CEA, revient d'une visite à Pékin. Les deux capitales insistent sur la sécurité de l'EPR, qui pourrait résister grâce à sa double coque, à des missiles et à la chute d'un airbus A 380. L'idée est d'avoir les mêmes exigences de sûreté en Chine et en Europe. #

30 mars 2011, « Sarkozy et Hu unis pour sauver la filière nucléaire »
Arnaud Leparmentier
Correspondant du Monde à New York depuis 2017

prove this commonplace, a general reflection commonly accepted by public opinion. Moreover, the pronominal anaphora [6, 7, 11] does not seem to be able to cover all aspects of anaphors beyond a text. It should be noted that this work attempts to identify the specificity of anaphors, in other words, the semantic substitution of a repetition of a notion or concept according to languages for information monitoring.

And in Chinese? What happens around rehearsals and anaphors?

To consider this, we decided it would be preferable to continue this research from the CLRG parallel corpus where the textual data are less eclectic and more comparable, using the repeated segment and poly-cooccurrency tools.

It should be recalled that parallel corpora are made up of source texts and translations, and make it possible to approach translinguistic phenomena with much greater security, while working on materials made up of translations into several languages, in order to better understand the problems of comparison.

5 English–Chinese Parallel Repeated Segments

Now, we will pay particular attention to the specificity of the Chinese language. In general, Chinese words do not know any conjugation, declination, time, inflexion or desinence (borrowed from French désinence, from Latin dēsinentia), gender, or number, with the exception of certain exceptions used in specific circumstances for numbers only. Chinese is an insulating language, a language where words are invariable.

Another specificity of the Chinese language is its spaceless writing, also called "*scriptio continua,*" "*scriptura continua*" or "*scripta continua.*" In order to be able to carry out automated analyses, it is essential to segment Chinese texts using automatic segmentors, the only ones capable of providing fast and homogeneous work that human beings are not able to perform satisfactorily. The segmentor cuts words in a string. To do this, it uses a lexical database containing all the words of the Chinese language. The problem lies in the fact that the same word can in some cases lead to several possible segmentations.

Let us now look at the most repeated segments in each of the components obtained from our CLRG corpus (Table 1). Immediately, it can be noticed that in Chinese, the

Table 1 Most repeated segments in CLRG

English			Chinese		
Segment length	Repeated segment	Frequency	Segment length	Repeated segment	Frequency
2	of the	8159	2	中国 的	1252
2	in the	5877	2	温室 气体	1016
2	to the	2790	2	中 的	897
2	and the	2558	2	的 是	888
2	climate change	2468	2	气候变化 的	830
2	on the	2143	2	的 问题	828
2	the world	1853	2	可 再生能源	803
2	to be	1839	2	新 的	758
2	for the	1720	2	的 影响	752
2	is a	1586	2	我们 的	733
2	at the	1560	2	上 的	693
2	that the	1530	2	可 持续	684
2	from the	1413	2	更 多	659
2	will be	1365	2	的 国家	630

content terms are classified at the top of the table, while in English, it is function words or syntactic words that are the most repeated. The meaning of the repeated segments of the English section reveals little interesting information. Of the function words or syntactic words that are the most repeated, only one theme related to our research is present: climate change. However, the segments repeated in Chinese reveal the real themes of our research: greenhouse gases, climate change, renewable energies, and new/new. We can say that two types of repetitions occur: grammatical words for English and content words for Chinese. It should be noted that the tool used in this work makes it possible to "clean" the extremely frequent repeated segments according to the language studied. It may be obvious to some that in French, function words are the most common. It would therefore be interesting to use this feature to obtain comparable results according to the languages studied.

It should be noted that this phenomenon also occurred in the comparable ENRG corpus (Table 2). The three-colored table shows the 16 most repeated segments of ENRG. As with the parallel corpus, our first observation is a particularly prominent thematic repetition for the Chinese subcorpus (ENRG_CN). For example, the frequency of the reduce emissions segment is 12,554, placed as the most repeated segment in ENRG_CN, word (the writing form) absent at the top of the table for the other two subcorpuses. However, these forms exist, albeit ranked much lower in the results of repeated segments. The other subthemes, represented by the repeated

Table 2 Most repeated segments in ENRG

ENRG-FR			ENRG-US			ENRG-CN		
Segment length	Repeated segment	Frequency	Segment length	Repeated segments	Frequency	Segment length	Repeated segment	Frequency
2	de la	23,101	2	of the	19,529	2	减 排	12,554
2	de l	19,787	2	in the	13,169	2	低 碳	10,730
2	à l	7426	2	to the	6361	2	新 能源	8753
2	à la	7244	2	on It	5339	1	气候 变化	8181
2	dans le	4975	2	and the	4943	2	风 电	7658
2	et de	4189	2	for the	4932	2	不 是	6658
2	dans les	3972	2	that the	4537	2	这 一	6409
2	sur le	3961	2	at the	3794	2	光 伏	6081
2	dans la	3720	2	from the	3326	2	大 的	5995
2	que les	3600	2	to be	3277	2	的 是	5911
2	a été	3328	2	by the	3185	2	的 一	5628
2	sue les	3205	2	New York	3137	2	也 是	5338
2	sur la	3092	2	in a	3095	2	就 是	5322
2	y a	2887	2	of a	2898	2	这 是	5124
2	dans l	2870	2	he said	2881	2	不 能	5014
2	que le	2763	2	with the	2684	2	环境 保护	4866

sequences directly related to energy and climate change, such as low carbon, wind energy, photovoltaics, etc., are also highlighted in Table 2.

For the French and American subcorpus, only grammatical words or function words appear in the most repeated segments. This phenomenon is essentially due to the anaphora mechanism or the deictic mechanism, which is not the same in French and American English. However, we note that in Chinese, these are key terms that are repeated, while in English and French, they are often named entities (proper names, toponyms, etc.).

6 Recurrences in Chinese Through the Prism of Their Poly-cooccurrences Experience

Let us look at the parallel poly-cooccurrences resulting from CLRG with the form "EPR." Only three forms are correlated in English. However, in the Chinese part, we obtained many more forms, such as "cost, reactor, Areva, CGN, etc." This phenomenon of repetitions in Chinese is frequent in this corpus (Fig. 9).

Chinese recurrences, deictic mechanism
To go further, we have chosen three terms that are sometimes difficult to transform into an anaphora in order to better observe this phenomenon of repetition.

These three pairs of words are "emission, 排, Pái," "problem, 问题, Wèntí," and "carbon, 碳, Tàn," and in this diagram (Fig. 10), we notice that there are always more repetitions in Chinese than in English.

For the neology "Jiǎn pái, 减排, reducing CO_2 emissions," it should be noted that in English the simple form "emission" covers much broader semantic fields; one would therefore have expected more repetitions in the English section.

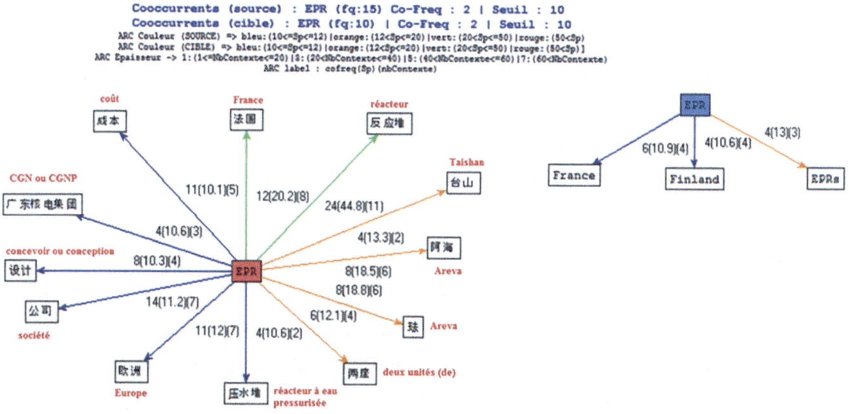

Fig. 9 Parallel poly-cooccurrences 2006 to 2014

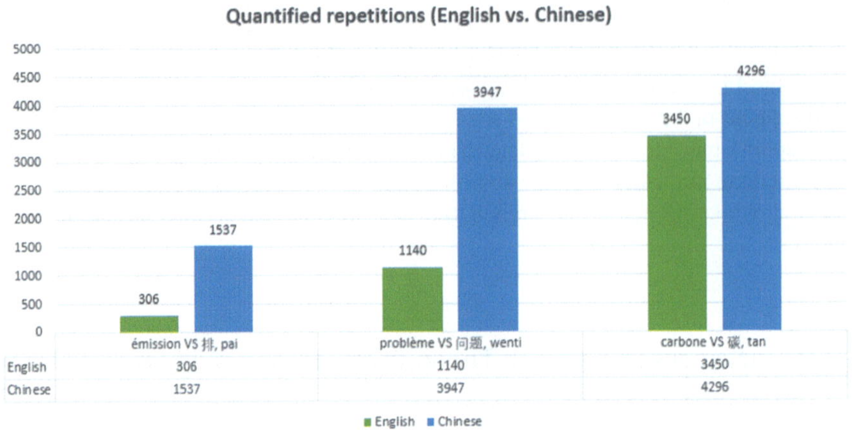

Fig. 10 Quantified repetitions (English vs. Chinese)

It is therefore possible to say that there are more repetitions in Chinese, fewer in English, and fewer still in French. This phenomenon is undoubtedly due to the anaphora mechanism or the deictic mechanism, which is different in French and English.

7 Conclusion

In conclusion, anaphors and recurrences are a fundamental asset for information mining and contribute to the effectiveness of multilingual investigations. The exploration work highlights the way in which the main analyzed concepts are perceived and formulated in each field and in each language. Over time, the information resulting from our analyses makes it possible to highlight linguistic facts and anaphora issues in the excavation of multilingual information [12]. Empirically, the semantic observations made through the repeated segments and poly-cooccurring networks in the Chinese texts allowed us to establish that anaphoric deictics, a strong tendency for the correct use of French and English, are a stylistic figure very rarely practiced by Chinese speakers—a fact which had not been previously highlighted. One of our objectives was to establish, evaluate, and prove this common place, a general reflection commonly accepted by public opinion. More generally, if, in Chinese, repetitions are more frequent, this should make it possible to keep the prominent referent in memory, facilitating the tasks of automatic anaphora resorption. Our study of particular corpora leads to a general linguistic fact that differentiates Chinese and the two languages of European origin. It should be noted that the BERT model [2], the one that has allowed better performance in deep learning and information retrieval recently, is based on repeated segments. This work would lead to contributions in various fields [8], such as multilingual semantics [9], the automation of monitoring, search engines, and generalized monitoring in all sectors.

References

1. Charolles M (1991) L'anaphore: problèmes de définition et de classification. Verbum XIV 2-3-4:203–215
2. Devlin J, Chang M, Lee K, Toutanova K (2018) BERT: pre-training of deep bidirectional transformers for language understanding. arXiv:1810.04805v2 [cs.CL] (Submitted on 11 Oct 2018 (v1), last revised 24 May 2019)
3. Habert B, Zweigenbaum P (2002) Problèmes épistémologiques: Régler les règles. TAL, Association pour le traitement automatique des langues 43(3):83–105
4. Martinez W (2012) Au-delà de la cooccurrence binaire... Poly-cooccurrences et trames de cooccurrence, Corpus [En ligne], 11 | 2012, mis en ligne le 18 juin 2013, https://corpus.revues.org/2262 (consulté le 25/02/2016).
5. Mayaffre D (2015) L'anaphore rhétorique. Pratiques n° 165–166. URL: https://journals.openedition.org/pratiques/2418
6. Mitkov R (1997) Factors in anaphora resolution: they are not the only things that matter. A case study based on two different approaches. In: Proceedings of the ACL workshop on operational factors in practical, robust anaphora resolution for unrestricted texts, pp 14–21
7. Mitkov R, Evans R, Orasan C (2002) A new, fully automatic version of Mitkov's knowledge-poor pronoun resolution method. In: Gelbukh A (ed) Computational linguistics and intelligent text processing. CICLing 2002. Lecture Notes in Computer Science, vol 2276. Springer, Berlin, Heidelberg
8. Rehm G, Stein D, Sasaki F, Witt A (eds) (2018) Language technologies for a multilingual Europe, translation and multilingual natural language. Language Science Press, Berlin, TC3 III, 1–4
9. Sailer M, Markanonatou S (eds) (2018) Multiword expressions: insights from a multi-lingual perspective. Language Science Press, Berlin, 374 p
10. Salem A (1987) Pratique des segments répétés, Essai de statistique textuelle. Histoire & mesures 2-3-4:224–227
11. Soon W, Ng H, Lim D (2001) A machine Learning approach to coreference of noun phrases. Comput Linguist 27(4):521–541
12. Zweigenbaum P, Sharoff S, Rapp R (2018) A multilingual dataset for evaluating parallel sentence extraction from comparable corpora. In: Language Resources and Evaluation Conference, Miyazaki, Japan, May 2018

Applications

Looking for *Topics*: A Brief Review

Ludovic Lebart

Abstract This paper presents a brief review of several endeavors to identify latent variables (axes or clusters). When dealing with textual data, these latent variables (clusters or axes) are sometimes designated *ex ante* by the term "topic". The first attempts to identify interpretable latent variables dates back to factor analysis at the beginning of last century. Recent years have witnessed a series of algorithmic attempts such as non-negative matrix factorization (NMF) or latent Dirichlet allocation (LDA). In the meantime, latent variables are also identified through several hybridizations and synergies of principal axes methods and clustering techniques. A single medium-sized classical corpus (Shakespeare's 154 Sonnets) will serve as a benchmark to sketch and compare in a compact way some characteristic features of several methods.

1 Introduction

There is a profusion of new disciplines around the industrial applications involving texts, with subsequent proliferations of tools and disparities of terminologies. There are also disparities in the attitude toward the texts, sometimes influenced by the availability and the user-friendliness of software. The problems entailed by huge sets of newsgroup or tweets are quite different from those encountered when dealing with literature, political discourses and psychological surveys. Because they are well known, translated in almost every language, deeply studied and commented, we will use Shakespeare's Sonnets as a benchmark to briefly compare the ability of several techniques to recognize topics in a corpus.

L. Lebart (✉)
Telecom-ParisTech, Paris, France
e-mail: ludovic@lebart.fr

D. F. Iezzi et al. (eds.), *Text Analytics*, Studies in Classification, Data Analysis, and Knowledge Organization,
https://doi.org/10.1007/978-3-030-52680-1_17

2 An Outline of the Contents of Shakespeare's Sonnets

The 154 sonnets of William Shakespeare deal with themes such as love, friendship, effects of time, beauty, treason, lust and death [19]. Note that the definition of topics in text mining is a pragmatic one and may also recover the concepts of theme and motif.

2.1 Theme, Topic, Motif

Usually, a topic is an objective explanation of the subject matter, whereas a theme represents the deeper underlying message. A motif is simply a recurring idea or pattern used to reinforce the main theme. Schematically, topics answer the questions: "What's the story about? Who? What? How?" and themes answer: "Why was the story written?". Topics in literature are easier to identify than themes.

Three main contiguous series of sonnets are generally recognized as three dominant themes:

Sonnets 1 to 17: (*Procreation*). These sonnets celebrate the beauty of a young man who is urged by the poet to marry so as to perpetuate that beauty.
Sonnets 18 to 126: (*Young man*). This longest sequence concerns the same young man (not definitively identified), the destructive effect of time, the force of love, friendship and poetry.
Sonnets 127 to 154: (*Dark Lady*). These sonnets are mostly addressed to a dark haired woman, not without some irony and cynicism (the two last sonnets 153 and 154 are specific epigrams in an ancient style; they should deserve in fact a specific category).

2.2 Eight Themes Derived from Expert Commentaries

The themes *Young man* and *Dark lady* could contain five sub-themes. While the first theme (*Procreation*) remains untouched, the new *Young man* and *Dark lady* themes will comprise only those sonnets which were not assigned to the five new categories below (*absence, storm, rivalry, death and eternal poetry*).

The partition of sonnets given in Table 1 is inspired by the works of Alden [1] and Paterson [2] but not explicitly mentioned by these authors. Figure 1 shows however that, after a blind correspondence analysis ignoring these themes, most of their locations are statistically significant on the principal plane of visualization.

Evidently, the following attempts to find topics into the corpus of sonnets will ignore that a priori partition into themes. We do not expect either to retrieve automatically these themes.

Table 1 List of eight a priori themes/topics with the corresponding sonnets numbers

Procreation	1–17
Young Man	20–25, 33–38, 40–42, 46, 47, 49, 53–55, 59–60, 62–70, 75–77, 88–106, 108–112, 115–125
Dark Lady	127–136, 139, 140, 143–146, 153, 154
Absence	26–32, 39, 43–45, 48, 50–52, 56–58, 61, 113–114
Storm	141, 142, 147–152
Rivalry	78–87
Death	71–74
Etern_poetry	18, 19, 81

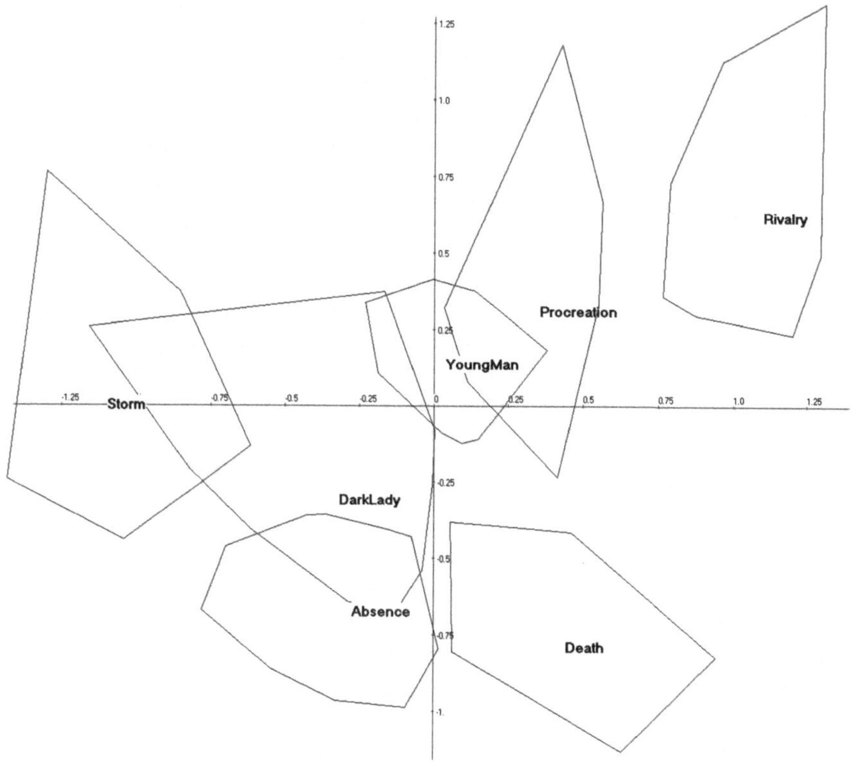

Fig. 1 Locations of seven themes/topics in the principal plane of the correspondence analysis of the lexical table (154 sonnets × 173 words, [min. frequency = 10]) as supplementary categorical variables. Conservative bootstrap confidence ellipses [drawing with replacement of sonnets] show the significant distances between several pairs of a priori themes. Note that the theme "Eternal Poetry", too much overlapping with others, is missing in this graphical display

However, the knowledge of these themes issued from literary criticism can provide us with a template for reading and interpreting the results more easily. Note that statistical tools mainly based on frequencies detect almost indifferently topics, themes or motifs.

3 Six Selected Methods for Topic Research

Among the six procedures selected in the present paper, four (RFA, FCA, LOA, LSA) make use of the singular values decomposition (SVD). The remaining two methods (NMF and LDA), less geometrical, involve a specific model and more complex algorithms.

Rotated factor analysis (RFA) is historically the first attempt to identify unobserved "latent factors" (Thurstone [3], after the pioneering papers of Spearman [4], and of Garnet [5]). RFA involves SVD in one of the most popular algorithms known as principal factor analysis. In this case, the *topics* are the words characterizing each kept factors. Initially, conceived for numerical values, it could be adapted to sparse frequency tables [**R** packages "psych" and "GPArotation"].

Fragmented correspondence analysis (FCA), in the vein of *ALCESTE* methodology [6], is based on the CA of fragments of texts (in our case seven consecutive lines, i.e.: half a sonnet), cf. Lebart [7]. The principal axes of CA serve to cluster these fragments (hybrid clustering using Hierarchical Classification—Ward criterion—and k-means). At the end of the process, the *topics* are defined by the series of words that characterize each cluster (software "DtmVic").

Logarithmic analysis (LOA) [8] is similar to spectral mapping [9] thanks to a difference of weighting. Both methods, like CA, comply with the principle of distributional equivalence (stability of the results vis-à-vis fusions of similar columns or rows). Applied to contingency or frequency tables, LOA often produces results similar to those of CA, with less sensitivity toward outliers as a consequence of the logarithmic shrinkage. A clustering of sonnets (similar to that of FCA) is then performed. The *topics* are then the words characterizing each cluster (software: "DtmVic").

Latent semantic analysis or indexing (LSA) [10], which is basically a SVD of the matrix of Tf.Idf coefficients (Term frequency × Inverse of document frequency). A clustering of sonnets (similar to that of FCA and LOA) is then performed. The *topics* are then the words characterizing each cluster (**R** package "lsa" [Fridolin Wild] and "DtmVic").

In the domain of text analysis, the two following methods belong more specifically to the field of "topic modeling".

Non-negative matrix factorization (NMF) starts with an equation that reminds singular values decomposition (SVD): Decomposition of a data matrix **A** as the product of two matrices of lower rank, **B** and **C**: $\mathbf{A} = \mathbf{B}\ \mathbf{C}$. The marked difference lies in a constraint of positivity of the coefficients of **B** and **C** (those of **A** being already supposed >0) [11, 12], (see Gaujoux [13]). In the topic modeling context, the main output of NMF is a set of topics characterized by list of words (software "scikit-learn" [Python] by Grisel O., Buitinck L., Yau C.K; In: Pedregosa et al. [14]).

Latent Dirichlet allocation (LDA) [15, 16] is a generative statistical model (involving unobserved topics, words and document) devised to uncover the underlying semantic structure of a collection of texts (documents, supposed to be a mixture of a small number of topics). The method is based on a hierarchical Bayesian analysis of the texts (package **R**: "topicmodels", and software "scikit-learn" [Python]).

At this stage, we have limited our investigation to six techniques out of a great number of approaches likely to identify topics. Among these approaches let us mention the direct use of CA without fragmentation of the texts, the techniques of clustering (used in FCA and LOA) which contain many more methods and variants, the already mentioned *Alceste* methodology [6]. The present piece of research evidently needs to be extended. In fact, each method involves also a series of parameters (threshold of frequency for the words, preprocessing options such as lemmatization/stop words, size of fragments or context units, number of iterations). The following experiment limited to six methods will be tersely summarized. A thorough investigation would need many more pages.

4 Excerpts from the List of 49 Topics (Limited to Two Topics Per Method)

The number of topics detected by each of the six selected methods varies between six and ten. Only two topics are printed below for each method.

4.1 Rotated Factor Analysis (Rotation Oblimin) (2 Topics Out of 6)

RFA1 eyes see bright lies best form say days.
RFA2 beauty false old face black now truth seem.

4.2 Fragmented Correspondence Analysis (FCA) (2 Topics Out of 7)

FCA1 beauty truth muse age youth praise old eyes glass long seen lies false time days.
FCA2 night day bright see look sight.

4.3 Logarithmic Analysis (Spectral Mapping) (2 Topics Out of 8)

LOA1 summer away youth sweet state hand seen age rich beauty time hold nature death.

LOA2 pen decay men live earth verse muse once life hours make give gentle death.

4.4 Latent Semantic Analysis (2 Topics Out of 8)

LSA1 time heart beauty more one eyes eye now myself art still sweet world.

LSA2 end grace leave words lie spirit change shame self could ever decay write.

4.5 NMF Topics (2 Topics Out of 10)

NMF0 love true new hate sweet dear say prove lest things best like ill let know fair soul.

NMF1 beauty fair praise art eyes old days truth sweet false summer nature brow black live.

4.6 Latent Dirichlet Allocation LDA (2 Topics Out of 10)

LDA0 summer worse praise nature making time like increase flower let copy rich year die.

LDA1 sing sweets summer hear love music eyes bear single confounds prove shade eternal.

5 A Synthesis of Produced Topics

How to compare the complete lists of topics, since neither the order of topics, nor the order of words within a topic are meaningful? We deal here with real "bags of words" exemplified by the excerpts of lines in Sect. 4. We will add the eight a priori themes defined in Table 1. Each a priori theme corresponds to a subset of sonnets. That subset will be described by its characteristics words. We can then perform a clustering of these 57 topics/themes (49 + 8). The technique of additive trees [17, 18] seems to be the most powerful tool for synthesizing in compact form these 57 topics/themes (Fig. 2). Let us recall one important property of additive trees: The real distance between two points can be read directly on the tree as the shortest path between the two points.

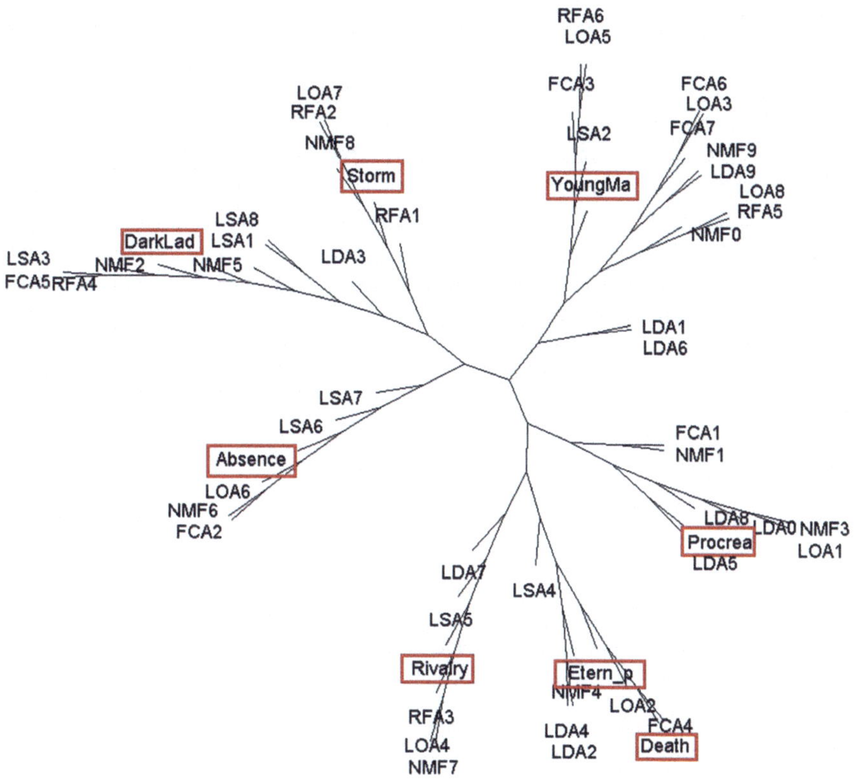

Fig. 2 Additive tree describing the links between the 49 topics provided by the 6 selected methods and the 8 a priori themes. The identifiers are those of Sect. 4 for the 6 selected methods. The 3 first letters indicate the method, followed by the index of the produced topic. The distance between two topics is the chi-square distance between their lexical profiles. Threshold of frequencies for words: 2. The boxed identifiers of the a priori themes are those (possibly shortened) of Table 1

Ideally, we expect to find a tree with as many branches as there are real topics in the corpus, each branch of the additive tree being characterized by seven labels: Six labels corresponding to the six methods briefly described above, plus one label corresponding to one a priori theme. Such situation occurs when each method has uncovered the same real topics. The observed configuration is not that good, but we can distinguish between six and nine main branches, which is probably the order of magnitude of the number of different topics. We note also that several different methods often participate in the same branch, which suggest that that branch correspond to a real topic discovered by almost all the six methods. Let us mention that a similar additive tree performed on the 49 topics (not involving the eight a priori themes) produces approximately the same branches. Thus, the eight a priori themes can be considered here as illustrative elements, serving only as potential identifiers of the branches.

It is remarkable that the eight a priori themes (boxed labels) are well distributed over the whole of Fig. 2. All the main branches have as a counterpart one of the a priori themes. The only exception is the branch located in the upper right part of Fig. 2. As an example of interpretation of Fig. 2, the branch in the lower center part of Fig. 2: [NMF7, LOA4, RFA3, LDA7, LSA5] is clearly linked to the a priori topic named *Rivalry* (see Sect. 2.2) (concurrence of five methods out of six). Most of the branches of the additive tree could be interpreted likewise. The upper right branch identified by none of the a priori themes may represent an unforeseen topic. More research and an expertise in Elizabethan poetry are required to confirm that we are dealing here with an undetected new theme. To conclude, we can only observe that each of the involved method, be it ancient or modern, may contribute to detect topics... and that exploratory tools are essential to visualize the complexity of the process and assess the obtained results.

References

1. Alden RM (1913) Sonnets and a lover's complaint. Macmillan, New York
2. Paterson D (2010) Reading Shakespeare sonnets. Faber & Faber Ltd., London
3. Thurstone LL (1947) Multiple factor analysis. The University of Chicago Press, Chicago
4. Spearman C (1904) General intelligence, objectively determined and measured. Am J Psychol 15:201–293
5. Garnett J-C (1919) General ability, cleverness and purpose. Br J Psychol 9:345–366
6. Reinert M (1986) Un logiciel d'analyse lexicale: [ALCESTE]. Cahiers de l'Analyse des Données 4:471–484
7. Lebart L (2012) Articulation entre exploration et inférence. In: Dister A, Longree D, Purnelle G (eds) JADT_2012. Presse Universitaire de Liège
8. Kazmierczak J-B (1985) Analyse logarithmique: deux exemples d'application. Rev Stat Appliquée 33(1):13–24
9. Lewi PJ (1976) Spectral mapping, a technique for classifying biological activity profiles of chemical compounds. Arzneimittel-Forschung [Drug Res] 26:1295–1300
10. Deerwester S, Dumais ST, Furnas GW, Landauer TK, Harshman R (1990) Indexing by latent semantic analysis. J Am Soc Inf Sci 41(6):391–407
11. Berry MW, Browne M, Langville AN, Pauca VP, Plemmons RJ (2007) Algorithms and applications for approximate nonnegative matrix factorization. Comput Stat Data Anal 52(1):155–173
12. Lee DD, Seung HS (1999) Learning the parts of objects by non-negative matrix factorization. Nature 401:788–791
13. Gaujoux R et al (2010) A flexible R package for nonnegative matrix factorization. BMC Bioinform 11(1):367
14. Pedregosa F, Varoquaux G, Gramfort A, Michel V, Thirion B, Grisel O, Blondel M, Prettenhofer P, Weiss R, Dubourg V, Vanderplas J, Passos A, Cournapeau D, Brucher M, Perrot M, Duchesnay E (2011) Scikit-learn: machine learning in python. J Mach Learn Res 12:2825–2830
15. Blei D, Ng A, Jordan M (2003) Latent Dirichlet allocation. J Mach Learn Res 3:993–1022
16. Griffiths TL, Steyvers M, Tenenbaum JB (2007) Topics in semantic representation. Psychol Rev 114(2):211–244

17. Huson DH, Bryant D (2006) Application of phylogenetic networks in evolutionary studies. Mol Biol Evol 23(2):254–267. Software available from www.splitstree.org
18. Sattath S, Tversky A (1977) Additive similarity trees. Psychometrika 42(3):319–345
19. Shakespeare W (1901) Poems and sonnets: Booklover's edition. In: The University Society and Israel Gollancz (ed). University Society Press, New York. Shakespeare Online, Dec 2017

Where Are the Social Sciences Going to? The Case of the EU-Funded SSH Research Projects

Matteo Gerli

Abstract This article investigates how the emergence of a European research funding affects the directions and processes of scientific knowledge production. It does so by examining the extended abstracts of the EU-funded research projects (2007–2013) realized within the broad domain of the Social Sciences and Humanities (SSH). By using some automated content analysis techniques, the essay tries to shed light on the interplay between the policy layer and the SSH field, taking into account the programmatic and legal structures inherent the 7FP, on one side, the national scientific systems and the disciplinary structure of the SSH field, on the other side.

1 Introduction: The EU "Clouded-Sky" Research Funding and Its Role in Affecting SSH Thematic Foci

The European Research Policy plays a strategic role for thousands of researchers and research institutions which operate within the EU borders. Thanks to the concomitant decrease in national public funds for scientific activities (see, e.g., [52, 53]), the European research agenda has dramatically increased its appeal among scholars and consequently its ability to have an impact on the directions and processes of scientific knowledge production. Indeed, starting from the 90s, the European Commission has endowed itself with a new means to combine and manage, on the basis of medium to long-term planning cycles, the whole set of scientific and technological initiatives financed by the EU budget: the so-called *framework programme* [5, 7, 28, 40, 45, 46]. Shortly, the underlying logic implies the programmatic intersection between research activities and other European policies, so that the promotion of scientific excellence complements the need to foster the creation of *cross-national* and *cross-disciplinary* collaborations for producing a type of knowledge "close" to the end users and to

M. Gerli (✉)
Department of Arts, University of Bologna, via Barberia 4, 40123 Bologna, Italy
e-mail: matteo.gerli@unibo.it

D. F. Iezzi et al. (eds.), *Text Analytics*, Studies in Classification, Data Analysis, and Knowledge Organization,
https://doi.org/10.1007/978-3-030-52680-1_18

the European institutions' interests, among which it is of primary importance the realization of the *European Research Area* [21, 22].

To be more precise, the emergence of the framework programme (from now on FP) can be interpreted as the result of the still ongoing Europeanization process of the pre-existing national research policy, which gave rise to a distinct governance apparatus dedicated to problem-solving and specialized in the production of binding rules for all member countries [17] in the field of science and innovation. In other words, it can be interpreted as a specific case of the more general *process of regionalization* of production and circulation of scientific knowledge [29, 30]. In particular, the decisions adopted at the European level and held in the FP can affect a wide range of aspects concerning scientific production and dissemination, acting on the "structure of opportunities" that characterized the scientific field at a specific time. Potentially, the whole scientific and technological production chain can be influenced, from the advancement of early stage careers to the consolidation of high-status disciplines and researchers [31, 32], passing through the social and economic exploitation of scientific results, the promotion of international cooperation and the training and mobility of scholars [1, 46].

Among the various activities that are planned and carried out on a multiannual basis, this article deals with that part of the FP—the most prominent in budgeting terms—that explicitly address scientific research via thematic calls for international and multidisciplinary projects. In particular, what is at issue here is the question on *how* and *how much* the thematic foci of scientific production are affected, if they are, by the so-called EU clouded-sky or "mission-oriented" research funding. Indeed, these programmes are defined by a top-down procedure and then "used by scholars to generate detailed research proposals", while retaining a space for diverging disciplinary interpretations, "just as clouds are perceived subjectively to represent imagined, different shapes" [46: 108].

In the next pages, I will focus on the EU-funded SSH (*Social Sciences and Humanities*) research projects in order to analyse the different "shapes" that social and human scientists have imagined and created in response to the European institutions research "demand". To be more precise, I will not examine the broad intention and the technical procedures underlying the European SSH multiannual programmes (for this, see [34, 46]), nor the difference in resource and structures that influence the way different countries participate to the European research programmes [24, 34]. Instead, I will focus on how different national scientific systems and disciplines deal with the thematic challenges posed by the FPs. Indeed, while the programmatic and legal structures inherent the FPs are an essential part of the process of Europeanization of scientific production and circulation, previous studies highlighted that, to understand how scientific knowledge is produced and circulates at a *inter* or *supra*-national level, one needs to take into account both *national scientific systems* and *disciplines* as a major institutional forces that affect scholarly contents [20, 23, 29]. Furthermore, the most convenient aspect of this approach is that it is fruitful for "systematic comparisons between developments in various disciplines and countries" ([23]: 5, see also [38]).

In its exploratory character, this study does not have strong hypothesis to be tested. Nevertheless, according to the theoretical premises, it aims to give an original perspective through which observing the relationship between the field of social sciences and the public policy field of the EU [24]. In particular, three main research questions are addressed: first, to what extent are the thematic foci of scientific production affected by the European Commission research demand? Second, are there any difference between disciplines and countries? Third, are there any pattern of linkage between scientific specialities and between national scientific systems? Before entering into specifics of the methodology and empirical results, however, some further theoretical premises are needed.

2 The Socio-political Dimensions of Scientific Research

The general thesis behind this article is that a mutually structuring relationship exists between scientific production and the EU integration [2]: on one side, the progress of integration process influenced (and still influences) research activities through the promotion of particular forms of knowledge and research questions (as far as I am concerned, mainly through the realization of cross-national and cross-disciplinary research projects), on the other side, the resulting knowledge contributes to conditioning the practices that take place in the European and national institutions, according to a circular process that brings the innovations to feedback into the system that expresses them.

Social Sciences and Humanities, which are less directly involved in the production of knowledge with a clear practical usability, are by no means unconcerned with this kind of phenomenon. At this regard, the *Journal of European Integration*[1] recently published a special issue on the relationship between social sciences and European integration, hosting some important articles that have highlighted the existence of several "crossroads" between the EU and the scientific community "itineraries": Rosamond [44], for instance, observed how certain theories on the political and economic integration (in particular that of the Hungarian Béla Balassa, from the economics side, and the neofunctionalism, from the political science side) had been informing the "strategic narrative" adopted by the European Commission during the 60s and 70s to legitimize its newly formed institutional role and its economic policy position, according to a quite peculiar two-ways traffic of influences process, being the economic integration theorized while it was happening. Deem [18] pointed out the existence of a relationship between the birth of a new field on higher education studies, the simultaneous evolution of national university systems and the launch of the so-called *Bologna process* at European level. Vauchez analysed, through a sociogenetic approach, the historical process through which the *acquis communautaire* «has been formulated, stretched, criticized, revised and finally naturalized as the most rigorous and objective measure of Europe against other possible methods» [51: 196]. This

[1] *Journal of European Integration*, 37/2015.

was possible thanks to the work of those who have been defined "methodological entrepreneurs", that is European officials who have politically invested and succeeded in establishing Europe's cognitive and technical equipment "dialoguing" with the scientific sphere.

Looking beyond such individual cases, what is really relevant to my purpose is the underlying idea about the possibility of studying science production from a sociological point of view, basically by rejecting what was traditionally regarded as an *internal/external division* [2], and thus admitting that even scientific and academic concepts can be formulated in conjunction with political-economic ambitions and practical problems such as those expressed by the EU and the stakeholders involved in the policy-making process.[2] This does not necessarily mean that science is equal to politics [13], what it does mean is that, in order to understand scientific production, one needs to recognize that "non-scientific" resources (financial or material resources, ideas and beliefs, symbolic resources, political or normative resources, etc.), may overstep scientific boundaries and be used for the production of new knowledge.

Bourdieu (see, e.g., [11, 12, 8, 9, 10]) described this phenomenon through the concept of "fields interrelations". In a few words, the social word is composed of multiple semi-autonomous *fields*, basically microcosms characterized by different *stakes*, *rules of the game* and particular *resources* which one needs to possess in order to get access to the game itself and its specific advantages. He conceptualized these spheres as partially independent, by which he means that, even though each field develops its own institutions, hierarchies, problems, tacit or explicit rules, they necessarily interact and affect each other. This is particularly true for cultural fields (art, cinema, religion, science, journalism, etc.), since they tend to be structurally dependent and subordinated to political and economic fields. Going straight to the point, this is to say that, if one is dealing with a sociological analysis of a cultural product (e.g. a text), one cannot just consider its formal characteristics, nor be limited to its context of production. Instead, one should use a "relational approach", taking into account both the *internal* features of the product and its *external* determinants.

There are several different ways through which one can observe *field-to-field relations* (see, e.g.: [27, 50, 54]). For the purpose of this article, I found particularly suitable the idea of "boundary objects" firstly developed by Star and Griesemer [48] (see also [36, 41]), namely objects that inhabit at the same time different intersecting social words, satisfying the informational and practical requirements to each of them. They are "objects which are both plastic enough to adapt to local needs and the constraints of the several parties employing them, yet robust enough to maintain a common identity across sites" [48: 393]. In other words, they allow the maximization of the fields (relative) autonomy, while guaranteeing their convergence towards common goals and thus the coherence of the whole collective project. From this

[2]The relationship between science, scientific knowledge and society constitutes the object of study of a multidisciplinary field of research called *Science and Technology Studies* (STS). For more details about the inner theoretical and methodological debate, see: Ancarani [3], Kaiser [33], Callon [16], Bucchi [14], Sismondo [47].

specific perspective, one can imagine an "EU project call", that is a call financed by the EU and managed by the European Commission [4, 46], as a set of coherent and standardized boundary objects that connect the European policy field with the SSH field, and the social and human scientists, from different countries, institutions and disciplines, to each other. The specifics of this approach are illustrated in the following paragraph.

3 The Corpus and Methodology

In engaging with this broad issue, this article will try to contribute further to the understanding of the topic by examining the *extended abstracts* of the SSH EU-funded research projects, as a part of the research activities financed by the 7FP (2007–2013) of the EU. By using some automatic text analysis tools, it will thus statistically explore the contents of such documents not per se, but in connection with some variables, which help to define the broader context of production.[3]

Unlike the studies discussed earlier [2], which are mainly based on micro-sociological observation, my investigation covers a macro-sociological analysis of a quite large *corpus* made of 5.847 graphic forms, equal to 70.215 occurrences. It is an *ad hoc* constructed *corpus*: it contains 248 texts (extended abstracts and projects titles), which were collected from the digital database CORDIS, the main institutional source of information related to the research projects financed by the EU. As a whole, the sample involves 249 research projects out of 251 realized by the SSH in 2007–2013.

The choice to focus on the projects extended abstracts and titles is not accidental but depends on their strict relevance to my research objectives. In fact, they can be effectively used to identify the structure of scientific ideas and research questions following the researchers in their lexical choices and strategies [15, 26, 55]. Basically, they represent an effective "shortcut" through which empirically one can observe the way in which the research groups that had been awarded a grant "actualized" the inputs they received from the European Commission.

As for the methodology, both simple and multivariate analysis are performed with the IRaMuTeQ software [6, 37]. In particular, the lexicographical analysis is used for a first exploration of the *corpus*, to identify and format texts units, turn texts into text segments and classify words by their frequency. The multivariate analysis, instead, is performed to detect the associations between textual data and some supplementary variables related to both the policy layer, as expressed in the calls (the boundary objects, in Star and Griesemer's view), and the SSH field. To be more precise, the research design consists of three main steps: first, mapping of the thematic-discursive

[3] A preliminary survey has been conducted on two peculiar scientific products realized by researchers operating within the SSH domain: *final reports* and *policy briefs* [25]. However, due to the lack of complete information and the differences between the aforementioned documents in terms of length and contents, I opted for an analysis of the *extended abstracts*, since they ensure greater representativeness and thus reliability.

structure of the European research projects through a descending hierarchical analysis (DHA) [42, 43]; second, analysis of the relation among the thematic clusters that arises from the DHA; third, analysis of the relations between the thematic clusters and the supplementary variables. The last two steps were realized through a factorial correspondence analysis (FCA) [19, 37, 39].

As for the supplementary variables, the analysis takes into account what in the 7FP was defined as *thematic area* (TA). The FP7 included eight different macro-thematic areas (Table 1), each one representing a specific thematic target that researchers were asked to follow in setting their project proposals and activities, even though with a certain degree of autonomy [46].[4]

From the scientific side, instead, the analysis includes the country and the geographic area (GA) in which the coordinating institution of each project was located (as a proxies of the national scientific systems) and the scientific discipline of the principal investigator as well. As for the latter, the following fourteen disciplines were detected from the principal investigators' institutional websites (Table 2).

4 The Main Findings

The DHA allowed the identification of eight clusters (Table 3). Following Reinert [43], they can be interpreted as "lexical words", namely specific semantic structures which, in our case, refer to different and even competing scientific topics and representations covered by the SSH between 2007 and 2013. In the right column, the first fifteen characteristic words are listed, accompanied by a "thematic label" and the percentage of text segments included in each cluster (left column), which can be interpreted as a proxy of their relative weight.

With regard to the association between the thematic areas and the detected clusters, Fig. 1 shows a clear functional homology between what had been scheduled in the FP7 by the European Commission and what researchers realized. Indeed, each cluster (horizontal axis) corresponds with a thematic area (verticals bars), clearly and strongly supporting the thesis of adaptability [48], to the extent that it can be argued that there is a quite clear "overlap" between the thematic areas defined in the project calls and the contents and discourses produced by the SSH scientists.

Figure 2, resulting from the FCA shows the relationship between the eight clusters in the form of a factorial plan. Briefly, the first four factors summarize together 67.5% of the total inertia: the first factor (21.9%), on the left plan, marks a clear opposition between what can be interpreted as *Science planning-impact* (negative half-plane) and *Societal issues* (positive half-plane), the latter referring to the whole set of clusters not directly related to science and technology issues (C1, C2, C3, C4 and C5). The second factor (17.5%), on the left plane, highlights a significant opposition between what can be named as the *Economic pole* (positive half-plane), by virtue of the

[4]For more details: Decision No. 1982/2006/EC of the European Parliament and of the Council of 18 December 2006.

Table 1 Thematic areas (TAs) by number of projects realized

TA	Definition	*N*	%
1	**Growth, employment and competitiveness in a knowledge society**	**42**	**16.9**
1.1	Changing role of knowledge throughout the economy	16	6.5
1.2	Structural changes in the European knowledge economy and society	15	6.0
1.3	Strengthening policy coherence and coordination in Europe	11	4.4
2	**Combining economic, social and environmental objectives**	**44**	**17.7**
2.1	Socio-economic development trajectories	30	12.1
2.2	Regional, territorial and social cohesion	14	5.6
3	**Major trends in society and their implications**	**38**	**15.3**
3.1	Demographic changes	10	4.1
3.2	Societal trends and lifestyles	19	7.6
3.3	Cultural interactions in an international perspective	9	3.6
4	**Europe in the world**	**41**	**16.5**
4.1	Interactions and interdependences between world regions	16	6.4
4.2	Conflicts, peace and human rights	15	6.0
4.3	Europe's changing role in the world	10	4.1
5	**The citizen in the European Union**	**37**	**14.9**
5.1	Participation and citizenship in Europe	19	7.6
5.2	Diversities and commonalities in Europe	18	7.3
6	**Socio-economic and scientific indicators**	**15**	**6.0**
6.1	How indicators are used in policy?	1	0.4
6.2	Developing better indicators for policy	7	2.8
6.3	Provision of underlying official statistics	4	1.6
6.4	Development of indicators for the evaluation of policies	3	1.2
7	**Foresight activities**	**12**	**4.8**
7.1	Wide socio-economic foresight on key challenges	5	2.0
7.4	Blue sky research on emerging issues affecting European S&T	6	2.4
7.5	Mutual learning and cooperation	1	0.4
8	**Strategic activities**	**20**	**8.0**
8.1	Emerging needs	4	1.6
8.2	Horizontal measures to support international cooperation	4	1.6
8.3	Measures to support dissemination of research	5	2.0
8.4	Transnational cooperation among National Contact Points	3	1.2
8.5	Evaluation and impact assessment studies	1	0.4
8.6	Monitoring European Trends in Research in the SSH	3	1.2
	Total	249	100.0

Table 2 Disciplines by number of coordinated projects

Scientific discipline	*N*	%
Economics and business (EB)	67	26.9
Sociology (S)	41	16.5
Politics and international studies (PIS)	33	13.3
Regional and urban sciences (RUS)	29	11.6
Project management (PM)	17	6.8
Education sciences (ES)	11	4.4
Law (L)	11	4.4
History (H)	10	4
Anthropology (A)	8	3.2
Science and technology studies (STS)	6	2.4
Media and communication (MC)	5	2
Statistics (ST)	5	2
Language sciences (LS)	3	1.2
Psychology (P)	3	1.2
Total	249	100

association between economic (C1) and, at a lower level, sustainable development (C3) issues, and the *Cultural-educational pole* (negative half-plane), by virtue of the association between educational (C4) and (inter-)cultural (C5) issues. In-between these two poles, the clusters C1 (*Conflict solution and political stability*) acts as a "thematic bridge" that links together the more economically oriented discourse with the more culturally oriented. From a semantic perspective, this axis is very relevant because it reproduces the real political conflict (in the middle of the plane) between powerful social institutions and interests (positive half-plane) and less favoured parts of society (negative half-plane).[5]

The third factor (14.3%), on the right plan, marks a polarization between what can be named as *External issues* (positive half-plane) of the EU, by virtue of the significant association with *Conflict solution and political stability* (C1), and *Internal issues* (negative half-plane) of the EU, by virtue of the association with *Education, care services and social inclusion* (C4) and, at a lower level, *Innovation, competitiveness and economic growth* (C2). Finally, the fourth factor (13.8%), on the right plan, shows an opposition between *Science for evaluation* (positive half-plane), by virtue of the strong association with *Research methods and tools* (C8), and *Science for cooperation* (negative half-plane), by virtue of the association with *SSH transnational and multidisciplinary collaborations* (C6).

Through the graphical representation of the supplementary variables in the factorial plane (Figs. 3 and 4), it is possible to clarify further the association between the policy layer and the structural characteristics of the SSH field, as they result from the national scientific system (country) in which the coordinating institution of each

[5] A similar interpretation can be found in Kropp [35].

Table 3 First fifteen characteristic words of the detected clusters

Cluster (Topic)	Characteristic words
C1: Conflict solution and political stability (16.5%)	Conflict, security, region, medium, power, role, stability, evolve, actor, play, accountability, change, multilateralism, pace, regional
C2: Innovation, competitiveness and economic growth (12.8%)	Investment, growth, firm, innovation, capital, productivity, market, intangible, performance, sector, economy, competitiveness, venture, service, private
C3: Sustainable development and lifestyle (8.2%)	Sustainable, environmental, lifestyle, sustainability, user, consumer, carbon, element, consumption, transition, pressure, liveable, scenario, model, energy
C4: Education, care services and social inclusion (7.1%)	Young, person, education, care, exclusion, childhood, youth, labour market, adult, educational, homelessness, child, combat, inequality, migrant
C5: Intercultural relations, civic engagement and political participation (14.9%)	Citizenship, citizen, political, individual, identity, cultural, democratic, tolerance, diversity, context, racism, collective, difference, everyday, immigrant
C6: SSH transnational and multidisciplinary collaborations (11%)	Partner, consortium, network, research, institute, collaboration, national contact points, humanity, university, trans, SSH, centre, research, team, advisory board
C7: Projects outcomes and their dissemination (15.5%)	Dissemination, stakeholder, policy makers, policy, workshop, research, result, dialogue, findings, communication, community, disseminate, academic, achieve, expert
C8: Research methods and tools (14%)	Datum, survey, method, empirical, tool, indicator, interview, qualitative, collection, theoretical, analysis, measurement, case, quantitative, patent

project was located and the scientific discipline of the relative principal investigator. In particular, Fig. 3 shows the cross-tabulation of the two factors as to the variables "geographic area" and "country" (only the eight most important countries, in terms of funded projects, are highlighted). In particular, what emerges from the analysis is the absence of very specific characterizations as to the scientific priorities and contents, since all the cases under investigation tend to be attracted by the centre of the axes (the medium profile), except for Sweden (SE), the most prone to the *Cultural pole*, Spain (ES), the most prone to the *Economic pole*, and the Eastern European countries (*EE*), the most associated with the *Internal issues*.

It is also worth to be pointed out the position of the Southern European countries (*SE*) with regard to the *External issues* (presumably because of their strategic Mediterranean position, at the southern European border).

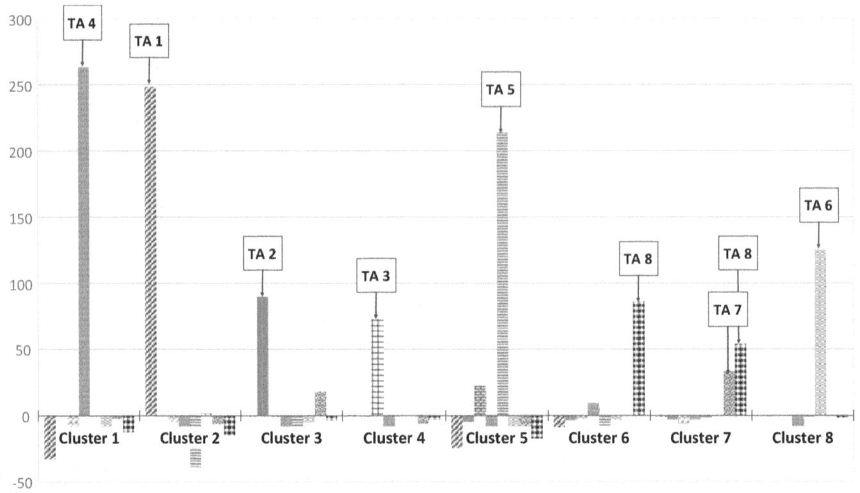

Fig. 1 Association (Chi^2) between thematic areas (TA) and the detected clusters. Legend: See Tables 1 and 3 for the labels of thematic areas and thematic clusters

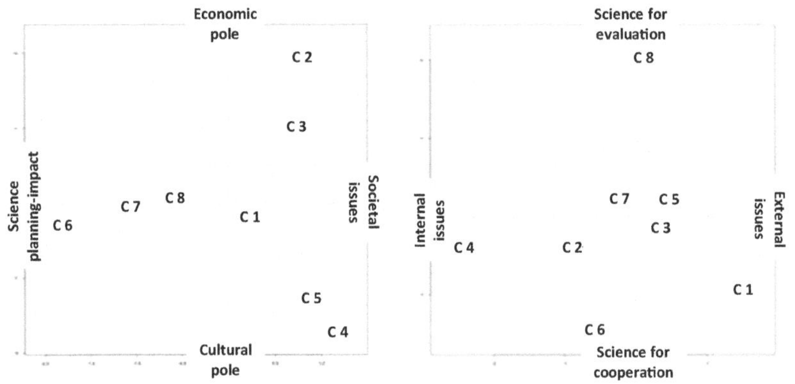

Fig. 2 Thematic map of the research activities realized by the SSH (2007–13). Legend: See Table 3 for clusters labels and their characteristic words

From the disciplinary perspective, instead, a significant division of scientific work emerges from the analysis, as shown by the position of the labels with respect to the four quarters that each factorial plane is composed of (roughly, each quarter can be considered as a cluster of disciplines, except for those which are very close to the centre of the factorial plane). In particular, on the left plan, the widest distance in terms of scientific interests and vocabulary occurs between Economics and business (EB), on one side, and Language sciences (LS) and Education sciences (ES), on the other; but also between the last two disciplines, which are very close to each other, and Project Management (PM). On the right plan, instead, the widest distance

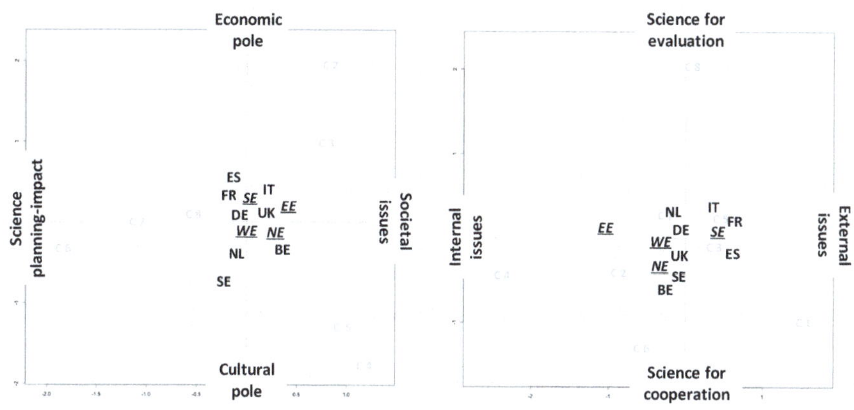

Fig. 3 Association between factors and supplementary variables (country and geographic area highlighted). Legend: *NE* (Northern Europe); *SE* (Southern Europe); *WE* (Western Europe); *EE* (Eastern Europe); *IT* (Italy); *FR* (France); *UK* (United Kingdom); *SE* (Sweden); *DE* (German); *BE* (Belgium); *NL* (Holland); *ES* (Spain)

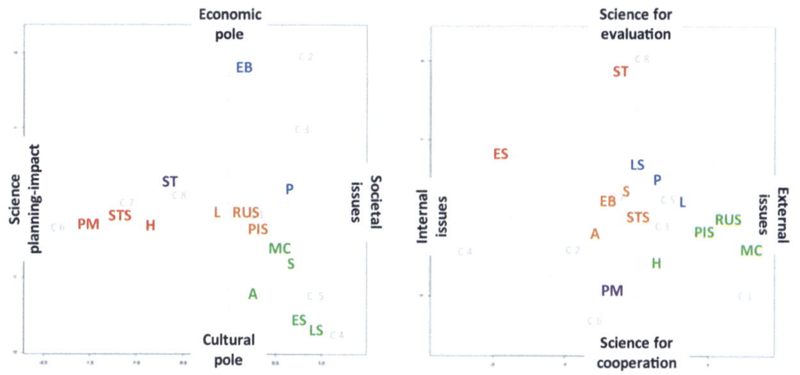

Fig. 4 Association between factors and supplementary variables (scientific disciplines highlighted). Legend: Sociology (S); Anthropology (A); Media and communication (MC); Science and technology studies (STS); Economics and business (EB); Statistics (ST); History (H); Law (L); Education sciences (ES); Language sciences (LS); Politics and international relations (PIS); Psychology (P); Regional and urban sciences (RUS); Project management (PM)

emerges between Media and communication (MC) and Education sciences (ES), and between Statistics (ST) and Project management (PM).

On the left plan, Sociology (S), Anthropology (A), Media and communication (MC), Education Sciences (ES) and Language sciences (LS) are placed on the same quarter, revealing the existence of a (partially) common way to approach the challenges posed by the European research projects. Statistics (ST) is close to Science and technology studies (STS) and Project management (PM), while Regional and urban sciences (RUS), a subfield concerned with problems that are common to different disciplines and approaches, such as spatial development, urban planning, migration, energy economics and evaluation, economic geography, is very close to the centre of the axes together with Law (L) and Politics and international relations (PIS). Finally, History (H) holds an unexpected position very close to STS and PM. However, this is due to the fact that some of the ten projects led by historians were more devoted to coordinate, bring together and thus valorize existing organizations and research institutions on historical topics than to produce new pieces of knowledge (e.g. this was the case of projects such as IDEAs, MELA, T-AP and RICHES).

On the right plan (Fig. 4), MC, RUS and PIS are very close to each other and share the same quarter with History (H); ST and ES shar the same quarter, since Statistics (ST) is used to evaluate performances of (internal) activities such as education; similarly, LS, P and L reveal similar interests and vocabulary, even though they are very close to the centre of the axes. Finally, STS, EB, S and A are the less characterized by the four semantic dimensions that arose from the FCA analysis and representation.

5 Conclusion: The SSH and the "European Spirit"

The findings presented herein indicate a close relationship between the programmatic framework, defined by the European Commission, and the SSH thematic foci, supporting the thesis of an interrelation between the policy field and the scientific field. In particular, looking at the scientific contents as a whole, it emerges clearly the ambition that the European institutions cultivate and try to achieve by "instilling" in the SSH that "European spirit" which is embodied, at least from the Lisbon Strategy [21] onwards, in the search for a balance between economic growth, sustainable development and social cohesion. The research topics that emerge from the analysis (Table 3 and Fig. 1) clearly demonstrate the existence of an "osmosis" process between the policy layer and the activities realized by the SSH scholars. However, this kind of evidence does not allow to understand how much the European research policy is able to affect the scientific practices, being necessary a different approach for this kind of investigation (e.g. a more qualitative case study approach).

As to the supplementary variables, the evidence (Figs. 3 and 4) shows that the high *porosity* of national borders is counterbalanced by a more rigid and differentiated division of scientific work between disciplines and specializations, which essentially correspond to different "models" of *disciplinary Europeanization*. In particular, some disciplines appear more mobile, flexible and "welcoming" than other with regard to the research foci and disciplinary proximity relationships. This is, for instance, the case of Economics and Business (EB) that appears very close to the *Economic pole* and separated from the other disciplines, on the left plan, while it is close to the centre of the axes together with Sociology (S), Anthropology (A) and Science and technology studies (STS), on the right plane. Similarly, Regional and urban sciences (RUS) shares the barycentre with Law (L) and Politics and international studies (PIS), on the left plane, while it is very prone to the *External issues* pole together with PIS and Media and communication (MC). By contrast, Project management (PM) and Education sciences (ES) seem to be the less flexible in terms of research topics and interests. Although the study of the process that generates this structure goes beyond the purposes of this first exploratory investigation, it still seems appropriate to point out with Ennis [20] how this may depend on various factors, including the sharing of cognitive and value orientations, the similarity in terms of methods and theories, the disciplinary prestige and therefore the costs (in symbolic terms) to be paid to allow heterodox themes and problems entering the disciplinary field. Furthermore, one also has to consider that disciplines are not "unified totalities", but dynamic social field that are driven by a mixture of internal and external forces [49].

Besides that, the SSH semantic space shows a clear cleavage between the economic discourse and the cultural discourse, which points out the existence of a lack of interaction and perhaps a dangerous separation between these two spheres, as it is also demonstrated, in terms of distance, by the analysis of the disciplinary structure of the SSH field (Figs. 2 and 4). Something similar appears even in relation to the other semantic poles. If this cannot be considered as a general rule, it is clear that, for what concerns the "mission-oriented" projects promoted by the EU, what is at stake for a linguist, an education scholar, an anthropologist and even a sociologist is far from what concerns an economist, a business management scholar or a statistician. In other words, the aforementioned "European spirit" manifests itself in a fragmented way, and it is not clear to what extent the "bridge" disciplines, such as Politics and international relations (PIS), Science and technology studies (STS) and Regional and urban sciences (RUS) are able to effectively contribute in reassembling the "mosaic".

In conclusion, what is the main lesson that can be drawn from this exercise of scientific ideas mapping? From a sociological perspective, one can argue that, in connecting the social sciences field with the policy field, the European research projects act as a magnifier of the interplay between the programmatic and legal structures inherent the 7FP and the disciplinary structures that the SSH inherited from their own history. From the analysis conducted, it is not clear enough whether the former prevail on the latter or vice versa; however, it is evident that, in its attempt to "dismantle" the boundaries between the European policy-making and science and between scientific themes and disciplines, the European Commission finds an "obstacle" in the resistance that the partially autonomous SSH field exerts over the policy lines and priorities held in the FPs.

References

1. Andrée D (2009) Priority-setting in the European research framework programme. Vinnova, Stockholm
2. Adler-Nissen R, Kropp K (2015) A sociology of knowledge approach to European integration: four analytical principles. J Eur Integr 37(2):155–173
3. Ancarani V (1996) La scienza decostruita. Teorie sociologiche della conoscenza scientifica. FrancoAngeli, Milano

4. Andrée D (2008) A rough guide to the FP7 work programmes (EU's seventh framework programme for research). Who can be involved in the preparation? What to do, when and how? Vinnova, Stockholm
5. Banchoff T (2002) The politics of the European research area. ACES working paper, vol 3, pp 1–25
6. Bolasco S (2013) L'analisi automatica dei testi. Fare ricerca con il text mining. Carocci, Roma
7. Borras S (2000) Science, technology and innovation in European politics. Research paper of the Roskilde University, vol 5, pp 1–30
8. Bourdieu P (1994) Raisons pratiques. Sur la théorie de l'action. Seuil, Paris
9. Bourdieu P (1995) Champ politique, champ des sciences sociales, champ journalistique. 5° Cahiers de recherche GPS (Groupe de recherche sur la socialisation) de l'Université Lumière Lyon
10. Bourdieu P (2001) Science de la science et réflexivité. Cours du Collège de France 2000/2001. Raisons d'agir, Paris
11. Bourdieu P (1975) The specificity of scientific field and the social condition of the progress of reason. Soc Sci Inf 6:19–47
12. Bourdieu P (1992) Les règles de l'art. Seuil, Paris
13. Breslau D (1998) In search of the unequivocal: the political economy of measurement in US labor market policy. Praeger, London
14. Bucchi M (2004) Science in society. An introduction to social studies of science. Routledge, London
15. Callon M, Law J, Rip A (eds) (1986) Mapping the dynamics of science and technology: sociology of science in the real world. Macmillan, London
16. Callon M (2001) Four models for the dynamics of science. In: Jasanoff S, Markle GE, Petersen JC, Pinch T (eds) Handbook of science and technology studies, revised. Sage, Thousand Oaks, pp 29–63
17. Cowles MG, Caporaso J, Risse T (eds) (2001) Transforming Europe: Europeanization and domestic change. Cornell University Press, Ithaca
18. Deem R (2015) What is the nature of the relationship between changes in European higher education and social science research on higher education and (why) does it matter? J Eur Integr 37(2):263–279
19. Di Franco G (2006) Corrispondenze multiple e altre tecniche multivariate per variabili categoriali. FrancoAngeli, Milano
20. Ennis GJ (1992) The social organization of sociological knowledge: modeling the intersection of specialties. Am Sociol Rev 57(2):259–265
21. European Commission (2000) Towards a European research area. COM(00)6 final of 18.01.2000
22. European Commission (2007) The European research area: new perspectives. COM(07)161 final of 04.04.2007
23. Fleck C, Duller M, Karády V (2018) Shaping human science disciplines. Institutional development in Europe and beyond. Palgrave Mcmillan, Cham
24. Gerli M (2017) Il campo sociale dei progetti di ricerca europei. Il caso delle SSH. Studi Culturali 1:127–150
25. Gerli M (2018) Looking through the lens of social sciences: the European Union in the EU-funded projects reporting. In: Iezzi DF, Celardo L, Misuraca M (eds) JADT'18. Proceedings of the 14th international conference on statistical analysis of textual data. Universitalia, Roma, pp 300–311
26. Gerli M, Santoro M (2018) Gramsciology. Studiare gli studi gramsciani nel mondo a distanza. Studi Culturali 3:439–465
27. Gorski PS (2013) Bourdieusian theory and historical analysis: maps, mechanisms and methods. In: Gorski PS (ed) Bourdieu and historical analysis. Duke University Press, Durham, USA, pp 327–366
28. Guzzetti L (1995) A brief history of European union research policy. Publications Office of the European Communities, Luxembourg

29. Heilbron J (2013) The social sciences as an emerging global field. Curr Sociol 62(5):685–703
30. Heilbron J, Bancourt T, Timans R (2013) Understanding social the sciences and humanities in Europe. Serendipities 2(1):1–9
31. Hoening B (2017) Competing for status: dynamics of scientific disciplines in the European transnational field. Serendipities 2(1):90–106
32. Hoening B (2017) Europe's new scientific elite: social mechanisms of science in the European research area. Routledge, Abingdon and New York
33. Kaiser D (1998) A Mannheim for all seasons: Bloor, Merton, and the roots of the sociology of scientific knowledge. Sci Context 11(1):51–87
34. Kastrinos N (2010) Policies for co-ordination in the European research area: a view from the social sciences and humanities. Sci Public Policy 37(4):297–310
35. Kropp K (2013) Social sciences in the field of power—the case of Danish social science. Soc Sci Inf 52(3):425–449
36. Lamont M, Molnár V (2002) The study of boundaries in the social sciences. Ann Rev Sociol 28:167–195
37. Lebart L, Salem A, Berry L (1998) Exploring textual data. Kluwer Academic, New York
38. Ienna G, Santoro M (2016) Per una scienza delle relazioni internazionali in campo culturale. Studi Culturali 1:69–82
39. Loubère L, Ratinaud P (2014) Documentation IRaMuTeQ (0.6 alpha 3 – version 0.1). Available on: www.iramuteq.org/documentation
40. Menéndez LS, Borrás S (2000) Explaining changes and continuity in EU technology policy: the politics of ideas. In: Dresner S, Gilbert N (eds) Changing European research system. Ashgate Press, Aldershot, pp 28–51
41. Quinn A (2014) Theoretically accounting for television formats in the new. International Division of Cultural Labour. In: Miller T (ed) The Routledge companion to global popular culture. Routledge, New York, pp 23–35
42. Ratinaud P, Marchand P (2012) Application de la méthode ALCESTE à de "gros" corpus et stabilité des "mondes lexicaux": analyse du "CableGate" avec IraMuTeQ. Actes des 11eme Journées internationales d'Analyse statistique des Données Textuelles. Liège, Belgique, pp 835–844
43. Reinert M (1987) Classification descendante hiérarchique et analyse lexicale par contexte: application au corpus des poésies d'Arthur Rimbaud. Bulletin de Méthodologie Sociologique 13:53–90
44. Rosamond B (2015) Performing theory/theorizing performance in emergent supranational governance: the live knowledge archive of European integration and the early European commission. J Eur Integr 37(2):175–191
45. Ruberti A, André G (1995) Uno spazio europeo della scienza. Riflessioni sulla politica europea della ricerca. Giunti, Firenze
46. Schögler RY, König T (2017) Thematic research funding in the European Union: what is expected from social scientific knowledge-making? Serendipities 2(1):107–130
47. Sismondo S (2010) An introduction to science and technology studies, 2nd edn. Wiley-Blackwell, Oxford
48. Star SL, Griesemer JR (1989) Institutional ecology, 'translations' and boundary objects: amateurs and professionals in Berkeley's museum of vertebrate zoology, 1907–39. Soc Stud Sci 19(3):387–420
49. Steinmetz G (2017) Field theory and interdisciplinary: history and sociology in Germany and France during the twentieth century. Comp Stud Soc Hist 59(2):477–514
50. Swartz D (1997) Culture & power. The sociology of Pierre Bourdieu. University of Chicago Press, Chicago
51. Vauchez A (2015) Methodological Europeanism at the cradle: Eur-lex, the acquis and the making of Europe's cognitive equipment. J Eur Integr 37(2):193–210
52. Vincent-Lancrin S (2006) What is changing in academic research? Trends and futures scenarios. Eur J Educ 41(2):169–202

53. Vincent-Lancrin S (2009) Finance and provision in higher education: a shift from public to private? In: Higher education to 2030, vol 2. Centre for Education Research and Innovation: OECD, pp 259–284
54. Wang Y (2016) Homology and isomorphism: Bourdieu in conversation with new institutionalism. Br J Sociol 67(2):348–370
55. Whittaker J (1989) Creativity and conformity in science: titles, keywords and co-word analysis. Soc Stud Sci 19(3):473–496

Topic Modeling of Twitter Conversations: The Case of the National University of Colombia

Eliana Sanandres, Raimundo Abello, and Camilo Madariaga

Abstract Topic modeling provides a useful method for finding symbolic representations of ongoing social events, which represent claims about the shape of social reality, its causes and the responsibility for action such causes imply. It has received special attention among social researchers in the last decade. During this time, Twitter has acted as the most common platform for people to share narratives about social events. This study proposes Latent Dirichlet Allocation (LDA) based topic modeling of Twitter conversations to determine the topics shared on Twitter about the financial crisis in the National University of Colombia. We downloaded all tweets that included the hashtag *#CrisisUNAL* (UNAL is the Spanish acronym for National University of Colombia) using the Twitter API interface and analyzed over 45,000 tweets published between 2011 and 2015. The results illustrate the strength of topic modeling for analyzing large text corpora and provide a way to study new emerging information that people share on Twitter.

1 Introduction

This chapter presents a way to analyze large amounts of textual data from Twitter conversations in an efficient and effective way. Specifically, we explain how to capture the narratives that people share on Twitter about social events, reduce their

E. Sanandres (✉)
Department of History and Social Sciences, Universidad del Norte, Barranquilla, Colombia
e-mail: esanandres@uninorte.edu.co

R. Abello
Directorate of Research, Development and Innovation, Universidad del Norte, Barranquilla, Colombia
e-mail: rabello@uninorte.edu.co

C. Madariaga
Department of Psychology, Universidad del Norte, Barranquilla, Colombia
e-mail: cmadaria@uninorte.edu.co

D. F. Iezzi et al. (eds.), *Text Analytics*, Studies in Classification, Data Analysis, and Knowledge Organization,
https://doi.org/10.1007/978-3-030-52680-1_19

complexity and provide plausible explanations. This research concern has received special attention among social researchers in the last decade [5, 15, 20, 26, 28, 32].

We propose an Latent Dirichlet Allocation (LDA) based topic model to address this challenge. Topic modeling is a useful approach to analyze large amounts of textual data by identifying the latent topics, which index discursive environments or frames that define patterns of association between a focal issue and other constructs [8]. These patterns of association are to be interpreted as symbolic representations of ongoing social events representing claims about the shape of social reality, its causes, and the responsibility for action such causes imply [2]. In this study, we use topic modeling to illustrate the symbolic representations of a social event on Twitter, in this case, the financial crisis in the National University of Colombia.

Over the last decade, Colombian academics and representatives of the government have recognized that the limitations of their budgets are the major constraint in the response of public universities to the increasing demands of society. To face this problem, the government proposed to reform the entire system of higher education [22]. The intention was to find new sources of money for higher education, enable more people to attend college, encourage transparency and good governance in the education sector, and improve the quality of higher education. One of the most controversial changes proposed by the government was the opening of the education sector to private investment by for-profit companies [10]. This was immediately rejected by public universities, who claimed that the proposed reform would lead to a full-scale privatization of the system of higher education [29].

At the National University of Colombia, the largest public higher education institution in Colombia, some students and professors claimed that such reform offered no clear solution to the financial crisis in the university. They explained that the university had been using a funding model with its sources of support mixed between the state and external resources, claiming that since 2004 this model had borne dwindling state support and ever-increasing costs to be covered by external resources. They showed that government transfers had decreased from 70% in 2004 to 64% in 2013, while the external resources produced from activities such as tuition fees, non-formal education courses and academic extension services, among others, had increased from 30 to 36% in the same period [25]. This statement reopened the debate on the financial crisis in the National University of Colombia and became a Twitter trending topic with the hashtag *#CrisisUnal* (UNAL is the Spanish acronym for the name of the National University). We analyzed over 45,000 tweets about the financial crisis in the National University of Colombia identified with the hashtag *#CrisisUnal* and illustrated the strength of topic modeling for the analysis of large text corpora as a way of studying narratives shared on Twitter.

It may be asked why should we devote attention to Twitter conversations on this topic? However, it should be considered that Twitter conversations offer clues about what the university's population is thinking and doing regarding the crisis. The use of Twitter in the National University of Colombia does not differ from other universities around the world. It is commonly used to share and communicate university news; to invite the academic community to engage in social activities [3], to spread updates about conferences, courses and scholarships, keeping users updated and fostering academic involvement, and it has also become a platform to publicize the academic offer, as well as to divulge cultural programs and university services [7, 12].

Among other social networks used in universities, Twitter stands out because it facilitates the dissemination of information, stimulates didactic interactions and energizes communication within the university's community [9]. Recently, Prieto and Rubio [27] have emphasized that Twitter's efficacy for the production and transmission of real-time information enables the generation of dialogs and the construction of virtual learning communities. In addition, it allows the analysis of communication networks which are created between users who share messages identified under a specific hashtag.

A central advantage of using Twitter for analyzes is that it covers topics in real time, producing a large amount of data that can be used to look at people's perceptions and narratives regarding to particular events. Twitter also provides a practical way to examine collective experiences related to a topical event, to study behaviors and attitudes where social desirability bias may occur in official surveys, and to collect large amounts of data with a limited budget [17]. Twitter conversations also illustrate the views of the reading public and show dominant viewpoints, which emerge quickly and are difficult to change [31].

2 Method

2.1 Data Collection and Pre-processing

The data collected for topic modeling corresponds to the published tweets between 2011 and 2015 that contained any reference to the financial crisis in the National University of Colombia identified with the hashtag *#CrisisUNAL*. Our collection formed a corpus of over 45,000 tweets.

The corpus was analyzed using topic modeling, a technique that facilitates the identification of specific subjects or latent topics that structure a document corpus [4, 23]. Its application implies the use of generative probabilistic algorithms, which enable the analysis of the probability distribution of the words that shape the original texts. Using this information, frequency patterns and correlations are determined for discovering the latent topics underlying the texts [30]. Some of these algorithms are: Probabilistic latent semantic analysis, Latent Dirichlet Allocation (LDA), Pachinko allocation, hierarchical LDA, among others.

In this study, we used LDA, which assumes that there is a set of topics to be found in a collection of documents. The intuition behind LDA is that documents exhibit multiple topics which are formally defined as a distribution of words over a fixed vocabulary [4]. To run the LDA model, we prepared the data in three steps:

Step 1: Converting the corpus into a matrix

Consider that the corpus size is 45,000 documents. The first step was to convert this corpus into a matrix D (document-term), with $d \times t$ dimensions, where each document or tweet is defined by a set of weighted frequencies of term i on document j. While building this matrix, these settings were made:

- All the text was standardized using lowercase
- Numeric characters and punctuation marks were removed
- All the words were reduced to their roots
- All the empty words or '*stopwords*' were removed
- Words with minimum 3 and maximum 30 characters were included
- URLs were removed

Step 2: Filtering the terms which do not provide information

In this step, term frequency—inverse document frequency (Tf-Idf) was used in order to filter the terms. This measure expresses how relevant a word is to a document, and it is used as a weighting factor in data mining; it allows filtering the words of the corpus that do not offer full information according to their frequency [18]. For its application, the average frequency of the terms was obtained and only those with a Tf-Idf value equal to or greater than the average was included. We only included terms with a Tf-Idf value of at least 0.1, which is a bit less than the median, to make sure that the most frequent terms were omitted.

Step 3: Determining the number of topics

The number of topics was defined by using the logarithmic probability ratio or log-likelihood (LL), a statistical test that automatically identifies a number k of topics and defines the likelihood that the corpus documents are generated by that number of topics. This criterion has been previously used in different topic modeling applications [13, 28, 32]. In accordance with this test, an optimal number of topics will be the one that is most likely to generate all the information in the corpus. The LDA model was then applied for each value of k, the maximum logarithmic probability values were identified and the harmonic mean of these values was calculated. We selected $k = 12$ as having the highest log-likelihood value (LL $= -198{,}000$).

2.2 Statistical Analysis

Once the data was prepared, we run the analysis in two steps:

Step 1: Running LDA

In this step, the LDA model was applied with the k value previously identified. This algorithm identifies the topics in two phases:

- Phase 1: Each word is assigned to a temporary topic. For the assignment, a belonging criterion to a Dirichlet distribution is followed, which allows a word that appears more than twice in different documents to be assigned to several subjects simultaneously.
- Phase 2 [Iterative]: Once the word is assigned, the algorithm reviews and updates the assignment of the topics, running one by one all the words in each document. For each word, the topic assignment is determined by two main criteria:
  - Criteria 1: How prevalent the word is across the topics?
  - Criteria 2: How prevalent the topics are on a document?

Step 2: Labeling the identified topics

Based on the researcher's criteria, each topic label was defined according to the content of its documents.

The aforementioned steps were performed using the R software version 3.2.2 with the next additional packages: 'SocialMediaLab' [14], 'topicmodels' [16], 'slam' [19], 'rmpfr' [21], 'tm' [11], 'stringr' [24].

3 Results

Table 1 displays the 12-topic solution and lists the 10 highest-ranking terms for each topic. We call attention to four sets of topics: Six topics concerned with social protest, three topics on educational reform, two topics calling for investment, and one topic emphasizing the role of the National University of Colombia in the Colombian peace process. To more easily interpret the topics, after reviewing the list of terms, we examined those tweets that exhibited each topic with the highest probability.

3.1 Protest Topics

Protest topics are the focus of the Twitter conversations on the financial crisis in the National University of Colombia. Topic 1 covers the protests of the education workers. The most highly ranked terms were '*sintraunal*' (the labor union covering all workers at public universities), '*protest*', '*strike*', '*campus*', '*riot*', '*gas*', '*blocked*' and '*wall*'. The tweets in which this topic was strongly represented locate protests in national and international contexts with terms like '*nation*' and '*clacso*' (Latin American Council of Social Sciences), indicating that the protests were a matter of concern in Colombia and in Latin America. Topic 3 also refers to the protests of the education workers. Some of the top words are '*sintraunal*', '*gases*', '*wall*', and '*block*'. This topic frequently exhibits tweets that show negative aspects of protests, such as '*confrontation*', '*death*' and '*bombs*'.

Table 1 12-topic solution

Topic 1	Topic 2	Topic 3	Topic 4	Topic 5	Topic 6
sintraunal	agricultural	sintraunal	agrarian	solidarity	no to the
protest	strikes	gases	protests	no to the	reform
strike	graffiti	wall	movement	reform	universities
campus	hate	block	mobilization	justice	listen
riot	block	undefined	participation	march	sciences
gas	bombs	bombs	people	respect	confrontation
blocked	terrorists	hood	bombs	charge	media
wall	crash	criticism	poor	help	classrooms
nation	delinquents	death	assembly	block	abandoned
clacso	guevara	confrontation	disturbance	upedagogica	mobilization
				studying	block
Topic 7	**Topic 8**	**Topic 9**	**Topic 10**	**Topic 11**	**Topic 12**
defend	no to the	stamp	intimidation	peace	revolutionary
university	reform	demand	blocked	process	victory
improvement	propose	support	abandoned	mobilization	popular
campus	threat	public	public	research	campus
crisis	oblivion	university	eviction	studying	strike
infrastructure	save	strike	strike	participation	eviction
cement	closed	resources	che	talks	denounce
hospital	blocked	deserve	graffiti	intellectuals	deserve
architecture	abnormality	financial	protest	solidarity	abandonment
sociology	upedagogica	pride	worker	civil	took
	uncertainty				

Topics 2 and 4 refer to the agricultural sector protests. While Topic 4 is related to the '*mobilization*' of '*people*' to take part in these '*protests*', Topic 2 emphasizes the participation of '*terrorists*' and '*delinquents*' in '*agricultural*' '*strikes*'. In this context, social protest is associated with the Argentine Marxist revolutionary Ernesto Che Guevara. Che is also mentioned in Topic 10, which deals with the protests of the working class and the intimidation of protesters. The most highly ranked terms in this topic are '*intimidation*', '*blocked*', '*abandoned*', '*public*', '*eviction*', '*worker*', '*strike*', '*che*', '*graffiti*' and '*protest*'. Finally, Topic 12 covers the revolutionary cause of social protest and includes the words '*revolutionary*', '*victory*', '*popular*', '*campus*', and '*strike*'.

With words such as '*protest*', '*agrarian*', '*confrontation*', and '*mobilization*', the aforementioned topics refer to the protest for the defense of education, but also for a fair and peaceful life for peasants and working class citizens. Its purpose, as mentioned in one of the tweets, aims to build a National University '*democratic and pluralistic, that is, the engine of change and transformation of Colombian society*'. One of the most frequent forms of protest is the strike, mentioned in messages which summon '*everyone to the #nationaluniversitystrike*'; there were also proclamations such as '*the countryside without peasants exists without existing,#Colombia resists*'; '*the students of #unal are tired of the lack of budget,the ruin of buildings and corrupt administrations*'; '*#unal students demand an open and deliberate* *campus*'; and '*at #unal all we hear is long live the national strike*'. The marches are also frequent in messages as '*torch march #unal #torches*'; '*torch march in defense of the university's budget*'; or '*feminist march in this moment at #unal*'. Other forms of protest are also mentioned, for instance, '*pupitrazo*', '*besatón*' and '*abrazatón*', where students gather massively to protest symbolically (carrying their chairs, with hugs or kisses, respectively). Finally, some tweets point to the presence of guerrilla groups in protest demonstrations, for example: '*Bolivarian movement today at #unal*'; '*hooded with farc-ep* (farc-ep stands for Fuerzas Revolucionarias de Colombia - Ejercito del Pueblo, a former guerrilla group in the Colombian conflict) *flags line up in the #unal central square*'; or '*bracelets were distributed today by the bolivarian movement hooded around the world-class campus*'.

3.2 Anti-reform Topics

Three topics deal with the reform of higher education proposed by the government. According to the terms included in Topic 5, public universities reject this reform and called for justice and respect; terms in this topic include '*solidarity*', '*no to the reform*', '*justice*', '*march*', and '*respect*'; tweets representing this topic show strong solidarity among public universities. Topic 8 is also related to the rejection of the planned educational reform to save public education; this includes terms like '*no to the reform*', '*propose*', '*threat*', '*oblivion*', and '*save*'. In the same way, Topic 6 indicates that public universities reject the reform of higher education, mobilize to denounce the government's abandonment and demand to be listened to. Some of the words in this topic are: '*no to the reform*', '*universities*', '*listen*', '*sciences*', '*confrontation*', '*media*', '*classrooms*', '*abandoned*', '*mobilization*' and '*block*'.

Topic 5 particularly calls for solidarity with the fight for the defense of the university and the interests of working class members. In this context, it is frequent the use of words such as '*solidarity*', '*truth*', '*march*', '*noalaley30*' (no to the 30 law, which states with the reform), '*volunteer*', among others. This same purpose operates as a justification for different manifestations of protest, including stones throwing and blockages. Among the tweets that address this topic is: '*#unal solidarity*'; '*strength and solidarity with the partners of the #unal*'; '*#MeAndMyIllusionThatSomeDay worker's fights join together,unity and solidarity with @ramocolombia and #unal workers*'; '*I invite you to give RT just as a small sample of solidarity with people who work as a few others for the #unal #workdignityatunal*'; or '*solidarity with the #unal students injured yesterday during protests against the #TLC*'.

3.3 Investment Topics

Topics 7 and 9 cover demands for investment to face the crisis. Topic 7 calls for infrastructure investment. Many tweets in which this topic is prominent focus on the '*infrastructure crisis*' of the '*campus*' buildings, in particular, the '*sociology*' and '*architecture*' buildings and the university's '*hospital*'. The top terms in this topic include '*defend*', '*university*', '*improvement*', '*campus*', '*crisis*', '*infrastructure*', '*cement*', '*hospital*', '*architecture*' and '*sociology*'. Topic 9 plays a similar role in investment demands focusing on the pro-National University of Colombia '*stamp*', created to acquire '*financial resources*' to improve the university facilities. Some tweets containing this topic highlight the role of the University as a national '*pride*'. The top-ranked terms include '*stamp*', '*demand*', '*support*', '*public*', '*university*', '*resources*', '*financial*', '*strike*', '*deserve*' and '*pride*'.

Topic 7 refers to the infrastructure crisis in the National University and presents it as a reflection of the public university crisis in Colombia. Some of the most representative tweets for this topic are: '*one thing is to say that the public university collapses, but it is the last straw to have to say that literally #unal collapses*'; '*nice to take classes outside the cloister to the streets but I didn't think it was the way students in #unal do today,due to lack of buildings*'; '*my university is so good that it rains even inside the buildings #unal*'; '*financial crisis due to government abandonment forces #unal to sell services to survive*'; and '*hopefully Mr. rector @mantillaignacio and @juanmansantos watch the news about the state structure of the #unal*'.

Associated with topic 7, is also the hashtag '*#siestampillaunal*'; which refers to the Pro Stamps National University Law and other Colombian state universities, created as a parafiscal contribution in 2013 for the strengthening of state universities, in particular, for the construction, adaptation and modernization of universities infrastructure [6]. This became a national trend and generated optimism among users, while receiving widespread support through messages such as: '*we do not want that the day to pay attention to the #unal infrastructure is because a student was injured #siestampillaunal*', '*from #unal in valle del cauca inviting support #siestampillaunal for infrastructure of the first public research university*', '*#siestampillaunal because #unal has excellent quality professors and students,but they require infrastructure to work and study*'. Nevertheless, some messages show skepticism about the scope of this law, for example: '*the stamp does not solve serious problems of infrastructure and democracy*'.

3.4 Peace Topic

Topic 11 represents the integration of the crisis in the National University of Colombia into a broader frame of national concern associated with the Colombian peace process. The top-ranked terms are '*peace*', '*process*', '*mobilization*', '*research*', '*studying*', '*participation*', '*talks*', '*intellectuals*', '*solidarity*', and '*civil*'. Tweets in which this topic was strongly represented are related to the role of the university as facilitator in peace talks among the government, rebel groups involved in the Colombia's internal armed conflict (which began more than 50 years ago and is currently in negotiation, in a process known as the Colombian peace process), intellectuals, and representatives of civil society.

Words like '*resistance*', '*solidarity*', '*reinventacolombia*' (reinventing Colombia) and '*procesodepaz*' (peace process) come together in a frame of reference that indicates a common interest in building a university for peace. In this context, messages of resistance to the intervention of groups trying to destroy the university are frequent, for example: '*a real cry of protest in the #unal arises when students face the hooded who destroy it*'. It is also common to find signs of solidarity with the interests of the university, for example '*#yomepongolaruana* (I wear the ruana) *and as a #unal student I put on the gown for a university hospital that wants to demolish us @angelamrobledo @manecolombia*'; '*The fight belongs to everyone... support us #unal #udea #itm #uptc #univalle #unipamplona #up #unicauca*'. Within this environment of resistance to external intervention and solidarity for the defense of the interests of the university, interest in building a university for peace is common in messages such as: '*#unal to preserve the peace campus,no to violence: the path is the peace*'; '*#unal accompanies peace process from the academy*'; '*we are preparing the fight to advance in the construction of a university for peace @feucolombia @manecolombia #unal*'; '*It is always desirable to end the war*'.

4 Conclusions

Producing an interpretable way to study Twitter conversations efficiently and effectively is only the beginning. The solution of this issue presents meaningful categories to address the analytic question that motivated the study: *How was the financial crisis in the National University of Colombia framed on Twitter?* The 12-topic solution showed that it was framed through four categories: protest, anti-reform, investment and peace.

Each topic constitutes a frame, in that it includes terms calling attention to particular ways in which the crisis under study may arouse controversy: protest frames emphasize public displays, demonstrations and the civil disobedience of the working class; anti-reform frames refer to the rejection of the proposed reform for the higher education in public universities; investment frames focus on investment demands to face the crisis; and the peace frame draws attention to the role the National University of Colombia played in acting as a facilitator in the Colombian peace process. Each of these frames represents a discursive environment for the financial crisis, which broadcasts not just the structural characteristics of the crisis (investment demands and education reform), but also symbolic representations of ongoing social events (workers protests and peace process), which can be seen as claims about ongoing social processes and demands of reparation.

The discussion about the financial crisis in the National University of Colombia on Twitter became a trending topic since 2011, when the government proposed to reform the law 30 of 1992 that would modify the financing scheme of higher education in Colombia. At that time, the contrary reactions and protests from the National University of Colombia did not wait, especially in the campus located in Bogota. The speech focused on denouncing the financing problems, the infrastructure crisis and the labor conflict that had been affecting the University for years and that, for some members of the academic community and officials of the institution, did not have a clear solution in the reform proposed. In fact, the directives of the institution were blunt in stating that the de-financing of the university, the infrastructure crisis and the labor conflict constituted 'structural problems' that did not compromise their academic activity nor their commitments to society [1].

However, the generalized discourse through Twitter has progressively displaced those representations based on the structural conditions, giving priority to symbolic representations that the actors built on these events as threats to their academic activity and their mission to contribute to the resignification of the nation project.

These results provide substantive insight into Twitter conversations about the financial crisis in the National University of Colombia. Using LDA to discover topics allowed us to locate two narratives: One focused on the structural characteristics of the crisis and the other concerned with symbolic representations of ongoing social events surrounding that crisis. For cultural sociologists, this is only the beginning of the analysis. A topic model allows a starting point to be found, which in this case is the structure of Twitter data. Used properly, with appropriate validation, topic models are valuable complements to other interpretive approaches, offering new ways to extract topics and make sense of online data.

References

1. Agencia de Noticias UN (2013) La U.N. no está en crisis. Accessed https://agenciadenoticias.unal.edu.co/detalle/article/la-un-no-esta-en-crisis.html. 18 Dec 2015
2. Alexander J (2004) Toward a theory of cultural trauma. In Alexander J, Eyerman R, Giesen B, Smelser N, Sztompka P (eds) Cultural trauma and collective identity. University of California Press, California
3. Atkinson C (2009) The backchannel: how audiences are using Twitter and social media and changing presentations forever. New Riders, Berkeley
4. Blei D (2012) Probabilistic topic models. Commun ACM 55(4):77–84
5. Chandelier M, Steuckardt A, Mathevet R, Diwersy S, Gimenez O (2018) Content analysis of newspaper coverage of wolf recolonization in France using structural topic modeling. Biol Cons 220:254–261
6. Congreso de Colombia (2013) LEY No. 1697 20 DIC 2013 Por la cual se crea la estampilla pro Universidad Nacional de Colombia y demás universidades estatales de Colombia, Bogotá
7. Curioso W, AlvaradoVásquez E, CalderónAnyosa R (2011) Usando Twitter para promover la educación continua y la investigación en salud en el Perú. Rev Peru Med Exp Salud Publica 28(1):163−164
8. DiMaggio P, Nag M, Blei D (2013) Exploiting affinities between topic modeling and the sociological perspective on culture: application to newspaper coverage of US government arts funding. Poetics 41(6):570–606
9. Ebner M, Lienhardt C, Rohs M, Meyer I (2010) Microblogs in higher education—a chance to facilitate informal and processoriented learning? Comput Educ 55(1):92–100
10. El Espectador (2011) Universidades con ánimo de lucro, apuesta del gobierno, Mar 10
11. Feinererm I, Hornik K (2018) Package 'tm'. Available via: https://cran.r-project.org/web/packages/tm/tm.pdf. Accessed 3 Dec 2019
12. Fields E (2010) A unique Twitter use for reference services. Library Hi Tech News 27(6/7):14–15

13. Fligstein N, Brundage JS, Schultz M (2014) Why the federal reserve failed to see the financial crisis of 2008: the role of "Macroeconomics" as a sense making and cultural frame. IRLE working paper no 111–14
14. Graham T, Ackland R (2016) Package 'SocialMediaLab'. Available via: https://vosonlab.net/papers/SocialMediaLab/SocialMediaLab.pdf. Accessed 20 Aug 2019
15. Griffiths T, Steyvers M (2004) Finding scientific topics. Proc Nat Acad Sci 5228–5235
16. Grün B, Hornik K, Blei D, Lafferty J, Phan X, Matsumoto M, Nishimura T, Cokus S (2019). Package "topicmodels". Available via: https://cran.r-project.org/web/packages/topicmodels/topicmodels.pdf. Accessed 10 Dec 2019
17. Himelboim I, McCreery S, Smith M (2013) Birds of a feather tweet together: integrating network and content analyses to examine cross-ideology exposure on Twitter. J Comput Mediated Commun 18(2):154–174
18. Hornik K, Grün B (2011) Topicmodels: an R package for fitting topic models. J Stat Softw 40(13):1–30
19. Hornik K, Meyer D, Buchta C (2019) Package "slam". Available via: https://cran.r-project.org/web/packages/slam/slam.pdf. Accessed 10 Dec 2019
20. Kovanović V, Joksimović S, Gašević D, Siemens G, Hatala M (2015) What public media reveals about MOOCs: a systematic analysis of news reports. Br J Educ Technol 46(3):510–527
21. Maechler M (2019) Package "Rmpfr". Available via: https://cran.r-project.org/web/packages/Rmpfr/Rmpfr.pdf. Accessed 10 Dec 2019
22. Ministry of National Education (2010) Proposal for the education reform in Colombia, Apr 12
23. Mohr J, Bogdanov P (2013) Topic models: what they are and why they matter. Poetics 41(6):545–569
24. Müller K, Wickham H, James D, Falcon S (2019) Package 'RSQLite'. Available via: https://cran.r-project.org/web/packages/RSQLite/RSQLite.pdf. Accessed 10 Dec 2019
25. National University of Colombia (2014) Estadísticas e indicadores de la Universidad Nacional de Colombia, 19. ISSN 2357-5646
26. Newman D, Block S (2006) Probabilistic topic decomposition of an eighteenth-century American newspaper. J Assoc Inf Sci Technol 57(6):753–767
27. Prieto J, Rubio R (2018) Comunicación cultural de masas. El caso del Instituto Cervantes y su diplomacia digital a través de Twitter. Commun Soc 31(3):239–259
28. Sanandres E, Otálora J (2015) Application of topic modeling for trauma studies: the case of Chevron in Ecuador. Invest Desarrollo 23(2):228–255
29. Semana (2011) Reforma a la Ley 30: por qué sí, por qué no, Apr 1
30. Steyvers M, Griffiths T (2007) Probabilistic topic models. In: Handbook of latent semantic analysis, vol 427, no 7, pp 424–440
31. Xiong F, Liu Y (2014) Opinion formation on social media: an empirical approach. Chaos Interdisc J Nonlinear Sci 24(1):013130
32. Yang T, Torget A, Mihalcea R (2011) Topic modeling on historical newspapers. In: Zervanou K, Lendvai P (eds) LaTeCH '11 Proceedings of the 5th ACL-HLT workshop on language technology for cultural heritage, social sciences, and humanities, pp 96–104

Analyzing Occupational Safety Culture Through Mass Media Monitoring

Livia Celardo, Rita Vallerotonda, Daniele De Santis, Claudio Scarici, and Antonio Leva

Abstract In the last years, a group of researchers within the National Institute for Insurance against Accidents at Work (INAIL) has launched a pilot project about mass media monitoring in order to find out how the press deal with the culture of safety and health at work. To monitor mass media, the institute has created a relational database of news concerning occupational injuries and diseases, which was filled with information obtained from the newspaper articles about work-related accidents and incidents, including the text itself of the articles. In keeping with that, the ultimate objective is to identify the major lines for awareness-raising actions on safety and health at work. The hypothesis is that, for different kind of accidents, a different language is used by journalists to narrate the events. To verify it, news have been preprocessed, and several analyses have been implemented on these data, using automatic text analysis techniques; our purpose is to find language distinctions connected to groups of similar injuries. The identification of various ways in reporting the events, in fact, could provide new elements to describe safety knowledge, also establishing collaborations with journalists in order to enhance the communication and raise people attention toward workers' safety. At the end, the results obtained have confirmed the starting hypothesis, and they have also provided useful tools to understand the media communication related to safety in workplaces.

1 Introduction

The study described here grew out of the collaboration between the Department of Social Sciences and Economics of Sapienza University of Rome and the Headquarters for Research of INAIL (Italian National Institute for Insurance against Accidents at

L. Celardo (✉)
Tor Vergata University of Rome, Rome, Italy
e-mail: livia.celardo@uniroma2.it

R. Vallerotonda · D. De Santis · C. Scarici · A. Leva
INAIL Research—Headquarters for Research of the Italian National Institute for Insurance against Accidents at Work, Rome, Italy

D. F. Iezzi et al. (eds.), *Text Analytics*, Studies in Classification, Data Analysis, and Knowledge Organization,
https://doi.org/10.1007/978-3-030-52680-1_20

Work) where since 2012 a team of researchers has developed the idea of monitoring the mass media in view of prevention against accidents at work [5, 8]. The idea behind this project, raised within the Institute and developed through this collaboration, is based on the hypothesis that there is an information asymmetry between real events and media news, in terms of both qualitative and quantitative aspects.

In the interest of accident prevention, the institute decided to start a mass media monitoring in order to identify and correct the critical issues present in media communication. In the field of health and safety at work, the archiving and reasoned organization of news from the press or Web can make a contribution to the knowledge of the phenomenon, as in other domains. For this reason, operators, societies, and private associations have already undertaken experiences of news archives. These archives, often limited to specific territories or sectors, have so far been predominantly used for the purpose of quantification of the phenomenon. However, it is important to know how the topic is treated and, consequently, the level of safety culture within the descriptions.

With this in mind, the institute has realized the so-called "Repertorio Notizie SSL" (*News Repository on Occupational Safety and Health*), that is a relational database of media news related to occupational injuries and diseases. In a first phase of this project, 1,858 news articles regarding 580 different accidents were collected; for each injury, not only the news texts but also several variables were identified.

The objective is to observe the culture of occupational safety and health communicated by mass media agencies in order to identify new elements for increasing prevention against accidents at work. Actually, it is well known that media communication affects people's perception, understanding, feelings, and values [11], and then, the way news are communicated can affect workers' behavior, encouraging or not more responsible attitudes. Other scholars have already used newspaper texts in order to analyze ideologies and opinions. To give some examples, Pollak et al. [12] have used news reports about the 2007 Kenyan election and post-election crisis in order to provide ideological differences between local and international press coverage, Balahur and Steinberger [1] explored sentiment analysis in newspaper articles, establishing guidelines for positive/negative opinions, and Fortuna et al. [7] investigated patterns in media through the observation of terms choices in different topics.

In this study, we focus on the hypothesis that there are some asymmetries in the language used to describe the injuries depending on the characteristics of the event. To test it, we performed on the repository data several automatic text analysis procedures.

The article is structured as follow: In section no. 2, the News Repository is presented; in section no. 3, data are presented and the methodology is exposed; in section no. 4, the results of the analyses are shown; in section no. 5, conclusions are drawn.

2 The Database

News Repository on Occupational Safety and Health (NeRO) is a tool created to allow analyses of news contents and texts related to occupational diseases and injuries. It arose from a previous collaboration with the University of Bologna, concerning the implementation of a database about agricultural accidents, in a prevention perspective. Actually, for agriculture, the phenomenon is not entirely covered by official databases, which do not deal with events occurred to hobbyists, assistants, and family helpers. We then decided to extend the information gathering to all economic sectors and create a new database release containing additional technical specifications, guessing that NeRO utility does not lie so much in identifying accidents not recorded by official databases, as rather in detecting how (and if) news is presented. In fact, our strategic objective is to increase public awareness and safety culture through a different approach which will be also based on the study of news articles, their composition, and communication dynamics. So, the first operational purpose is to understand:

- which kind of terms are used in news articles about accidents at work or occupational diseases;
- what inspires a title;
- how the same news is treated by different sources/media;
- how the news text could be interpreted in different ways due to who communicates the news itself;
- whether or not some specific aspects of the events are considered by media.

Our study plans to analyze the cultural characteristics of mass media communication regarding occupational safety and health (OSH), observing the attitude of mass media (and journalists) toward the subject and the way users perceive the news depending on which words are used.

2.1 Portrayal of the Tool

NeRO is an ad hoc relational database, centered on the gathering of newspaper articles regarding accidents at work, but it is also arranged to gather news on near misses, occupational diseases, and incidents from all kind of sources (press, television, or radio). It involves several digital interconnected tables related to different entities (Table 1), which contain structured (i.e., based on appropriate classifications) and unstructured (i.e., textual) information.

From the conceptual scheme of entities to be entered, attributes, and relations, a first release was produced in ANSI C++ language. It ran on all the major platforms (Windows, OS X, and Linux), but it did not provide some features that would have been useful for data entry and analyses. We achieved an access release, thereby overcoming those problems. Information retrieval regards events happened

Table 1 Entities of the database NeRO, description of the information available for each entity

Entity	Description
Event	Specific accident or occupational disease drawn from the news. It constitutes the central entity of the database, as well as the unit of analysis
News	Newspaper article related to one or more *Events*
Person	Person involved in a work accident or occupational disease
Injury	Type of injury
Disease	Type of disease
Legal personality	Company where an "Injury" took place, or a "Disease" was developed
Activity	Economic activity of the company
Location	Geographical location of the "Event"

in Italy, and it could contain both online and directly consulting newspapers, since we exploited Google Alert Service (using some suitable keywords) and a daily-newspaper subscription ("la Repubblica"). The reference unit is the event (right now, we are restricting events to accidents), and different aspects and information are linked to it: one or more articles about it, one or more workers injured, and so on. The data-entry interface consists of a series of thematic screens, starting from the opening one, which covers the list of already recorded events. These screens allow to enter the following data, step by step:

- [Screen "Event"] Text containing event report (written down taking worthwhile information from all the articles about that event), date of the event, venue, company where accident occurred (if appropriate), economic sector (according to ATECO 2007 classification);
- [Screens "News"] Texts of each article about the event, its source (newspaper name or press affiliation), article title, Web URL, date of the article to be compared to the event one;
- [Screens "Worker" and Sub-screens "Accident" and "Harms, disorders or diseases"] Injured worker's biographical data, information about accident (using all the variables of ESAW classification, European Statistics on Accidents at Work), type of injury, physical implication or resulting disease (according ICD 10 classification).

3 Methodology and Data

The data collection is resulted from a pilot phase of the project, and it was created reporting articles about occupational accidents occurred in five Italian regions (Apulia, Lazio, Campania, Sicily, and Emilia-Romagna). The news have been selected from the Web sites of the national and local newspapers using the *Google*

Table 2 Dataset variables, extracted from the NeRO repository

Labels	Meaning
Year	When the accident happened
Newspaper	In which journal the article was published
Article	Text of the news
Event	Accidents listed by sequential numeration
Area	Area of worker's residence
Economic activity	Economic activity where the worker is employed by the NACE Rev. 2 classification
Place	Where the injury took place
Gender	Worker's gender
Nationality	Worker's nationality
Fatality	Fatal accident or not
In-itinere	Commuting accident or not
Road	Road accident or not
Motivation	Cause of the injury

Alert tool with the following keywords (both singular and plural forms): *work accident*; *work injury*; *white death*; *on-the-job fatality*.[1]

At the end of the data collection, the repository was composed of 1858 articles regarding 580 different work accidents. From the repository, we extracted and merged the tables related to events and news, creating a single dataset composed of structured and unstructured information (Table 2).

On articles and events data, we involved different analyses. Initially, we analyzed some event characteristics—economic activity, fatality, ongoing and road accident variables—comparing their distributions with official data, in order to examine if mass media give the same relevance to all the accidents. After that, we focused our attention on the news, analyzing them through several text mining methods [2, 6, 10]. In order to analyze the communication features of mass media concerning our topic, we explored by way of textual analysis many aspects of the language used by the journalists to narrate the events. We started calculating some lexical indexes, i.e., the type token ratio and the hapax percentage, in order to have an overall view on the corpus from a linguistic perspective. Then, we focused our attention on the bi-grams, applying on the corpus a sequence analysis (also known as Markovian analysis), in order to investigate sequences of words. We did it using the T-LAB software [14]. This method allows to calculate for all the predecessors and successors of each lexical unit the transition probabilities between pairs of words (also known as Markov chains). Next, we implemented a content analysis for discovering the subjects treated in news using the Reinert's method [13], and it is a descendant hierarchical classification algorithm implemented within the IRAMUTEQ software

[1]For further information about the news collection procedure, see: INAIL [8], *Il monitoraggio dei mass media in materia di salute e sicurezza: Strumenti per la raccolta e l'analisi delle informazioni.*

that, using the words co-occurrences matrix, produces groups of similar lexical units. Finally, we implemented on the term-document matrix a co-clustering procedure [3, 4], the aim was to identify if there are language specificities [9] connected to some event characteristics. The results will show how the language used by journalists depends on some event characteristics, and then, to better understand the differences between news texts, we analyzed the specificities related to the modalities of one specific variable, regarding the fatality of the accidents.

4 Main Results and Discussion

As previously mentioned, in the first part of the analysis, we considered the events connected to the news. We focused on four variables, comparing the data from the repository with the INAIL official statistics of accidents reports (Figs. 1, 2 and 3).

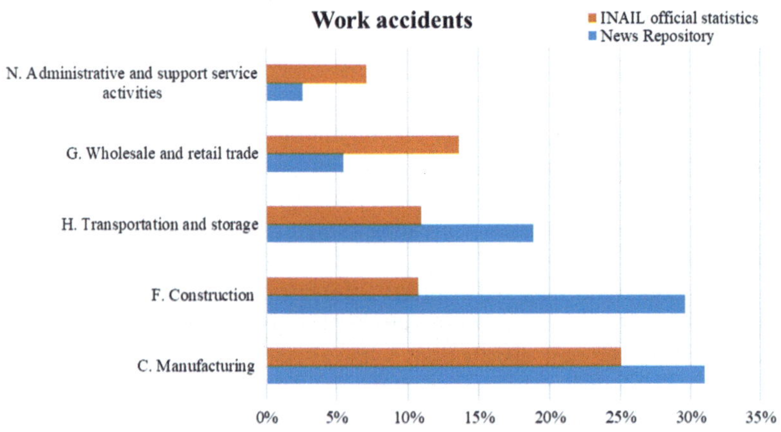

Fig. 1 Work accidents by economic activity, first five sectors. *Source* Our elaboration

Fig. 2 Work accidents by mortality. *Source* Our elaboration

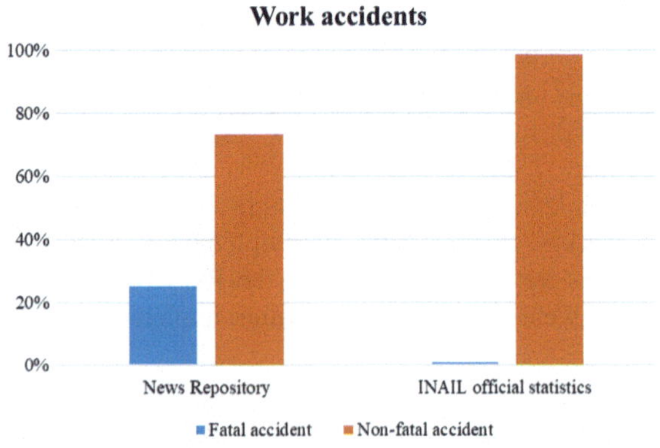

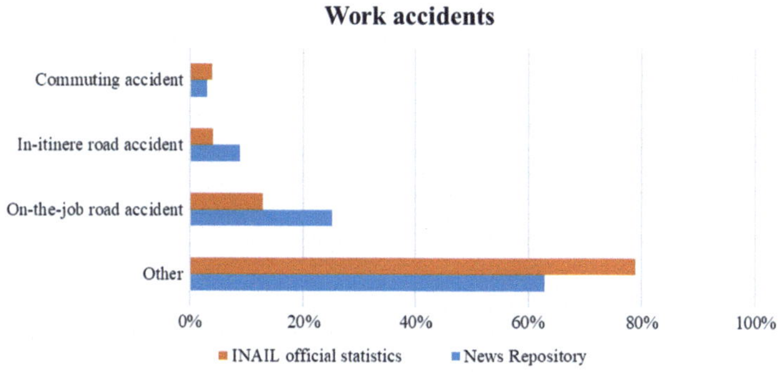

Fig. 3 Work accidents by situation. *Source* Our elaboration

The figures show that there is an asymmetry between accidents reports and the events described by mass media. Through the comparison, we observed that newspapers tend to focus on certain accidents rather than others: road accidents, on-the-job fatalities, and those economic activities perceived more dangerous for workers—like construction or manufacturing—are associated with a higher level of journalists' attention.

In the second part of our analysis, we worked on the news texts; from the collection of the 1858 articles, we obtained a corpus—since now called NeRO corpus—of medium dimension, composed of 326,938 occurrences and 14,057 types (Table 3). In order to check whether it was possible to statistically process data, two lexical indicators were calculated: the type-token ratio (TTR: 4.3%) and the hapax percentage (HP: 33.89%). According to the large size of the corpus, both lexical indicators highlighted its richness and indicated the possibility to proceed with the analysis.

After lemmatization, hapax, and stop-words removal (articles, numbers, auxiliary verbs, conjunctions, prepositions and pronouns), the number of types has fallen to 7074. Then, we removed also the words not recognized by the dictionary, obtaining a set of 4183 keywords.

The next step was to apply a sequence analysis to the corpus. We selected several words in order to analyze their predecessors and successors; we reported only the chains with transition probability higher than 0.1 (Table 4). In the results, it is not important to focus on predecessors rather than successors (and vice versa), but the objective of this analysis is the association, i.e., the closeness, between pairs of words, independently from the order.

With this technique, we investigated connections between some selected words and the others. Specifically, we concentrated on terms like fatality, tragic, accidental, to find if accidents have been communicated by media like something that is

Table 3 Main lexical indexes for the NeRO corpus

Documents	Occurrences	Types	No. of Hapax
1858	326,938	14,057	4763

Table 4 Sequence analysis, Markov chains

Terms	Predecessor	Successor	Probability
Fatality	Tragic		0.40
Tragic		Destiny	0.30
Accidentally	Equipment		0.20
Accidentally	To slip		0.20
Terrible	Misfortune		0.17
Tragic		Event	0.15
Distraction	Instant		0.13

predictable, or not. As shown in the table, there are many significant chains; in fact, the association of specific words in the texts—e.g., tragic fatality, tragic destiny, an instant of distraction—transmits the idea that work injuries are random, fortuitous, and most important, something we could not prevent.

To identify the main subjects of the corpus, we performed a content analysis using the Reinert's method for descending hierarchical classification. We implemented this method using as unit analysis both the segments and the texts. The implementation on the segments produced four overlapped clusters, while in the other case, we identified three well-separated groups. Then, using as unit analysis the texts, the algorithm detected three clusters of words (in order, the red, the blue, and the green ones), identifying distinct contents within the corpus (Fig. 4):

1. EVENT (56.5%). In this group, the largest one, we found all those terms that are connected to the accident descriptions;

Fig. 4 Result of cluster analysis, Reinert's method. *Source* Our elaboration

Table 5 Row clusters evaluation for the NeRO corpus, Calinski–Harabasz index

N. clusters	2	3	4	5	6	7	8	9
CH index	**32.11**	23.42	**26.83**	21.49	20.97	20.26	13.58	11.10

Bold indicate optimal number of clusters

2. ROAD (26.5%). This cluster identifies the focus given in the media news to road accidents by journalists. This group is very interesting, because it confirms the relevance given to injuries that took place on the road;
3. RITE (17%). In this, groups are contained with those terms related to the ritual aspects connected to events like accidents, where the emotive part is predominant.

As previously mentioned, with the preprocessing phase, we obtained a term-document matrix of dimension (4183 × 1858). The matrix was then weighted using the specificities method, so we performed the double *k*-means algorithm on several combinations for rows and columns—from (2 × 2) to (9 × 9). On these combinations, we calculated the Calinski–Harabasz index for identifying the optimal combination of clusters. For both documents and terms, the CH index suggested the solutions of two and four clusters (Table 5). Then, we analyzed the centroid matrix of the two combinations—(2 × 2) and (4 × 4)—finding that the second one was the most interpretable.

The centroid matrix calculated on the NeRO corpus (Table 6) shows that the first cluster of words is not related to any specificity with the document clusters; it means that this group contains the general terms, i.e., those words that are present in all the documents (Table 7). In fact, in this cluster, we found those terms that are connected to the description of the events, e.g., place, tragedy, worker, factory, to happen, installation, rescuer. The second cluster of terms presents a relationship with the third cluster of documents; the words characterizing this group are related to the description of road events (road, kilometer, onboard, motorway, crossroad, etc.), and the connected column cluster contains mostly news of non-fatal road accidents (Table 7). The third-row cluster shows an interconnection with the second cluster of documents; the terms belonging to this group (hospital, assistance, ambulance, doctor, trauma, to intervene, etc.) regards the narration of the rescue and the first aid, and the document cluster covers most of the news about non-fatal accidents.

Table 6 Centroid matrix (terms × documents)

	Cluster 1 (43%)	Cluster 2 (36%)	Cluster 3 (17%)	Cluster 4 (4%)
Cluster 1	0.0163	0.0062	0.0058	0.0073
Cluster 2	0.0235	0.0002	**0.2406**	0.0821
Cluster 3	0.0414	**0.1533**	0.0196	−0.0469
Cluster 4	0.0166	−0.0034	0.0116	**0.6378**

Bold indicate higher level of specificity in clusters

Table 7 Word clusters, first 25 words listed by frequency of occurrence

Cluster 1	Cluster 2	Cluster 3	Cluster 4
Resident	Accident	Man	Victim
Place	Firefighter	Year	To crush
Death	To occur	Hospital	Son/Daughter
Tragedy	Police	Laborer	Mortal
Province	Again	Assistance	Carriageway
To arrive	Road	To transport	Wife
Body	Kilometer	To intervene	To run over
Tragic	Wound	Ambulance	To kill
Operation	Wounded	Life	Unintentional
Strong	Young	Condition	Homicide
Worker	To involve	To injure	Biker
Service	Onboard	Injury	To arrest
House	Impact	Code	To track down
To cause	Motorway	Company	Automobile
Factory	Crossroad	Colleague	Omission
Serious	Direction	Verification	Minibus
Warehouse	Crash	To admit to hospital	Pirate
Minute	Lorry	Intervention	To stop up
To happen	Car	Cause	Escape
Civil	To extract	Trauma	Concert
Passing	Traffic	Safety	Scooter
Installation	Fiat	Doctor	Musical
Military	Driver	Construction site	Band
Family	Metal sheet	To reconstruct	Anterior
Rescuer	Chauffeur	To hit	To jump

The fourth cluster of words indicates an association with the fourth column cluster; the terms of this group concern mostly the death of the worker (to leave, mortal, to kill, homicide, etc.), and the documents connected to these words are mainly fatal road accidents. It is interesting to note that in this, last group of words reported those terms connected to the worker's family.

From the cluster analysis, we have seen that there are some interesting characterizations of the language used in newspapers. Some variables, like the fatality of the injury and the accident site, present a strong lexical differentiation among the modalities; this means that who is narrating the event, i.e., the journalist, uses a specific language to describe the accident, on the basis of these characteristics.

Finally, to better understand the lexical differences connected to the fatality of the event, we analyzed the specificities for this particular variable (Table 8).

Table 8 Analysis of the specificities for the variable "accident mortality"

Fatal accident—No	z = test-value	Fatal accident—Yes	z = test-value
Hospital	59.17	Tragedy	35.68
Serious	58.84	Family	27.17
To transfer	54.90	Useless	23.62
Dangerous	28.38	To leave	19.84
Rescue	24.13	Victim	18.68
Ambulance	24.09	Tragic	17.71
Leg	23.12	Friend	14.95
Injury	22.06	Band	14.89
Trauma	20.55	Condolence	12.65
Hand	18.84	Province	12.15
Fracture	16.70	Son	11.49
Helicopter	13.70	Wife	11.48
Bus	12.23	Escape	10.63
Crossroad	10.20	Mayor	9.11

Starting from the results showed in table, we can observe and confirm that there is a significant difference in the language utilized when the accident is fatal or not. The terms used in the case of a non-fatal event are related to the description of the injury, while in the case of a mortal accident, the situation is completely different: the words utilized refer to the emotional sphere of the event, so concepts like the family or the unpredictability are very often used to describe what was happened.

5 Conclusions

The project here presented showed how News Repository on OSH (NeRO) can contribute to analyze occupational safety and health, although in some institutions there are already databases dedicated to newspaper articles dealing with OSH. Actually, in addition to news texts, NeRO provides several systematized information, enabling to filter news according to various search criteria and, above all, to carry out a number of studies and organized analysis on textual data, too.

The analyses conducted in this first phase gave us many causes for reflection. What we found shows that if official statistics provide the quantification of the phenomenon, the mass media monitoring offers a different point of view connected to the construction of prevention culture in the occupational environment. The analyses implemented in this chapter showed that there are several asymmetries between accidents reports and media news, in terms of both quantitative and qualitative aspects. This knowledge acquisition has relevant implications, firstly because potential asymmetries could be reduced in order to improve prevention and safety in workplaces, maybe shrinking job-related accidents. If mass media have the power to create public opinion on specific issues, the importance of monitoring the information and encouraging truthful communication is very clear. This goal could be achieved by means of a strict collaboration with institutions and journalists, through the improvement of knowledge, capabilities, and awareness. The National Institute for Insurance against Accidents at Work is committed to improving safety knowledge and accidents prevention through the monitoring of the information.

Until now, a preliminary phase was carried on in order to identify the potentiality of the project; in the future, the monitoring will be systematic, and more advanced techniques will be implemented in order to have a continuous and constant update on mass media production.

References

1. Balahur A, Steinberger R (2009) Rethinking sentiment analysis in the news: from theory to practice and back. In: Proceeding of WOMSA
2. Bolasco S, De Mauro T (2013) L'analisi automatica dei testi: fare ricerca con il text mining. Carocci Editore.
3. Celardo L, Iezzi DF, Vichi M (2016) Multi-mode partitioning for text clustering to reduce dimensionality and noises. In Mayaffe D, Poudat C, Vanni L, Magri V, Follette P (eds) JADT 2016: statistical analysis of textual data. Les Press de Fac Imprimeur, pp 181–192
4. Celardo L (2017) Classifying textual data: a two-way approach. Working papers series-Ph.D. course in applied social sciences
5. Celardo L (2018) Opportunities of using big data in social sciences: work injuries through media analysis. Working papers series-Ph.D. course in applied social sciences
6. Feldman R, Sanger J (2007) The text mining handbook: advanced approaches in analyzing unstructured data. Cambridge University Press, Cambridge
7. Fortuna B, Galleguillos C, Cristianini N (2009) Detection of bias in media outlets with statistical learning methods. In: Text mining. Chapman and Hall/CRC, London, pp 57–80
8. INAIL (2015) Il monitoraggio dei mass media in materia di salute e sicurezza: Strumenti per la raccolta e l'analisi delle informazioni
9. Lafon P (1980) Sur la variabilité de la fréquence des formes dans un corpus. Mots 1(1):127–165
10. Lebart L, Salem A, Berry L (1997) Exploring textual data, vol 4. Springer Science & Business Media, Berlin
11. McLuhan M (1994) Understanding media: the extensions of man. MIT Press, Cambridge
12. Pollak S, Coesemans R, Daelemans W, Lavrač N (2011) Detecting contrast patterns in newspaper articles by combining discourse analysis and text mining. Q Publ Int Pragmatics Assoc 21(4):647–683
13. Reinert M (1983) Une méthode de classification descendante hiérarchique: application à l'analyse lexicale par contexte. Les cahiers de l'analyse des données 8(2):187–198
14. T-LAB (2017) User's manual T-LAB plus 2017. T-LAB. Retrieved 14 July 2019. https://tlab.it/en/download.php

What Volunteers Do? A Textual Analysis of Voluntary Activities in the Italian Context

Francesco Santelli, Giancarlo Ragozini, and Marco Musella

Abstract Over the past decades, the complex phenomenon of volunteering has been mainly analyzed within the field of economic literature with respect to its "economic value added," that is the capability of these activities to increase the productivity level of some specific goods or services. The paper adopts a different point of view, in that voluntary organizations are analyzed as places of innovation, where new jobs arise and people acquire new skills. Thus, volunteering can be understood as a "social innovation" factor. In order to gain a deeper insight into the types of voluntary works, we have used data coming from the Istat survey "Multiscopo, Aspetti della vita quotidiana" (Multipurposes survey, daily life aspects), released in the year 2013. In our textual analysis, we have utilized the information included in the open-ended question section provided by respondents regarding the description of the tasks performed individually as volunteers. After stemming, lemmatization, and cleaning, the data have been analyzed by means of community detection based on semantic network analysis, with the purpose of identifying job patterns, then by a clustering procedure on answers (Reinert approach to textual clustering), and lastly through correspondence analysis on generalized aggregated lexical tables (CA-GALT) with the aim to explore volunteers' profiles. In particular, we have singled out differences in gender, age, educational level, region of residence, and type of voluntary association.

F. Santelli (✉) · G. Ragozini · M. Musella
University of Naples Federico II, Naples, Italy
e-mail: fsantelli@units.it

G. Ragozini
e-mail: giancarloragozini@unina.it

M. Musella
e-mail: marcomusella@unina.it

D. F. Iezzi et al. (eds.), *Text Analytics*, Studies in Classification, Data Analysis, and Knowledge Organization,
https://doi.org/10.1007/978-3-030-52680-1_21

1 Introduction

Volunteer work differs from the traditional forms of work in several features. Nevertheless, most of the authors approaching the volunteering phenomenon are interested mainly in the economic value that this sector is able to add to the labor market [10, 15]. At the same time, these scholars consider voluntary work as a special type of employment within the economic theoretical framework. From this point of view, volunteering is assumed to be a peculiar sector of the production with a considerable number of divergent rules and different dynamics compared to the standard work patterns, but still able to provide goods and services to the community just like the other sectors. In this sense, it can be argued that volunteering is supposed to increase the overall economic value of the society.

In the current research, the analytical focus will be instead on how volunteering may be considered as a laboratory of social innovation embedded in the labor market. The main concept behind it is that volunteering is based on various guidelines and different principles [20]; therefore, it should be noted that from within the volunteering field, new professional profiles could be developed as well as pre-existent ones modified. Text mining approach will be performed on open-ended questions given by volunteers, assuming that their self-concepts is a consistent proxy of the volunteering world. The empirical statistical analysis will make use of two tools chosen for their capability to profile both groups of words and cluster of volunteers. The latter, in the Italian context, will be analyzed in parallel with the traditional categories applicable to the classic labor theory. It will be shown that most of the determinants of the segmentation of the professions [5], such as gender, age, or geographic area of origin, can be as well adopted in this framework.

2 Volunteering Context in Italy

Over the past thirty years, the history of the Italian welfare has been characterized by an evolution that has also impacted the so-called *social professions* [1].

What is more, the consolidation of the social assistant profession as a job position has been also endorsed from a legislative viewpoint with the establishment of the "Ordini degli assistenti sociali" (association of the social assistants), which took place at the end of the 80s. The legal constitution of this association has marked a shift from an informal (even if already remarkable as amount) framework of volunteering activities to an official legitimation of such professions. It is beyond the aims of this work to profile and describe, even from a historical perspective, the chronological phases that have come to comprise the whole ensemble of the social professions nowadays. However, it is clear that, over the last years as well as in recent years, there is a strong flow of innovation that has affected the sector of social professions, which has nonetheless been for years active in nonformal context. Because of this, such flow has invigorated, as a whole, from the volunteering world. It is possible to identify some peculiar macro-areas within which job opportunities have been more affected by changes within the field of volunteering activities:

1. First contact operators (i.e., those who take care of people's first needs).
   The post-Fordism Era and the subsequent need segmentation have raised the issue of a welfare system that cannot be easily standardized to cater all the population needs. People belonging to different economic, social, cultural, and demographic status are likely to ask for different and specific needs. Subsidies and official welfare institutions are not the only tools available in nowadays' society to answer the social demands of people who request them. This has consequently led to the birth of professions related specifically to first needs: hospitality, clothing, food drive, and first aid. Volunteers have put into action their capacities in such activities in an experimental fashion, but such professions are increasingly finding a place within an ever more traditional and standardized working framework.
2. Entertainers of socio-educational centers for children and elders.
   With regards to this area, there is an increase in the demand of new meeting places, especially in urban areas, where people usually spend their spare time. These places assume a particular relevance when old people and children use them as an opportunity to "learn," namely for educational purposes as well as for entertaining themselves. Consistently with this context, new professions related to the field of education and entertainment have emerged within the volunteering framework and are now finding legitimation even outside the abovementioned sector.
3. Management and administrative professions at different levels.
   The necessity to find resources, may they be financial or not, together with the need to organize voluntary activities, has put into question the emergence of new professions related to such field. People involved in this kind of activities ask for resources to official institutions, report, and account events and projects (accounting sector). These individuals do that in a perspective that is far from the market logic, with no profit allowed in this case. The people involved have expertise in management, governance, and accounting.
4. Managers and head managers involved in planning, management, and projects evaluation (highest profile).
   The professions in these category account for the highest professional profiles that have arisen from the field of volunteering. They act mainly as reference persons when it comes to the organization of group of volunteers, often organized as an association. These individuals are able to deal both with the legal field of the organization as well as the management point of view (executive field). They have skills in project evaluation and financial planning, and coordinate those subjects involved in the jobs presented previously in Sect. 3. These professionals are also focused on maintaining networks (short and/or long networks) around volunteers' organizations: They communicate with political, social, and economic entities that provide help and support to volunteers.

Given this preamble, statistical findings related to such theoretical framework will be presented in the next section, with the aim to understand how strong is the linkage between answers of volunteers and the theoretical proposal of macro-areas of activities from which the research has started.

3 Data and Statistical Approach

The research data have been elicited from the 2013 Istat Survey "Multiscopo, Aspetti della vita quotidiana" (Multipurposes survey, daily life aspects) [9]. This is a large annual sample survey that includes the resident population living in private households. The survey was carried out by interviewing a sample of about 20,000 households and approximately 50,000 people with P.A.P.I. technique. The main sections in the questionnaires concern education, work, family and social life, spare time, political and social participation, health, life style, and access to the services.

From the whole sample, we selected around 5000 respondents, who declared to be involved in volunteering activities. More specifically, we took into consideration the section of open-ended questions in which they specified what kind of activities they performed and whether they did it within an organization or autonomously. The main core of the statistical text mining procedure focused on those brief descriptions that deal with their own autonomous voluntary work. We analyzed the descriptions alongside the socio-demographic variables available: gender, age, geographic macro-area, and educational level. Bearing in mind the definition of *volunteering* [9, 18], several descriptions were excluded from the database, since they are not considered voluntary activities (e.g., people donating blood to AVIS organization, or people providing help to their family members). They have claimed to be volunteers, but we considered them as erroneous inclusions.

Therefore, following this preliminary cleaning procedure by which inappropriate or missing answers were deleted, the valid number of volunteers amounted to 4254 from the original 5000. Before carrying out the analysis, we performed a preliminary transformation of the original lexical data by removing punctuation and stop-words, and by stemming the words, i.e., deleting all the derivational and inflectional suffixes [12, 17]. Consequently, all the words that evolved from the same root were considered morphologically equal after the stemming process. To perform this task, we used the porter stemming algorithm through the software R implemented in the package tm [13].

After this initial analytical step consisting in cleaning processes, we have applied two different methodological approaches to single out groups of activities that could account for voluntary jobs, that is: firstly, semantic network analysis [7, 16] community detection, and secondly, the Reinert textual clustering method, called "Analysis of the Co-occurring Lexemes within Simple Statements of Text" [2, 14]. Moreover, with the aim to profile voluntary jobs with respect to socio-demographic dimensions, we used correspondence analysis on generalized aggregated lexical tables (CA-GALT) [11].

The former is an extension of social network analysis that treats the text as a graph structure: each word is defined as a *node*, and the ties between words are undirected links weighted by the count of co-occurrences (how many times do these words appear together in the same answers). Groups of terms corresponding to semantic clusters can be found through community detection algorithms [8]. Among several plausible community detection algorithm, we have chosen the fast greedy method, given that it is highly suitable to deal with undirected and weighted edges [3]. The key quantity in this framework is *Modularity*, which measures how good the division in a graph is and how separated the different vertex types are from each other, with respect to one division (either vertex types or figured out clusters). In each step, the algorithm merges together the two words that maximize modularity, and stops when no statistical gain in modularity is given when merging two close words up to that iteration. Modularity **Q** formula is:

$$Q = \sum_{i=1}^{c} (e_{ii} - a_i^2)$$

where c accounts for the number of identified communities, e_{ii} is the probability of edge e to fall into group i, and a_i^2 is the squared probability of a random edge a falls into community i. Such communities are groups of words that are likely to appear together and will be used as a tool to describe several sets of activities, which are thus identified within the volunteering framework.

The second method, the Reinert approach, which was applied for textual clustering, aims to identify classes (groups) of terms that are maximally associated among them within the class while being minimally associated with other classes. However, in contrast to the social network approach, it does not exploit the graph theory and its properties, but rather a different matrix representation of data. It first constructs a matrix that takes into account all the short answers of volunteers, the so-called elementary context units (ECUs) and lemmas, where each cell indicates the presence or absence, in a dichotomic fashion, of that word in the original statement. It is very similar to a term-document matrix, but at a smaller unit of textual analysis used as reference. Here, there is an example of the elementary context units matrix:

$$\mathrm{ECU}_{p,n} = \text{word} \overset{\text{ECU}}{\begin{bmatrix} 0\,0 \cdots 1\,1 \cdots 0 \\ 1\,1 \cdots 1\,1 \cdots 0 \\ \vdots\;\vdots \ddots \vdots\;\vdots \ddots \vdots \\ 1\,0 \cdots 0\,1 \cdots 1 \end{bmatrix}}$$

Subsequently, it makes partitions, iteratively, of ECUs with the goal of maximizing at the same time similarity in the same class and the difference between classes. The iterative, descending hierarchical classification process, comes to conclusion when a predetermined number of iterations results no longer in statistically significant divisions.

The procedure does not involve only the lemmas, but it involves texts as well. The second phase of the algorithm splits the rows of the ECUs matrix which are in this case study the volunteers' short answers. These partitions are useful to identify groups of words that can help in the interpretation of common patterns within the answers.

The idea behind using two different theoretical formalizations to achieve the same task—that is to detect group of words that, in turn, are expression of a specific group of activities—is that if findings are consistent among different clustering approaches (and different methodological choices), there is a strong hint that results are significant, and therefore, interpretation can be done with greater confidence.

On the other hand, the CA-GALT method allowed us to jointly analyze in a multiple correspondence framework both the lexical table and socio-demographic profiles, combining the document-term matrix and the matrix containing the individual characteristics. For CA-GALT analysis, the **X** matrix is a block matrix made up by two different matrices combined by columns: a **DT** document-term matrix and a **V** matrix with, for each individual, the observed variables:

$$X_{n,(p+k)} = [\mathrm{DT}_{n,p}|V_{n,k}] = \begin{bmatrix} \mathrm{dt}_{1,1} & \mathrm{dt}_{1,2} & \cdots & \mathrm{dt}_{1,p}| & v_{1,1} & \cdots & v_{1,k} \\ \mathrm{dt}_{2,1} & \mathrm{dt}_{2,2} & \cdots & \mathrm{dt}_{2,p}| & v_{2,1} & \cdots & v_{2,k} \\ \vdots & \vdots & \ddots & \vdots & \vdots & \ddots & \vdots \\ \mathrm{dt}_{n,1} & \mathrm{dt}_{n,2} & \cdots & \mathrm{dt}_{n,p}| & v_{n,1} & \cdots & v_{n,k} \end{bmatrix}$$

This is made to exploit the potential relationship between used lexicon and individual features; the final aim is to validate whether volunteering is consistent or not with the job segmentation determinants of traditional labor market.

4 Main Findings of the Analysis

After the preliminary transformations, the overall corpus shows a high degree of heterogeneity with 1649 different words, and a high level of sparsity, close to 100%, due to the large number of documents and their shortness. The term frequency distribution has a median equal to 2, and a $p_{0.75}$ percentile equal to 4. Given the sparsity, the analysis focuses on the most frequent words that profile and describe voluntary activities, taking into account exclusively those words that are above the $p_{0.90}$ percentile (frequency equal to 11), and ending up in a vocabulary consisting of 175 words. The most used are *organizz* (to organize, or organization) that appears 296 times, *assistent* (assistent) with 225 occurrences, *attiv* (activity) that occurs 215 times, then *assoc* (association), *aiut* (to help), and *volontar* (volunteer and derived words). Those terms can be considered pretty generic and could be related to several aspects inside the volunteers' community, without showing additional informative power to profile volunteers. They are followed by terms describing specific fields of intervention: sport, *fond* (fund), event, *bambin* (child/children), *anzian* (senior/old). Furthermore, some of them express just one semantic meaning and can be considered bi-grams [4]: *croce rossa* (red cross), *croce verde* (green cross), *croce bianca* (white cross), *protezione civile* (civil protection/defense), *vigili fuoco* (firefighters), *capo scout* (scoutmaster). We merge them in the following (the following what?).

Applying the semantic network and the community detection algorithm to these data, we found seven groups/communities. In Fig. 1, we plot the semantic network along with the communities, in which words are colored according to the community. It is possible to identify a set of "jobs" related to the typical charity organizations, mainly in a religious context: the care of old people and hospitalized people—*ospedal, malat, assistenz, ascolt, accud, cur, sostegn*—(orange), the education and animation of disadvantaged children, carried out mainly by religious organizations—*insegn, parrocc, scuol, orator, cateches, anim*—(purple), the food and cloth drive and its distribution to the poor—*cibo, vestiar, caritas, raccolt, aliment, mens, pover*—(green). Another large group is related to executives and officers of the (voluntary?) organizations and to cultural event organizers—*organizz, event, cultural, membr, consigl, dirigent, reunion*—(blue). Related to this large group, we also found musicians (black) characterized by *suon, band, musical*. Finally, the last important area

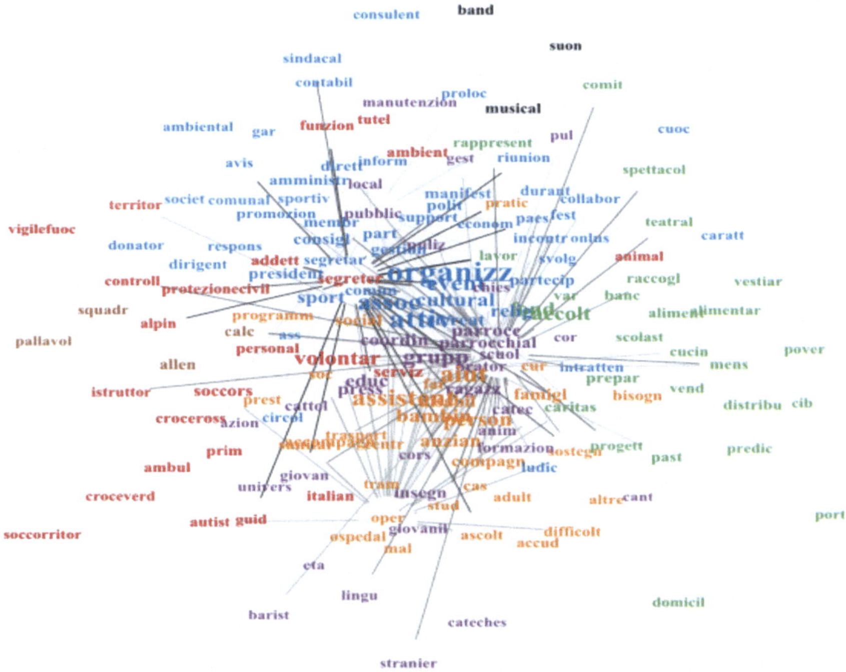

Fig. 1 Semantic network: different colors for different communities identified by fast greedy algorithm. Size of words and width of the edges are proportional to the weights

of the network is associated to the organized volunteers on the territory—*vigile-fuoc, protezionecivil, territor, cr oceross, soccors, ambul*—(red). The coaches are mixed with this group—*squadr, allen, calc, pallavol*—(brown). All these activities are mainly done in nonreligious organizations and are not directly related to charity aims.

The second method, the Reinert approach (clusters in Fig. 2) allowed us to find words (lemmas) that show the less amount of "overlapping" across clusters. Such overlapping quantity was measured in the ECUs matrix by means of χ^2 statistics. From this point of view, it reminds to a similar logic to the one of the correspondence analysis. The clustering partition has been done from a number of clusters going from 2 to 10, and the best result (in terms of χ^2) is given by a partition with seven identified clusters, with one cluster of texts classified as "NA"; it means that this cluster includes those documents that are too generic in terms of words. Volunteers belonging to this clusters likely have used lemmas such as "I help...," "I do volunteering...," and "I take care...," which are too common to be classified as a specific group of activities.

These results are pretty consistent with the seven communities identified in semantic SNA approach. And, even if with some small exceptions, interpretation follows the same pattern, starting from light blue, group of trainers (*atlet, sport, calcio, basket, allen*); in blue, group of people involved in first care and drivers (*ambulanz, guid, croce_ross, barell*); in light green, firefighters and technician. (*vigile_fuoco, tecnic*); in dark green, volunteers involved in taking care of people with difficulties mainly in religious context (*catechism, disabil, insegn, aiut, anzian*); in pink, people related to the employment context in the volunteering world and in sportive environments (*segretar, dirigent, istruttor*); in red, those who are involved in organization and administration tasks (*organizz, event, cultural, coordin, associaz, consigl*) and lastly, in orange, people organizing clothing and food driving (*raccolt, fond, aliment, vestiar*). In the next section, exploiting lexical CA findings, we will try to find an answer to the following question: which category of people belongs to each of those clusters of words that are expression of different activities in volunteering?

Analyzing categories and lexical CA (Fig. 3), it was/is possible to profile individuals according to their demographic status. In this context, a real clustering procedure was/is not performed, but rather, as for the classical correspondence analysis, the two spaces, units, and variables, are linked together taking into account that words close to a specific category are more likely to occur for people belonging to the given category. It is clear that there is a gender gap: men are related to sport activities, they play music in bands, they are drivers (mainly ambulance drivers) and are involved in administration tasks. Women are more involved in providing services to individuals (by taking care of children and old people), also carrying out food and cloth drive to the poor. Geographic differences emerged as well: volunteers from north-east and north-west describe their activities as *manutenzion, dirigent, addett, consigl*, showing a higher organization level. The volunteering performed in the south and the islands is more related to a female style of voluntary work, with a predisposition for religious organization and mainly aimed to assistance. Educational level and age have as well an impact: lower levels of education, crossed with age information, profile a group of old and less educated volunteers involved in religious volunteering. Higher

Fig. 2 Reinert approach to textual clustering identifies seven clusters of texts as best partition. A world cloud is depicted for each cluster

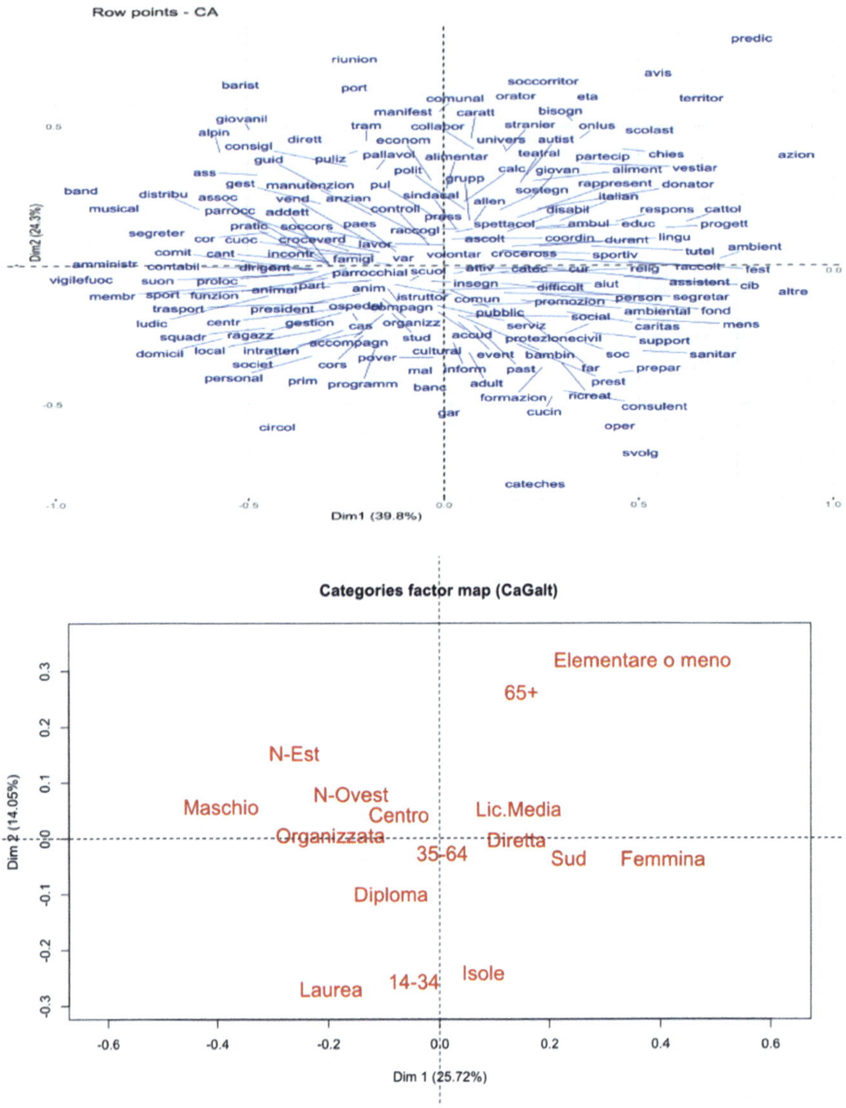

Fig. 3 CA-GALT for both terms (blue) and categories (red). Overlapping both factor maps is possible to profile cluster of individuals

educated people carry out mainly administrative tasks. The central age group (35–64) shows, on the other hand, an average profile close to the origin of axis, as well as people/volunteers from the center of Italy.

5 Discussion and Conclusion

As introduced in the first section, the aim of this paper/research is to present a general perspective about volunteering work in Italy, under the assumption that it is possible to study the subject matter in an analog way as labor market is studied in classic economic literature. Some authors already provided examples to show how the voluntary work actually follows the rule of supply and demand under given conditions (i.e., Wolff et al. [19]) and also volunteering companies make use of marketing strategies similarly to business companies [6]. The two different statistical tools presented in previous section give to the empirical analysis different hints and are somehow complementary. Communities in semantic network of (Fig. 1) are based on the connection level between words, without taking into account other previously known characteristics of individuals. Thus, the communities discovered are groups of words that define several activities and so clusters of jobs in some specific fields. The same task has been achieved using Reinert textual clustering procedure, leading to a classification of volunteers very similar to the one in the community detection. Clusters seems to be consistent, and this highlights how macro-areas of activities described in the second section can be discussed on a clear segmentation as a result of statistical analysis. In the second analysis, both spaces build in CA-GALT, individuals and categories, stress out how segmentation is clearly present in volunteering as in labor market, and words used (and so activities done) change with regards to gender, education, age, and macro-area, in an equivalent way as for standard jobs. As a matter of fact, the findings offer an overview about the relationships between words (as description of activities) and categories (socio-demographic variables). Summing up, both analyses not only highlight how volunteering is complex and heterogeneous; but also, that people involved are in some cases highly skilled, in that they often use some of the competencies trained in their professional life. Generally, they are able to describe their activities in a thorough way while explaining overtly the aim of their voluntary jobs. The text mining analysis presented in this paper revealed that some of the population's needs are not adequately satisfied, given the assumption that volunteers spend their time and use their skills to give something to individuals that strongly ask for services, in a framework similar to supply and demand mechanism. Furthermore, in order to have a more exhaustive overview for the enactment of future policies, the next step could be to focus on the opposite side. In other words, another survey should be carried out to investigate why people *ask for help* to volunteers. In doing so, such an investigation will probably provide a better understanding of those specific needs that are not fully satisfied as well as unveil to what extent the individuals of the Italian society obtain, in terms of assistance, especially from official welfare institutions.

References

1. Amati F, Musella M, Santoro M (2015) Per una teoria economica del volontariato, vol 1. G. Giappichelli Editore, Torino
2. Bholat D, Hansen S, Santos P, Schonhardt-Bailey C (2015) Text mining for central banks. Available at SSRN 2624811
3. Clauset A, Newman ME, Moore C (2004) Finding community structure in very large networks. Phys Rev E 70(6):066111
4. Collins M (1996) A new statistical parser based on bigram lexical dependencies. In: Proceedings of the 34th annual meeting on Association for Computational Linguistics, pp 184–191. Association for Computational Linguistics
5. Colombo A (2003) Razza, genere, classe. Le tre dimensioni del lavoro domestico in Italia. Polis 17(2):317–344
6. Dolnicar S, Randle M (2007) The international volunteering market: market segments and competitive relations. Int J Nonprofit Voluntary Sect Mark 12(4):350–370
7. Drieger P (2013) Semantic network analysis as a method for visual text analytics. Procedia Soc Behav Sci 79:4–17
8. Fortunato S (2010) Community detection in graphs. Phys Rep 486(3–5):75–174
9. Indagine Istat Multiscopo sulle famiglie: aspetti della vita quotidiana (2013) Retrieved from https://www.istat.it/it/archivio/91926
10. Ironmonger D (2000) Measuring volunteering in economic terms. Volunteers and volunteering. The Federation Press, Sydney, pp 56–72
11. Kostov B, Bécue Bertaut M, Husson F (2015) Correspondence analysis on generalised aggregated lexical tables (CA-GALT) in the FactoMineR package. R J 7(1):109–117
12. Lovins J (1968) Development of a stemming algorithm. Mech Transl Comput Linguist 11(1–2):22–31
13. Meyer D, Hornik K, Feinerer I (2008) Text mining infrastructure in R. J Stat Softw 25(5):1–54
14. Reinert M (1990) Alceste une méthodologie d'analyse des données textuelles et une application: Aurelia De Gerard De Nerval. Bull Méthodol Sociol 26(1). https://doi.org/10.1177/075910639002600103
15. Salamon L, Sokolowski S, Haddock M (2011) Measuring the economic value of volunteer work globally: concepts, estimates, and a roadmap to the future. Ann Public Coop Econ 82(3):217–252
16. van Atteveldt W (2008) Semantic network analysis: techniques for extracting, representing, and querying media content. BookSurge Publishers, Charleston SC
17. Willett P (2006) The Porter stemming algorithm: then and now. Program 40(3):219–223. https://doi.org/10.1108/00330330610681295
18. Wilson J (2000) Volunteering. Ann Rev Sociol 26(1):215–240
19. Wolff N, Weisbrod BA, Bird EJ (1993) The supply of volunteer labor: the case of hospitals. Nonprofit Manage Leadersh 4(1):23–45
20. Zamagni S (2005) Gratuità e agire economico: il senso del volontariato. Working paper presented at Aiccon meeting, Bologna

Free Text Analysis in Electronic Clinical Documentation

Antonella Bitetto and Luigi Bollani

Abstract This study finds reason in the now wide availability of clinical documentation stored in electronic form to track the patient's health status during his care path. The diffusion of these practices make available many biomedical collections of electronic data, easily accessible at low cost that can be used for research purposes in the field of observational epidemiological studies, in analogy with what was historically already done in studies based on the reviewing of medical records. However, since these collections are not organized according to specific survey schemes, they sometimes do not allow the index events to be discriminated with the necessary reliability between one source and another automatically. This poses the problem of finding effective methods of critical rereading of texts able of solving this deficiency and, according to the possibility, of bringing the words or segments back into subdomains that can be analyzed statistically. It is proposed to address the problem, showing study criteria and an empirical experience, consistent with the needs of a biomedical context.

1 Introduction

The progressive process of dematerialization of clinical documentation (specialist outpatient assessments, emergency department consultation notes, diagnostic examination reports) and the implementation of health data flows for administrative purposes (for the payment of the services provided or for updating the registry of INPS etc.) have made available information that can also be used for purposes other than those for which the data are collected.

A. Bitetto
Azienda Socio Sanitaria Territoriale di Monza, Monza, Italy
e-mail: a.bitetto@asst-monza.it

A. Bitetto · L. Bollani (✉)
Università di Torino, Turin, Italy
e-mail: luigi.bollani@unito.it

D. F. Iezzi et al. (eds.), *Text Analytics*, Studies in Classification, Data Analysis, and Knowledge Organization,
https://doi.org/10.1007/978-3-030-52680-1_22

The use of electronic health information began in the second half of the 1980s for the transmission of computerized data between hospitals and local or governmental authorities for administrative and demographic purposes. This type of health data describes the "pathways" of the case treatment in terms of services provided (i.e., type of ward, days of hospitalization, care provided). Above all, they summarize the reason for access to one or more formally coded diagnoses according to international disease classification criteria (10th or 11th version of the World Health Organization).

Since the 2000s, the replacement of paper-based medical records with electronic ones has led to the dissemination of semi- or un-structured health documents, based on texts written in general practices and hospital settings. It is a corpus of texts that cover most of the medical field and allows to trace the path and the results of the specialist visits, the diagnostic exams and the prescribed treatments, first aid reports, and clinical diary during the hospitalization. The data are stored in dedicated servers and only available to authorized healthcare personnel.

It follows that are currently available different types of computerized health data (EHRs—"electronic health records"), classified into: (a) structured or registered using controlled clinical terminologies such as the International Classification of Diseases—10th revision (ICD10) or the systematic nomenclature of medicine—Clinical terms (SNOMED-CT); (b) semi-structured (e.g., laboratory tests and information on prescription) that follows a pattern that varies according to the conventions adopted locally; (c) unstructured (e.g., clinical text); (d) binary (e.g., image files such as Rx and TAC).

The systematic nature of these data collections, mostly organized by individual entities, makes them particularly valuable for epidemiological research purposes that uses observational-type designs: Over the years, numerous examples of research based on EHR have been produced both in the context of classical epidemiology, such as studies on the risks or outcomes of diseases, or the effectiveness and quality of care [14].

On the other hand, since these collections are organized for purposes other than those of scientific research, they must be "treated" before they can be analyzed using statistical methods. In the past, this was done by re-reading the medical records by subject matter experts [7].

The rapid growth of EHRs in medicine has placed emphasis on alternative methods to manual review to identify the subdomains of medical narrative texts that subsequently allow the application of the statistical predictive analysis models necessary for medical research. The pretreatment is especially necessary for unstructured health documents where the subjects' health status (symptoms, functioning disorders, exposure risks, or triggering events, sometimes the same diagnosis) is described in a short text that refers to a wide range of semantic aspects, inherent in the natural language as well as in the technical-scientific one. In this type of corpus, it is believed that the automatic analysis of the texts can solve the shortcomings due to the narrative language, tracing words, and segments back to statistically analyzable subdomains [5, 17].

When the data are collected systematically on the entire population concerned (there are no defects in coverage) and that the errors of charges are minimal, automatic analysis of information of EHR has allowed to investigate costs of treatments [16] and safety areas such as—between many—domestic accidents, allergic reactions, and drug side effects [4, 6].

Further, the EHRs texts are rich of information that can be the used to study the outcomes of chronic illnesses with high prevalence such as cardiovascular diseases, diabetes, or asthma because it facilitates the follow up across the time [11, 15]. In Italy over the last 15 years, similar studies have been developed: some of them as research financed by the European community but most of them as autonomous local research. The latter—thanks to the extensive sources of information available at low cost—have succeeded in developing examples of epidemiological research with extensive case studies even if developed in local contexts [9, 13, 19].

Some research has focused on mental disorders, the medical area chosen for the text analysis experience presented below. In this context, EHRs allows obtaining low-cost information on large sectors of the population that can include difficult case histories otherwise to be recruited: This is the case of subjects in the prodromal phase at high risk of developing psychosis [8] or self-defeating [10, 21].

Psychiatry is a field of medicine where clinical evaluation needs of a detailed analysis of signs, symptoms, and life situation: Psychiatric consultations take place essentially on the clinical interview with the patient. The assessment is supplemented with information from accredited sources where possible on triggering (e.g., alcohol or drug use) and complicating (e.g., self-harm, aggression) events. The result is a clinical text that describes the mental state according to the rules of psychiatric semeiotics which uses a language and a theoretical construct considerably variable according to the academic schools. In general, the information is supplemented by the narrative description of the triggering and complicating events. For these reasons, the classification of medical and even more psychiatric corpora into subdomains is a controversial topic that justifies research studies on methodologies for automatic text analysis. In our opinion for psychiatry, it is appropriate to choose a priori an interpretative model of words and lemma that allows defining subdomains in line with a validated and univocal theoretical construct.

In the case, report proposed a method for the treatment of unstructured psychiatric texts, according to criteria already partly used in previous experiences [3]. The method use knowledge engineering techniques based on rules established by subject matter experts according with Sebastiani [18]. Then, the associations between the subdomains were studied in order to verify the predictive ability of the methodology to identify the relevant subdomains in the event of a psychiatric emergency.

2 Materials and Methods

This section presents the methods used to treat the reports of psychiatric consultations carried out at the emergency department of a large Lombard University Hospital (1250 accredited beds).

Table 1 *Corpus* description

Number psychiatric consultations	1721
Number of total *corpus* words	177394
Number of words with recurrence >10	1679
Average number of words for report (±SD)	103.08 (±47.51)

2.1 *Corpus*

The *corpus* is monolingual—in Italian—composed of short texts written by the psychiatrist on duty at the end of the emergency consultation.

The reports are verified and then kept by the hospital information service, ISO 9001/2015 certificate, which provided the *corpus* of the data, in anonymous form.

1721 reports were analyzed, relating to the period 01/01/2012–31/12/2012 (Table 1).

2.2 *Linguistic Filtering Procedure*

The *corpus* has been subjected to a linguistic filtering treatment. From 177,349 words present in the original reports, punctuation, numbers, pronouns, articles, propositions, proper names—even drugs—and words with a recurrence of less than ten have been eliminated.

The result was a list of 1679 distinct words, which has been manually reviewed by an expert to select words or brief phrases capable of describing mental health problems/needs according to the structural model used by the HoNOS scale [12, 20].

This is a model for assessing the state of mental health setup for problems and not for diagnoses, which are hardly reported in first aid reports. The model distinguishes 12 "problems" related to the following concepts:

item H1—HYPERACTIVE, AGGRESSIVE BEHAVIORS; item H2—DELIBERATELY SELF-LEASING BEHAVIORS; item H3—PROBLEMS RELATED TO THE ASSUMPTION OF ALCOHOL OR DRUGS; item H4—COGNITIVE PROBLEMS; item H5—PHISICAL ILLNESS PROBLEMS; item H6—PROBLEMS LINKED TO HALLUCINATIONS AND DELUSIONS; item H7—PROBLEMS RELATED TO DEPRESSED MOOD; item H8—OTHER PROBLEMS FROM OTHER PSYCHIC SYMPTOMS; item H9—PROBLEMS IN SIGNIFICANT RELATIONSHIPS; item H10—PROBLEMS IN THE PERFORMANCE OF DAILY LIFE ACTIVITIES; item 11—PROBLEMS IN THE LIVING CONDITIONS; item H12—PROBLEMS IN WORK AND RECREATIONAL ACTIVITIES.

In this way, a *thesaurus* was created consisting of 214 keywords (short phrases or single words, expressed in form of lemmas). At every single lemma, it has been associated a category of problems according to HoNOS criteria.

Only H10, i.e., problems in the performance of daily life activities, was not considered, given the lack of words that can lead to information on this kind of disability.

Table 2 Description of filtering process

Number of words with repetition >10	495
Number of key words classified	214
Number of words analyzed such as short phrases, short locutions (2 or 3 words)	207
Number of words analyzed such as single words	81

This is due to the emergency context in which the consultation is carried out which does not normally take into consideration the information on the difficulties, disability to manage everyday life. On the other hand, in the thesaurus, an additional area compared to HoNOS was considered, related to the "refusal of treatments" which is an important causal factor of the emergency in psychiatry.

Subsequently, the filtering procedure provided for the elimination of the text segments in which a single classified keyword was denied. This for avoiding misclassification of the text.

At the end of this phase of the review of the filtering procedure, 1629 reports from the original database were classifiable in problem areas according to HoNOS scale (92% of consultations). This reclassified corpus represents the basis of the statistical analysis (Table 2).

2.3 *Statistical Analysis*

The different psychiatric consultation reports were automatically examined for the presence or absence of each classified keyword or short segment: this procedure led to the selection of m reports (in this case $m = 1629$), which cross n keywords (in this case $n = 214$); subsequently, the keywords were translated—on the basis of the expert *thesaurus*—into the HoNOS codes in order to facilitate the interpretation of problem areas and their associations. In addition to the HoNOS scale (12 items; 11 of them were considered, excluding as mentioned H10), we have included information on refusal or interruption of care, which is a relevant problem in emergency psychiatry, considering r items (in this case $r = 12$) in our final coding.

More in detail, we consider a *reports* × *keywords* binary matrix **B1** (m, n), in which the generic term t_{ij} is equal to one if the report i contains—at least once—the keywords j (t_{ij} is equal to zero in other cases). Then, after reducing the keywords to the HoNOS codes through the thesaurus (each keyword is connected with one and only one HoNOS item), another binary matrix **B2** (m, r) is built, where the generic term q_{ij} is equal to one if the report i contains—at least once—the HoNOS code j (q_{ij} is equal to zero in other cases).

In order to study the associations among the different problem areas, a correspondence analysis (according to Benzécri [2] was carried out on the **B2** matrix (*reports* × *HoNOS problem area*).

Furthermore, to clarify the relationship between the problem areas of HoNOS and the words of the *thesaurus*, another correspondence analysis was performed on the matrix **BB** $(m, r + n) =$ [**B2**, **B1**], obtained by juxtaposing the matrixes **B2** and **B1**. In this analysis, keywords were considered as additional elements.

3 Results

Table 3 shows the frequency distribution of the reclassified problem areas according to HoNOS criteria.

As expected, emergency psychiatric consultations primarily define mental distress through detailed descriptions of the observed symptoms (H1–H8). The description of precipitating environmental factors appears more rarely (H9, H11, H12).

Among the symptoms, those most frequently found are the depressed mood (H7) and the class that collects all unspecified clinical manifestations "other psychic symptoms" (H8).

Also the description of secondary problems to physical symptoms (H5) is very frequent, as expected, given that the management of psychiatric emergencies takes place in the emergency room of a general hospital where the request for an opinion on subjects with concomitant problems of organic disease is higher than in a psychiatric clinic.

The occurrence of violent and hyperactive behavior (H1) also shows a high frequency, being one of the most typical reasons for the emergency intervention in psychiatry.

Figure 1 describes the results of the analysis of the correspondences on the categories that describe the clinical manifestation of the mental distress symptoms, the most common problem area in the texts.

Table 3 Item HoNOS and percentage of problem frequency found in the emergency psychiatric reports

Item HoNOS	% of problem frequency
H1	30.82%
H2	15.22%
H3	12.22%
H4	7.18%
H5	20.32%
H6	18.35%
H7	32.72%
H8	59.55%
H9	5.10%
H11	1.23%
H12	7.31%
Care refused	18.97%

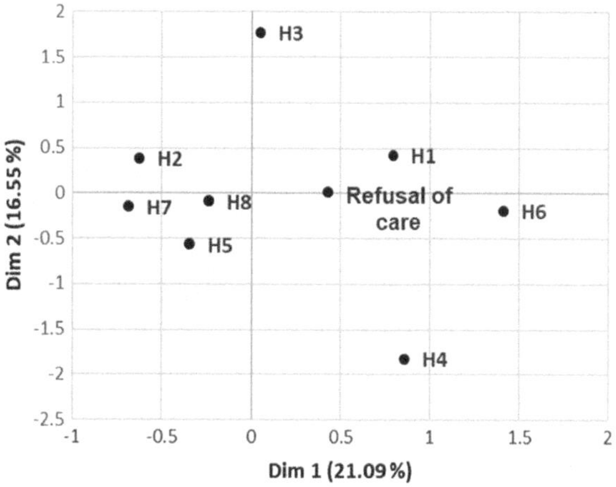

Fig. 1 Joint representation of the first eight items' HoNOS (mental and physical symptom)

The first factorial plane shown in the graph explains 37.64% of the total variance.

Compared to dimension 1, along the axis of the abscissa, the categories of symptoms are divided into two groups: On the right, we find the problems related to the "depressed mood" (H7) close to "other symptoms" (H8), of which anxiety is the most decrypted. Near in the map, there is also the category "physical symptoms" (H5), confirming the probable psychosomatic origin of a part of them. Finally, in the same group, we also find self-injurious and suicidal behaviors, which are often associated with problems of depression according to literature.

On high abscissa values, we find instead the class of "psychotic symptoms" (H6) and of "agitated behaviors" (H1) that result in relation to the "refusal of care."

The problems related to "alcohol and drug abuse" (H3) and those due to the "presence of cognitive problems" of neurological origin (H4) reach the extremes of dimension 2, identified by the ordinate axis.

The same analysis is represented in Fig. 2, also projecting the relevant words of the *thesaurus* used.

The keywords in Fig. 2 are distributed around their classification code, confirming and clarifying the results shown above.

Figure 3 introduces the effect of environmental factors in triggering an acute episode. In it, we expose a second analysis of the correspondences conducted on the categories of psychic and behavioral symptoms together with the precipitating environmental factors. In this case, the factorial first plane explains 30.33% of the total variance.

The distribution of clinical manifestations along the abscissa axis confirms the results of the analysis of the first subset of categories. In this case, it is possible to note the tendency of the class "problems related to the abuse of alcohol and drugs" (H3) to position itself toward the center of the graph in the vicinity of the "other symptoms" category (H8), with which it is possible that certain symptoms, mainly anxiety, are related.

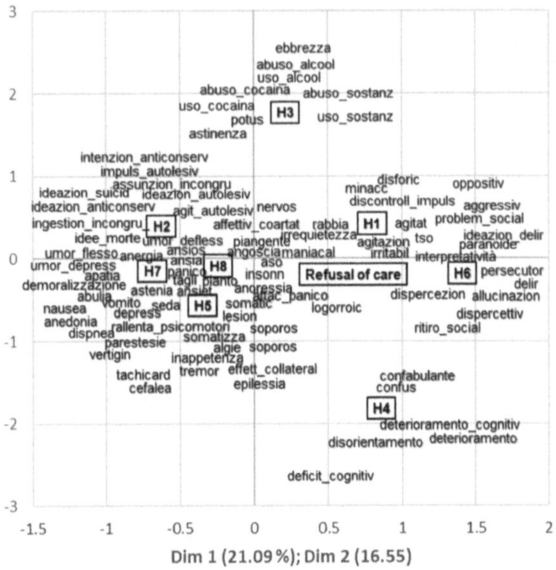

Fig. 2 Joint representation of the HoNOS items related to the first eight items and additional representation of the key words/phrases used to identify the different items

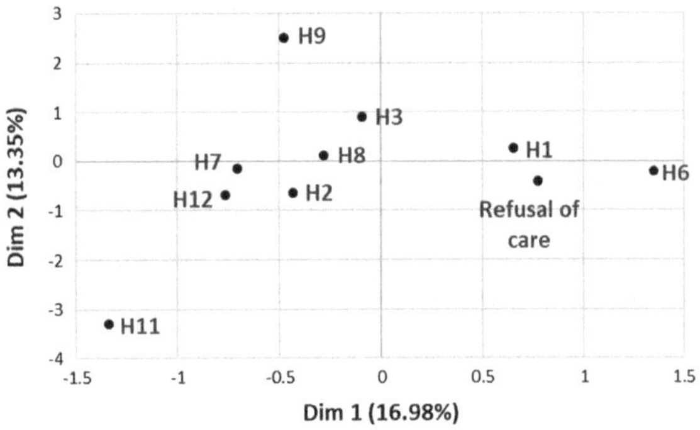

Fig. 3 Behavioral symptoms (H1, H2, H3), psychic symptoms (H6, H7, H8), and precipitating environmental factors (H9, H11, H12)

As regards environmental factors, the data show an association between "job problems" (H12), "depressive spectrum symptoms" (H7), and "deliberately self-injurious behaviors" (H2). It is possible that the emergency department represents a first point of access for a user with reactive forms, even serious ones, secondary to occupational stress factors (burnout, reactive depressions). The other categories related to environmental problems (H9 and H11) are located at the extremes of dimension 2, showing a certain degree of independence with respect to the occurrence of behavioral and psychic symptoms.

4 Conclusions

The empirical experience of semiautomatic textual analysis on emergency psychiatric reports confirms its usefulness in investigating complex phenomena such as clinical manifestations and risk factors in psychiatric emergency. It is a context in which it is difficult to conduct a traditional study due to the complexity of clinical management and the poor collaboration of many clients. In healthcare settings of this type, the only systematic sources of information are those required by law such as the compilation of the medical records. For this reason, it is important to explore methods in order to make the texts contained in these types of documents scientifically analyzable as experimented in this case report.

On the other hand, there are some problems related to the quality of the information which, as collected for other purposes, present an excess of information with respect to some areas (mainly symptoms and conduct disorders) while they are lacking in others, such as the degree of disability of the subject or precipitating environmental factors. It is possible that these deficiencies can be overcome by acquiring information from other clinical sources as some researchers have done [8]. More complex studies require the presence of knowledge in the engineering of the neuronal network and the creation of research medical groups different than usual, implementing the costs.

The experimental analysis of free texts contained in electronic clinical documentation, reclassified according to standardized criteria, such as the HoNOS scale in psychiatry, seems able to provide an intelligible description of a framework of complex needs such as that in the area of acute mental distress. The problem of standardizing methods in the different phases of the investigation, from the ways in which information is collected and written in clinical reports to the creation of a *thesaurus* of standard words, remains a critical aspect that could invalidate reproducibility of the method in similar research contexts. The case report shows how it is possible to improve the misrepresentation where different technical language conventions are used, reclassifying the natural and technical words in a *thesaurus* created according to theoretically validated models.

The correspondence analysis proves to be a simple and useful method for exploring the relationships between the different subdimensions under consideration. The results are in line with what emerges from the sector literature review [1].

References

1. Anderson EL, Nordstrom K, Wilson MP, Peltzer-Jones JM, Zun L, Ng A, Allen MH (2017) American association for emergency psychiatry task force on medical clearance of adults part I: introduction, review and evidence-based guidelines. W J Emerg Med 18(2):235
2. Benzécri JP (1973) L'analyse des données, vol 2. Dunod, Paris
3. Bitetto A, Milani S, Giampieri E, Monzani E, Clerici M (2017) La consultazione psichiatrica in Pronto Soccorso come fonte informativa sui bisogni inespressi di salute mentale. Nuova rassegna studi psichiatrici, vol 15 (atti convegno SIEP)

4. Coloma PM, Schuemie MJ, Trifiro G, Gini R, Herings R, Hippisley-Cox J, Mazzaglia G, Giaquinto C, Corrao G, Pedersen L, van der Lei J, Sturkenboom M (2011) Combining electronic healthcare databases in Europe to allow for large-scale drug safety monitoring: the EU-ADR project. Pharmacoepidemiol Drug Safety 20(1):1–11
5. Denaxas S, Direk K, Gonzalez-Izquierdo A, Pikoula M, Cakiroglu A, Moore J, Hemingway H, Smeeth L (2017) Methods for enhancing the reproducibility of biomedical research findings using electronic health records. Bio Data Min 10(1):31
6. Ehrenberg A, Ehnfors M (1999) Patient problems, needs, and nursing diagnoses in Swedish nursing home records. Nurs Diagn 10(2):65–76
7. Friedman C, Shagina L, Lussier Y, Hripcsak G (2004) Automated encoding of clinical documents based on natural language processing. J Am Med Inform Assoc 11(5):392–402
8. Fusar-Poli P, Cappucciati M, De Micheli A, Rutigliano G, Bonoldi I, Tognin S, Ramella-Cravaro V, Castagnini A, McGuire P (2016) Diagnostic and prognostic significance of brief limited intermittent psychotic symptoms (BLIPS) in individuals at ultra high risk. Schizophr Bull 43(1):48–56
9. Gini R, Schuemie MJ, Mazzaglia G, Lapi F, Francesconi P, Pasqua A, Bianchini E, Montalbano C, Roberto G, Barletta V, Cricelli I, Cricelli C, Dal Co G, Bellentani M, Sturkenboom M, Klazinga N (2016) Automatic identification of type 2 diabetes, hypertension, ischaemic heart disease, heart failure and their levels of severity from Italian General Practitioners' electronic medical records: a validation study. BMJ Open 6(12):e012413
10. Hall GC (2009) Validation of death and suicide recording on the THIN UK primary care database. Pharmacoepidemiol Drug Saf 18(2):120–131
11. Ho ML, Lawrence N, van Walraven C, Manuel D, Keely E, Malcolm J, Reid RD, Forster AJ (2012) The accuracy of using integrated electronic health care data to identify patients with undiagnosed diabetes mellitus. J Eval Clin Pract 18(3):606–611
12. Lora A, Bai G, Bianchi S, Bolongar G, Civenti G, Erlicher A, Maresca G, Monzani E, Panetta B, Von Morgen D, Rossi F, Torri V, Morosini P (2001) La versione italiana della HoNOS (Health of the Nation Outcome Scales), una scala per la valutazione della gravità e dell'esito nei servizi di salute mentale. Epidemiol Psichiatr Soc 10(3):198–212
13. Migliardi A, Gilardi L, Fubini L, Bena A (2004) Descrizione degli incidenti domestici in Piemonte a partire dalle fonti informative correnti. Epidemiol Prev 28(1):20–26
14. Mitchell JB, Bubolz T, Paul JE, Pashos CL, Escarce JJ, Muhlbaier LH, Wiesman JM, Young WW, Epstein RS, Javitt JC (1994). Using Medicare claims for outcomes research. Med Care JS38–JS51
15. Persell SD, Dunne AP, Lloyd-Jones DM, Baker DW (2009) Electronic health record-based cardiac risk assessment and identification of unmet preventive needs. Med Care 47(4):418–424
16. Riley GF (2009) Administrative and claims records as sources of health care cost data. Med Care 47(7 Suppl 1):S51–5
17. Schuemie MJ, Sen E, Jong GW, van Soest EM, Sturkenboom MC, Kors JA (2012) Automating classification of free-text electronic health records for epidemiological studies. Pharmacoepidemiol Drug Saf 21:651–658
18. Sebastiani F (2002) Machine learning in automated text categorization. ACM Comput Surv (CSUR) 31(1):1–47
19. Vaona A, Del Zotti F, Girotto S, Marafetti C, Rigon G, Marcon A (2017) Data collection of patients with diabetes in family medicine: a study in north-eastern Italy. BMC Health Serv Res 17(1):565
20. Wing JK, Beevor AS, Curtis RH, Park SGB, Hadden J, Burns A (1998) Health of the Nation Outcome Scales (HoNOS): research and development. Br J Psychiatry 172(1):11–18
21. Zanus C, Battistutta S, Aliverti R, Montico M, Cremaschi S, Ronfani L, Monasta L, Carrozzi M (2017) Adolescent admissions to emergency departments for self-injurious thoughts and behaviors. PLoS One 12(1):e0170979

Educational Culture and Job Market: A Text Mining Approach

Barbara Cordella, Francesca Greco, Paolo Meoli, Vittorio Palermo, and Massimo Grasso

Abstract The paper explores teachers and students' training culture in clinical psychology in an Italian university in order to understand whether the educational context support the development of useful professional skills support the access of the newly graduated psychologists into the job market. To this aim, students and teachers' interview transcriptions ($n = 47$) were collected into two corpora. Both corpora underwent Emotional Text Mining (ETM), which is a text mining procedure allowing for the identification of the social mental functioning (culture) setting people's behaviors, sensations, expectation, attitudes, and communication. The results show 4 clusters and 3 factors for each corpus highlighting similarity in the representation of the training culture between student and teachers. The education and the professional training are perceived as two different processes with different goals. The text mining procedure turned out to be an enlightening tool to explore the university training culture: The research results were presented to the Master Coordinator who decided to plan a group activity for the students aiming to improve the educational process.

B. Cordella · F. Greco (✉) · P. Meoli · V. Palermo · M. Grasso
Sapienza University of Rome, Rome, Italy
e-mail: francesca.greco@uniroma1.it

B. Cordella
e-mail: barbara.cordella@uniroma1.it

P. Meoli
e-mail: paolomeoli3@libero.it

V. Palermo
e-mail: vittorio.palermo2511@gmail.com

M. Grasso
e-mail: massimo.grasso10@gmail.com

D. F. Iezzi et al. (eds.), *Text Analytics*, Studies in Classification, Data Analysis, and Knowledge Organization,
https://doi.org/10.1007/978-3-030-52680-1_23

1 Introduction

In psychology, the text mining is frequently used to analyze the organizational culture as it sets the interaction and the productivity in a specific environment. In this study, we present the analysis of the educational culture in an Italian university. According to the literature, there is a relationship between the faculty's training culture and the young professional ability in getting into the labor market (e.g., [1–4]).

According to McKinsey's study [5], 40% of youth unemployment does not depend on the economic cycle but on "structural causes." Among the others, education is one of the relevant factors because of the evaluation mismatch of the fitness of the young people's skills between the educational system and the productive one. While the educational institutions consider 70% of young people appropriately prepared, for the employers the rate lower to 42%. Nevertheless, education is a protection factor for poverty and quality of life, as stated by the ISTAT (National Institute of Statistic) [6]; in fact, graduates are less likely to become poor although the employability and the wage depend on the type of degree.

The youth unemployment is relevant issue in Italy as, in 2016, it was the European country with the highest rate of NEETs (Not in Employment, Education, or Training). While in Europe the rate of NEETs was on average 14.2%, in Italy almost 2.2 million of young people under 30 years old were unemployed (24.3%) and the rate grew to 32.2% within under 25. However, among NEETs only 24% were "inactive", that is, not interested in finding a job [6].

A recent report on Sapienza's Psychological Graduates [7] shows consistent data with the McKinsey's study: postgraduates psychologists find the first job after 16 months on average, and not as a psychologist in 32% of cases, 80% young graduates in psychology is employed after four years and the wage is 980 € per month on average, and only 60% of them find a job in working areas consistent with the psychological degree.

In the European countries, the academic educational system splits the training process in two: graduate and postgraduate courses, in which the theoretical background of the discipline is taught, and the post-university internship, in which the theoretical background is used in order to develop professional skills. This could explain for the mismatch between the educational system, focused mostly on the theoretical knowledge, and the productive system, based on the skills performance, and for the difficulty of young postgraduate in entering into the job market [8].

Moreover, the educational culture influences the success in acquiring useful professional skills for the job market. Several studies were performed in order to investigate students' representation in the psychology faculty in order to evaluate and improve the training process (e.g., [4, 9]). They show that the effectiveness of the education depends in part on the representation of the professional training characterizing the learning context.

In order to understand whether the educational culture at the university support students in developing their ability to get into the job market, we performed an

Emotional Text Mining (ETM) [10–12] of the interviews of students and teachers of the Master Degree in Clinical Psychology at Sapienza University of Rome.

2 The Emotional Text Mining

According to a semiotic approach to the analysis of textual data, Emotional Text Mining (ETM) [11, 12] is a procedure that, by means of its bottom-up logic, allows for a context-sensitive semiotic text mining approach on unstructured data. This has already been applied in different fields ranging from political debate, in order to profile social media users and to anticipate their political choices [13–16, 31], to brain structure [17], marketing [12], and to the impact of the law on society (e.g., [10, 11, 18]).

ETM is an unsupervised text mining procedure, based on a socio-constructivist approach and a psychodynamic model [19–22], which allows for the identification of the elements setting people's interactions, behavior, attitudes, expectations, and communication. We know that a person's behavior depends not only on their rationale thinking but also, and sometimes most of all, on their emotional and social way of mental functioning (e.g., [23, 32]). In other words, people consciously categorize reality and, at the same time, unconsciously symbolize it emotionally, in order to adapt to their social environment. The unconscious symbolization is social, as people generate it interactively and share the same emotional meanings through this interaction. Since communication and behavior are the outcome of this social mental functioning, it is possible to analyze the communication (text) to infer the social mental functioning (symbolic matrix) and understand people's behavior in different contexts.

Since the conscious process sets the manifest content of the communication, i.e., what is communicated, the unconscious process can be inferred through how it is communicated, namely, the words chosen to communicate and their association within the text. We consider that people emotionally symbolize an event, or an object, and socially share this symbolization. The words they choose to discuss an event, or object, are the product of the socially shared, unconscious mental functioning. That is, people sharing the same symbolization would use and associate some words in the same way. This could explain the lack of a random sampling in the literature, as all the participants share the same symbolizations. The researches based on an emotional textual analysis method often use a judgment sampling, in order to control for all the characteristics that could influence the local culture and define their number according to the available resources, choosing the textual analysis method according to the lexical indices [10, 11].

The emotional symbolization and the representation of an object [8, 10, 11], e.g., the educational culture, are interconnected, as the symbolization generates the representation [19] as well as the behavior [22]. Moreover, they imply different levels of generalization and awareness. While a person is aware of his/her behavior, she/he is not directly aware of the representation, nor is she/he aware of the symbolization, which is unconscious and socially shared. For this reason, ETM procedure allows for the detection of both the semantic and the semiotic aspects conveyed by the communication.

While the mental functioning proceeds from the semiotic level to the semantic one in generating the text, the statistical procedure simulates the inverse process of the mental functioning, from the semantic level to the semiotic one. For this reason, ETM performs a sequence of synthesis procedures, from the reduction of the type to lemma and the selection of the keywords to the clustering and the factorial analysis, in order to identify the semiotic level (the symbolic matrix), starting from the semantic one (the word co-occurrence) [11, 12].

In order to detect the associative links between the words and to infer the symbolic matrix determining their coexistence into the text, first we perform a bisecting k-means algorithm [24, 25], limited in the number of partitions, excluding all the text that does not have at least two keywords co-occurrence to classify the text. We have selected this clustering procedure as it is the most commonly used one in the semiotic approach [11, 12]. As in the literature, the identification of a reliable methodology for results evaluation is still controversial (e.g., [26]) and three clustering validation measures are considered in order to identify the optimal solution: the Calinski–Harabasz, the Davies–Bouldin, and the intraclass correlation coefficient (ICC) indices. Next, we perform a correspondence analysis [27] on the cluster per keywords matrix. While the cluster analysis allows for the detection of the representations, the correspondence analysis detects the symbolization [12, 28].

The interpretation process proceeds from the highest level of synthesis to the lowest one, simulating once again the mental functioning. Therefore, first we interpret the factorial space according to word polarization, in order to identify the symbolic matrix setting the communication. Then, we interpret the cluster according to their location in the factorial space and to the words characterizing the context units classified in the cluster, in order to identify the representation.

3 Methods

In order to explore the academic educational culture, we interviewed 30 students (13% of students) and 17 teachers (71% of teachers) of the Master of Science in Clinical Psychology of Sapienza University of Rome accordingly to their voluntary participation. Data on their experience of the training process were collected with an open-ended interview.

The transcripts were collected into two corpora: students' interviews resulted in a medium-size corpus (S) of 57.387 tokens, and teachers' interviews resulted in a small size corpus (T) of 28.746 tokens. In order to check whether it was possible to statistically process data, two lexical indicators were calculated: the type-token ratio and the hapax percentage ($TTR_S = 0.09$; $Hapax_S = 50.3\%$; $TTR_T = 0.147$; $Hapax_T = 53.8\%$). According to the size of the corpus, both lexical indicators highlight its richness and indicate the possibility to proceed with the analysis.

First, data were cleaned and preprocessed by the software T-Lab [29] and keywords were selected. In order to choose the keywords, we used the selection criteria proposed by Greco [10, 11]. In particular, we used stem as keywords instead of type, filtering out the lemma of the open-ended questions of the interviews. Then, on the context units per keywords matrix, we performed a cluster analysis with a bisecting *k*-means algorithm [24] limited to ten partitions, excluding all the context units that did not have at least two keywords co-occurrence. In order to choose the optimal solution, we calculated the Calinski–Harabasz, the Davies–Bouldin, and the intraclass correlation coefficient (ρ) indices. To finalize the analysis, a correspondence analysis on the keywords per clusters matrix was made [27] in order to explore the relationship between clusters and to identify the emotional categories setting participants' educational culture.

4 Main Results and Discussion

The results of the cluster analysis show that the keywords selected allow the classification on the average of 96% for both corpora. The cluster validation measures show that the optimal solution is four clusters for both corpora. The correspondence analysis detected three latent dimensions. In Figs. 1 and 2, we can appreciate the emotional map of the educational culture emerging from teachers and students' interviews and cluster location in the factorial space.

The teachers' corpus first factor (Table 1) represents the motivation in teaching, focusing on the group of students and their specific needs or on the Institutional generic scopes; the second factor focuses on the training outcome, the degree or the professional skills; and the third factor reflects the role of the academic professor that could represent oneself as a teacher or a professional. As regards the students

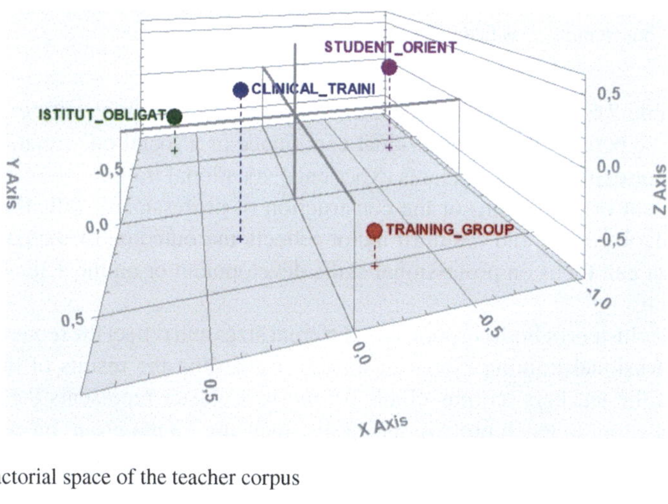

Fig. 1. Factorial space of the teacher corpus

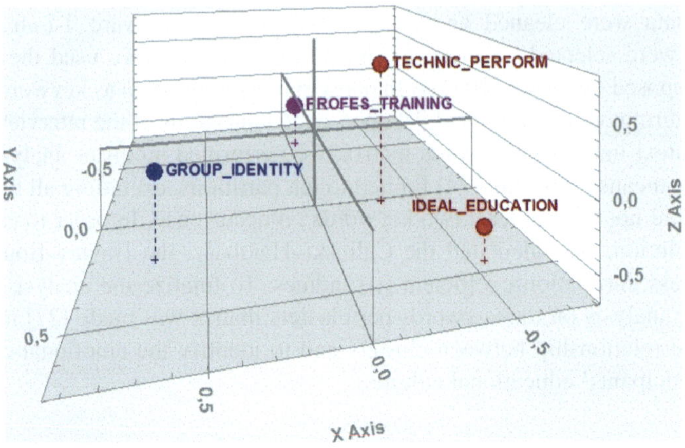

Fig. 2. Factorial space of the student's corpus

Table 1 Cluster coordinates on factors of the teachers' corpus (the percentage of explained inertia is reported between brackets beside each factor)

Cluster *N*	Label (CU in Cl %)	Factor 1 (44.1%) Motivation	Factor 2 (31.6%) Outcome	Factor 3 (24.3%) Role
1	Training group	Group	Competence	Teacher
	(22.3%)	−0.21	0.51	−0.50
2	Clinical training	Institution	Competence	Professional
	(33.7%)	0.33	0.23	0.39
3	Institutional obligations	Institution	Degree	Teacher
	(20.2%)	0.65	−0.66	−0.38
4	Student orientation	Group	Degree	Professional
	(23.8%)	−0.79	−0.39	0.16

CU in Cl = context units classified in the cluster

corpus (Table 2), the first factor represents the approach to university experience, which can be perceived as an individual experience or a social one (relational); the second factor explains how students experience vocational training, perceiving it as the fulfillment of obligations or the construction of professional skills that require personal involvement; and the third factor reflects the outcome of the educational training that can focus on professional skills development or on the achievement of qualifications.

The four clusters of both corpora have different sizes and reflect the representations of the professional training (Tables 1 and 2). Regarding the results of the cluster analysis of the teachers' corpus (Table 3), the first cluster represents the group of students as a tool to teach professional skills, focusing on the group process where relational dynamics are experienced; the second cluster focuses on clinical training,

Table 2 Cluster coordinates on factors of the students' corpus (the percentage of explained inertia is reported between brackets beside each factor)

Cluster *N*	Label (CU in Cl %)	Factor 1 (44.2%) Approach	Factor 2 (29.1%) Training	Factor 3 (26.7%) Outcome
1	Idealized education	Individual	Fulfilment	Skills
	(27.6%)	−0.56	0.45	−0.43
2	Professional training		Construction	Skills
	(20.8%)	−0.04	−0.63	−0.24
3	Group Identity	Relational	Fulfilment	
	(26.3%)	0.69	0.22	−0.01
4	Technical performance	Individual		Qualifications
	(25.3%)	−0.32	0.01	0.59

CU in Cl = context units classified in the cluster

Table 3 Teachers' cluster analysis results (the percentage of context units (*n*) classified in the cluster is reported between brackets beside the cluster)

Cluster 1 (22.3%)		Cluster 2 (33.7%)		Cluster 3 (20.2%)		Cluster 4 (23.8%)	
Training group		Clinical training		Institutional obligations		Student orientation	
Keywords	CU	Keywords	CU	Keywords	CU	Keywords	CU
Student	59	Psychology	94	School	29	Demand	42
To look for	43	Work	81	Person	28	Idea	40
Course	43	Clinical	54	Degree	19	Organization	33
Theory	32	To teach	36	University	18	To add	32
Lesson	21	Context	29	To find	17	Process	30
Modality	21	Problem	27	Specialization	16	Respect	29
Organization	20	Intervention	27	Important	16	To orientate	21
Intervention	19	Different	25	To enter	15	To talk	21
Report	17	Knowledge	22	To choose	14	Degree course	20
Template	16	Internal	22	Route	14	Training activities	18

teaching skills marketable in the job market; the third cluster focuses on the teachers' institutional obligations regardless of the students' training needs; and the fourth cluster represents students' orientation as a way to support students in managing their academic training regardless of professional skills.

As regards the students' corpus (Table 4), in the first cluster the good training involves students' adherence to lesson tasks regardless of critical thinking on the theoretical model proposed; in the second cluster, learning professional skills is strictly connected to the ability to get and respond to market demand; the third cluster reflects the relevance of belonging to a group of colleagues supporting the construction of

Table 4 Students' cluster analysis results (the percentage of context units (CU) classified in the cluster is reported between brackets beside the cluster)

Cluster 1 (27.6%)		Cluster 2 (20.8%)		Cluster 3 (26.3%)		Cluster 4 (25.3%)	
Idealized education		Professional education		Group identity		Technical performance	
Keyword	CU	Keyword	CU	Keyword	CU	Keyword	CU
Experience	116	To think	89	Choice	154	People	105
Bachelor	44	Exam	71	To study	153	To feel	91
Course	43	Psychology	65	To attend	104	To find	85
Profession	41	Follow	55	Relationship	102	Level	35
University	37	Reality	55	To like	98	To take	33
Possibility	35	To see	55	Colleagues	97	To success	30
To understand	33	To start	53	To talk	74	To live	26
Different	31	Bachelor	53	To organize	68	Way	23
Sense	30	Work	44	Demand	55	Thesis	20
To live	25	Interesting	44	To add	36	Laboratory	18

a professional identity that, unfortunately, seems unconnected to professional skills development; and the fourth cluster represents professional training as a process in which the degree achievement is the main goal, regardless of the job market demand.

Students and teachers seem to have similar representations of the training process: the academic need of building a network, highlighted by the students' cluster on *group identity*, and the teachers' cluster on *training group* and *student orientation*; the relevance of achieving a qualification, highlighted by the students' cluster on *technical performance* and the teachers' cluster on *institutional obligation*; and the development of professional skills marketable in the job market reflected by the teachers' cluster on *clinical training* and the students' cluster on *professional education* in line with what it was found by Carli et al. [9] and Paniccia et al. [4] by means of a similar methodology, the emotional textual analysis [30]. The awareness of the psychological demand of the labor market is an indicator of the professional training process's effectiveness. Nevertheless, students and teachers split the academic achievement from the development of professional skills. This could be a critical aspect, possibly explaining young graduates' difficulty in entering the job market, focusing more on academic context rather than on market demand. Therefore, during the training process, students do not develop the connection between professional training (what they are learning) and professional skills (what they are going to do in the future).

5 Conclusion

Although the study results could not be generalized, due to participants selection criteria and the methodology we used, they highlight the educational culture influencing the postgraduate psychologists' positioning in the job market. The results are in line with other researches performed in the field of clinical psychology education. In particular, Paniccia and colleagues' research (2009) highlights the element influencing students' satisfaction in their clinical psychology training. Students seem to be highly satisfied with the theoretical background and their relationship with other students, while they seem to be dissatisfied with their professional training and their job. Moreover, students' satisfaction relates to their educational culture. Thus, those who plan a hypothetical professional development are satisfied with their training, while those who feel their expectations of prestige, power, money, that the degree should confer, disconfirmed are more dissatisfied.

This research was performed in the same faculty of the Paniccia and colleagues one, and it seems to show similar results for what it concerns the students' educational culture. After ten years, the faculty is composed by fulfilling students, addressing passively their professional training, as well as by students who address actively their training, focusing on the construction of professional skills spendable in the labor market. In addition, this study highlights that teachers share the same educational culture of the students, which is in line with the theoretical framework. People interacting in the same social environment share the same local culture, i.e., the educational one. Therefore, it seems that the role is not an element determining an educational culture difference in this specific context.

In conclusion, the text mining approach to explore the educational culture seems to be a useful method to identify the elements hindering, or facilitating, the postgraduates psychologist's ability in entering into the job market and to plan useful training strategies [9]. Nonetheless, we think that the educational culture is only one element, among others, explaining for the difficulty faced by postgraduate professionals, especially psychologists. This issue appears to be much more complex, and it involves different factors within an economic context, which proposes precarious job and hinders a development-focused representation [1].

Despite these considerations, the research results were presented to the Master Coordinator and it was planned an intervention aiming to improve the educational process. A group of newly postgraduate psychologists conducted students' group of discussion, and they were supervised by three professors. The focus of the discussion was the relationship between their educational training and their professional plans, and the target population was the first-year students of the master.

Although the intervention is still in an experimental phase, it aims to involve students in the sense-making of their educational training according to their professional choice and plans. The possibility to define a setting in which it is possible to reflect with other students, about oneself choices, expectations, market constraints, training possibilities, seems to offer an opportunity to plan the educational training enhancing it.

References

1. Fanelli F, Terri F, Bagnato S, Pagano P, Potì S, Attanasio S, Carli R (2006) Atypical employment: cultural models, criticisms and development lines. Riv Psicol Clin. Retrieved from https://www.rivistadipsicologiaclinica.it/english/number3_07/Pagano.htm
2. Lombardo G (1994) Storia e modelli della formazione dello psicologo. Le teorie dell'intervento [History and models of the psychologist's training. The intervention theories]. Franco Angeli, Milano
3. Montesarchio G, Venuleo C (2006) Narrazione di un iter di gruppo. Intorno alla formazione in psicologia clinica [Narration of a group process. Around training in clinical psychology]. Franco Angeli, Milano
4. Paniccia RM, Giovagnoli F, Giuliano S, Terenzi V, Bonavita V, Bucci F, Dolcetti F, Scalabrella F, Carli R (2009) Cultura Locale e soddisfazione degli studenti di psicologia. Una indagine sul corso di laurea "intervento clinico" alla Facoltà di Psicologia 1 dell'Università di Roma "Sapienza" [Local culture and satisfaction of psychology students. A survey on the "clinical intervention" degree course at the Faculty of Psychology 1 of the University of Rome "Sapienza"]. Riv Psicol Clin. Retrieved from https://www.rivistadipsicologiaclinica.it/ojs/index.php/rpc/article/view/655/698
5. Castellano A, Kastorinis X, Lancellotti R, Marracino R, Villani LA (2014) Studio ergo Lavoro: come facilitare la transizione scuola lavoro per ridurre in modo strutturale la disoccupazione giovanile in italia [I Study Ergo I Work: how to facilitate the transition from school to work to structurally reduce youth unemployment in Italy]. McKinsey & Co.
6. ISTAT (2017) Rapporto annuale 2017 [Annual report 2017]. ISTAT
7. Servizi A (2017) L'inserimento occupazionale dei laureati in psicologia, dell'università La Sapienza di Roma [The employment of graduates in psychology from the La Sapienza University of Rome]. Direzione e studi analisi statistica—SAS
8. Carli R, Grasso M, Paniccia RM (2007) La formazione alla psicologia clinica. Pensare emozioni [The training in clinical psychology. Thinkink the emotions]. Franco Angeli, Milano
9. Carli R, Dolcetti F, Battisti N (2004) L'analisi emozionale del testo: un caso di verifica nella formazione professionale [The emotional textual analysis: a case professional training check]. In: Purnelle G, Fairon C, Dister A (eds) Le poids des mots. UCL Presses Universitaires de Louvain, Louvain-la-Neuve, pp 250–261
10. Cordella B, Greco F, Raso A (2014) Lavorare con Corpus di Piccole Dimensioni in Psicologia Clinica: Una Proposta per la Preparazione e l'Analisi dei Dati [Working with small size corpus in clinical psychology: a proposal for data preprocessing and analysis]. In: Nee E, Daube M, Valette M, Fleury S (eds) Actes JADT 2014, 12es Journées internationales d'Analyse Statistque des Données Textuelles. JADT.org, pp 173–184
11. Greco F (2016) Integrare la disabilità. Una metodologia interdisciplinare per leggere il cambiamento culturale [Including disability. An interdisciplinary methodology to analyse the cultural change]. Franco Angeli, Milano
12. Greco F, Polli A (2019) Emotional text mining: customer profiling in brand management. Int J Inf Manage. https://doi.org/10.1016/j.ijinfomgt.2019.04.007
13. Greco F, Polli A (2019) Vaccines in Italy: the emotional text mining of social media. Riv Italiana Econ Demografia Stat 73(1):89–98
14. Greco F, Alaimo L, Celardo L (2018) Brexit and Twitter: the voice of people. In: Iezzi DF, Celardo L, Misuraca M (eds) JADT'18: Proceedings of the 14th international conference on statistical analysis of textual data. Universitalia, Roma, pp 327–334
15. Greco F, Celardo L, Alaimo LM (2018) Brexit in Italy: text mining of social media. In: Abbruzzo A, Piacentino D, Chiodi M, Brentari E (eds) Book of short papers SIS 2018. Pearson, Milano, pp 767–772
16. Greco F, Maschietti D, Polli A (2017) Emotional text mining of social networks: the French pre-electoral sentiment on migration. RIEDS 71(2):125–136

17. Laricchiuta D, Greco F, Piras F, Cordella B, Cuturi D, Picerni E, Assogna F, Lai C, Spalletta G, Petrosini L (2018) "The grief that doesn't speak": text mining and brain structure. In: Iezzi DF, Celardo L, Misuraca M (eds) JADT'18: Proceedings of the 14th international conference on statistical analysis of textual data. Universitalia, Roma, pp 419–427
18. Greco F (2019) Il dibattito sulla migrazione in campagna elettorale: Confronto tra il caso francese e italiano [The debate on migration in the election campaign: comparison between the French and Italian cases]. Cult Studi Soc 4(2):205–213
19. Carli R (1990) Il processo di collusione nelle rappresentazioni sociali [The collusion process in social representations]. Riv Psicol Clin 4:282–296
20. Fornari F (1979) Fondamenti di una teoria psicoanalitica del linguaggio [Foundations of a psychoanalytic theory of the language]. Bollati Boringhieri, Torino, IT
21. Matte Blanco I (1975) The unconscious as infinite sets: an essay in bi-logic. Gerald Duckworth, London
22. Moscovici S (2005) Le rappresentazioni sociali [Social representations]. Il Mulino, Bologna, IT
23. Salvatore S, Freda MF (2011) Affect, unconscious and sensemaking: a psychodynamic, semiotic and dialogic model. New Ideas Psychol 29:119–135
24. Savaresi SM, Boley DL (2004) A comparative analysis on the bisecting K-means and the PDDP clustering algorithms. Intell Data Anal 8(4):345–362
25. Steinbach M, Karypis G, Kumar V (2000) A comparison of document clustering techniques. KDD workshop on text mining, vol 400, pp 525–526
26. Misuraca M, Spano M, Balbi S (2018) BMS: an improved Dunn index for document clustering validation. Commun Stat Theory Methods 1–14
27. Lebart L, Salem A (1994) Statistique Textuelle. Dunod, Paris
28. Grasso M, Cordella B, Pennella AR (2016) L'intervento in psicologia clinica. Nuova edizione [The intervention in clinical psychology, new edition]. Carocci, Roma
29. Lancia F (2017) User's manual: tools for text analysis. T-Lab version Plus 2017
30. Carli R, Paniccia RM, Giovagnoli F, Carbone A, Bucci F (2016) Emotional textual analysis. In: Jason LA, Glenwick DS (eds) Handbook of methodological approaches to community-based research: qualitative, quantitative, and mixed methods. Oxford University Press, New York, NY
31. Greco F,Polli A (2019) Anatomy of a government crisis. Political institutions, security, and consensus. In: Alaimo LS, Arcagni A, di Bella E, Maggino F, Trapani M (eds) Libro dei Contributi Brevi: AIQUAV 2019, VI Convegno Nazionale dell'Associazione Italiana, per gli Studi sulla Qualità della Vita, Benessere Collettivo e Scelte Individuali, Fiesole (FI), 12–14 Dicembre 2019 (pp 177–183). Genova: Genova University Press. ISBN: 978-88-94943-76-4
32. Carli R, Paniccia RM (2002) L'Analisi Emozionale del Testo: uno strumento psicologico per leggere testi e discorsi [Emotional textual analysis: a psychological tool to analyze textes and discourses]. Milano: FrancoAngeli

Author Index

D. F. Iezzi et al. (eds.), *Text Analytics*, Studies in Classification, Data Analysis, and Knowledge Organization,
https://doi.org/10.1007/978-3-030-52680-1

Subject Index

D. F. Iezzi et al. (eds.), *Text Analytics*, Studies in Classification, Data Analysis, and Knowledge Organization, https://doi.org/10.1007/978-3-030-52680-1

The manufacturer's authorised representative in the EU is Springer
Nature Customer Service Centre GmbH, Europaplatz 3, 69115 Heidelberg,
Germany. If you have any concerns regarding our products, please
contact ProductSafety@springernature.com

Printed and bound by CPI Group (UK) Ltd, Croydon, CR0 4YY
30/09/2026
02225654-0012